ELECTROCUTION OF AMERICA

Russ Allen

Glenmore
Books

Electrocution of America: Is Your Utility Company Out to Kill You?

Published by Glenmore Books

For further information, please contact:
Russ Allen
allen4568@sbcglobal.net

Book designed by:
Arbor Books
19 Spear Road, Suite 301
Ramsey, NJ 07446
www.arborbooks.com
www.hotshotexpertsandspeakers.com

Printed in the United States of America

Electrocution of America: Is Your Utility Company Out to Kill You?
Russ Allen

1. Title 2. Author
3. Environmental Politics/Electric Industry/Public Safety & Health

Library of Congress Control Number: 2006925133

ISBN-10: 0-9779685-0-2
ISBN-13: 9780977968503

This book is dedicated to the many voiceless souls who, without financial resources, are afflicted by the scourge of electrical poisoning in North America. This is also their story.

Acknowledgements

I want to give special thanks to Chris Baldwin, whose assistance in developing this manuscript was invaluable. My thanks also goes to the graphic artists at Arbor Books.

I also wish to express gratitude to attorneys Scott Lawrence, Barry Hammarback, and Dan Murray; they are contemporary St. Georges, battling powerful dragons. They might have chosen a more lucrative and unethical way of life and gone to work for the utilities. But then, they would not be the men they are.

Many thanks to Don Zipse, Larry Neubauer, Don Adams, Bill English, Andy Johnson, Doug Sutter, Terry Homan, Dick Schulte, and Mike Horeth. Those folk are dedicated to making the world a better place to live.

I'm forever indebted to my brother Bob. I still don't know how he managed to survive those dark days treating fifty or more sick, suffering cows. Patti Bowers and Scott Allen took that trip through hell along with him.

And last, my thanks to Jeff and Bill Van Geothem and Rick Seiltz.

Russ Allen
De Pere, Wisconsin
October, 2007

Table of Contents

Prologue
RENDERING DEATH

Death often came to our Wisconsin dairy farm early in the morning, in the guise of a crusty old man. From out of the Rendering Plant, or Mink Ranch as my brother Bob called it with his gallows humor, would come the executioner: white-haired, sixtyish, impassive, wearing a battered Green Bay Packers cap, a half-chewed cigar clenched between his teeth. The man sent chills down my spine. I always wondered but never asked what he thought about his vocation: destroying animals with a bullet to the brain. I was afraid that he might reveal himself a ghoul out of some horror film.

He sure looked the part. He would pull up in a spanking-new, silver 12-foot Ford truck—the business of destroying electrified cows was obviously booming—and park near the barn, cardboard coffee cup in one hand and blued rifle in the other, to shamble across to the birthing area where we had set up makeshift nursing quarters and quarantined our sick cows.

They were my last thought at night and first in the morning. During the worst of it, we had cows down every day. The Rendering Plant—what prankster had dreamt up the name?—refused to take live animals even though they were about to die from their exposure to electric current in the ground; cows had to be *rendered* stone-cold dead before the carcasses would be winched into the truck with a cable hooked round their necks, stacked on the lift gate, hoisted into the bed and hauled off. During peak electric usage, when our animals were violently ill, the Rendering Plant, which charged twenty bucks per execution—with designs for the corpse—had more work than it could handle. Hence, the need for itinerant executioners, men like the judges of old who rode a circuit on horseback, dispensing justice at the end of a rope.

It wounded my heart to watch emaciated, downed cows retch and twitch, kicking impotently in a vain struggle to rise, and be unable to help them, try as we might. Our veterinarian just shook his head and sighed. Immune systems completely shot, there was nothing left for him to do. The animals had grown so weak from mastitis and other bacterial infections that my brothers and I were forced to push them onto pallets and ferry them to the birthing area with the forklift. The look in their eyes was uncanny as they watched us; they seemed to understand what was about to happen.

The cruel irony of populating the birthing barn with dying creatures was not lost on me. Normally a site of happiness, of wet, milky smells, where pregnant Holsteins delivered beautiful black-and-white, wobbly-legged little calves, the area had become a charnel house. I could never bring myself to watch the killing, turning away at the last instant. But I could smell the sickly sweet odor of fresh blood, taste the acrid burn of cordite, and hear the animal's low grunt, its death rattle. And the short sharp crack of the executioner's .22. That round to a cow's head reverberated in my mind for hours afterward. It didn't matter how many times I'd heard it before; each gunshot tore away a chunk of my heart. The feeling of irreparable loss was always the same as the very first time.

Though I'd been born on the Allen family farm and worked it alongside my parents and ten siblings since I was five, I never thought of dairy farming as a job; it was a way of life. Our lives were simple and uncomplicated, sometimes hard, but always happy. We loved the wide, flat expanse of farmland, the cows and the cats, the fields and crops, corn and alfalfa, the green earth, the tall silver silos stretching away to the horizon, the infinite blue roof of Wisconsin sky, and one another. The herd, which we all fed, milked, and tended, was more to us than a commodity or property; our cows were part of the family. To lose them and our dairy business, which my father Alvin and mother Alice built over painstaking years of labor; to watch family members turn viciously against one another; to lose two beloved sisters to suicide and one brother to a heart attack before his time—and all for no better reason than corporate venality—the corruption, arrogance and avarice of big utilities which are dumping electricity into the ground and laying waste to the American heartland—were blows I couldn't tolerate without fighting back.

The battle has come close to bankrupting and destroying me. But, I still have a few arrows left in the quiver, and one of them is to expose the lies of Wisconsin Public Service, more aptly called Wisconsin Public *Dis*service. My story is bizarre and frightening, but I must shout it from the rooftops so that the desperate plight of small farmers, ordinary men and women without wealth or power or voice, may be made known. And the sinister methods of the utility, that newest breed of Robber Baron, held up for the world to see. Sunlight, a great disinfectant, may prove to be the best weapon.

But, it took years to understand what was occurring and why. In the meantime, we nearly went crazy. And broke. The animals' sickness was insidious, gradual, a "creeping menace" in the words of Barry Hammarback and Scott Lawrence, my lawyers. Those two are heroes who have been fighting utilities on behalf of farmers for over twenty years.

Cows' health and behavioral problems worsened over time regardless of treatment: they suffered sore hooves and diminished appetites, refused to drink because of shocks at the trough, became panicked and unruly upon nearing the metal milking parlor, kicked off milking devices, delivered sickly newborn calves which often died after several days. Had these things happened all at once, we might've been able to pinpoint the cause—electric current—which is well-conducted by the moisture and metal that form large parts of a cow's environment.

But that's the nature of the absorption of current by living things. Like AIDS, it kills without fanfare by destroying natural immunity. If a calf is born where AC current is flowing, her lifetime production capacity is greatly impaired. Should she develop pneumonia after birth—common after electrical poisoning—she is scarred for life.

Similar, mysterious illnesses have affected thousands of thoroughbred race horses in Kentucky in recent years. The rate of spontaneous abortions has been as high as 75% on some farms and higher-than-normal rates of stillbirths among foals bred in the past few years have been reported. The foals suffocated, suffered respiratory illnesses, dehydration and low white blood cell counts, indicating the inability to fight infection. More than 20 veterinarians and health professionals were called in to find out why. They *investigated and ruled out every possible cause but one: utility earth currents.* The financial losses to Kentucky's multibillion-dollar horse industry have been devastating, and it would be worthwhile and enlightening to investigate the situation there regarding the utility's stray current in the ground.

So, after years of declining milk production; sick and dying cows, heifers and calves, the death from childhood leukemia of my fifteen-month-old nephew, my quest *to discover what the hell was happening to us;* and swallowing a bellyful of lies from Wisconsin Public Service and its well-paid "experts," I finally decided to demand justice and sue this utility.

Money couldn't bring back the dead. But perhaps my people would rest easier when those responsible were exposed and convicted in a court of law. To be fair, WPS didn't single me out. It's an equal opportunity destroyer. So many family farms throughout the dairy states are being devastated as by an invasion of Huns, with electrical current constantly flowing through the soil of farms on its return to the substation, that courts all across the Midwest are starting to clog with lawsuits over the wanton destruction of livestock and businesses.

And it isn't just farms being poisoned. My farm is a ten-minute drive from Lambeau Field, home of the Green Bay Packers. It would be interesting to have an independent investigator—say, a Martin Graham, inventor of the guidance system for the Patriot Missile—go there and measure for earth currents. I'll wager Lambeau is shot through with electrical pollution. And if so, those currents may well be harming Packer players who practice as well as play there.

Stray electrical current, often incorrectly called stray voltage, is the latest link in an old chain of technological scourges. First, the corporations polluted our streams and rivers by pouring in toxic sludge. To cite a personal and morally disgusting example, in 1972, the state of Wisconsin passed a law prohibiting paper mills from dumping waste into the Fox River. It so happened that I chanced to speak with the man who installed a secret pipe beneath the mill, which led to the river whereby the dumping continued unabated during the night, in defiance of the law. Then they contaminated the air with noxious chemical fumes. And now they're torturing the very earth itself, running current through the soil where it debilitates and kills. And all in the name of the twin gods: Technology and Progress. There's a third god they worship but don't name, completing an unholy trinity: the god of Windfall Profits.

I'm only a simple dairy farmer, not a tree hugger nor a Greenpeace member; not a Naderite crusader or PETA agitator; I've never belonged to the Sierra Club or subscribed to the ozone hole hysteria; I don't believe that Martians built the pyramids, and I do believe that Lee Harvey Oswald acted alone. However, I also believe that Power corrupts and that greed begets more greed and a willingness to commit any vile act. Power thinks it can get away with anything. And utilities are nothing if not huge power centers.

Corporate greed isn't new. We have the recent examples of Exxon and Enron and Merck before us. Though they inflicted terrible damage on employees, as well as on the US economy and environment, at least those moral monstrosities were unmasked and suffered prosecution and harsh penalties.

The utilities are a different story. Flying beneath the radar, they keep largely out of the public and media eye. But stray current, the result of a badly designed distribution system (they are grounding the neutral to earth by their poles, using one wire for two purposes; see two poles on front cover), even today as I am writing this book the practice is still the same. WPS simply refuses to spend the money to revamp and refurbish them, preferring to pay out far cheaper cash settlements (recouped in the form of rate increases) in nuisance and damage suits—and has plagued farmers for several decades and is now beginning to affect homeowners, even in eastern states.

And so with their unlimited financial resources, the utilities hire "experts" from prestigious universities to support their preposterous claim that it isn't stray current destroying dairy farms but the ineptitude and incompetence of uneducated and paranoid farmers, who haven't learned to correctly feed and doctor our own animals. As though the dairy business was something new and foreign to Americans, who for decades have supplied milk and cheese and butter to the entire world. Paranoia and common sense came together as my world went insane.

No, the utility and its hired guns blame the farmer for falling milk production and dying cows. And when a thriving, profitable farm like ours begins experiencing the problems I've mentioned, *shortly after the electrical distribution system is reconfigured and its pathway rerouted through our property*, it must be more than a coincidence.

The tragedy is that we didn't discover the cause of our misfortunes much sooner. But who'd have thunk it? Who could've imagined a science-fiction disaster, a war of the worlds, on our own property? It was classic catch-22: we couldn't maintain the farm without electrical power but uncontrolled current flowing over earth rendered death to our cows and our livelihood. That old man with the .22-long rifle was, in a real sense, a hit man for the utility.

But before I tell you more about the utility's brazen acts and the lies of its paid stooges, what I did and said in the course of the fight, and the fantastic way events unfolded in a courtroom duel dominated by Barry and Scott through their knowledge and logic, I want to paint you a picture of Eden and how it all began for me, of what life was like before the Fall, before the demons descended on De Pere. "Guard, protect and cherish your land," said the artist Charles M. Russell. "For there is no afterlife for a place that started out as Heaven."

Guarding the land my parents left me is what I've attempted to do, what I'm currently doing; it's been a ferocious and bloody war. And a war, by the way, that still isn't over.

Chapter One
MIDWESTERN IDYLL

The land, this sweet-smelling *terra firma* in De Pere, these acres, this region, is as much a mother as the woman who gave me life. The land is in my blood and bones; I'm part and parcel of it. If this small sliver of Wisconsin had never existed, I would probably have imagined it. Like Antaeus of Greek mythology, who drew his strength from contact with Mother Earth, I've always drawn inspiration from the ground beneath my feet and the crops and animals growing from its soil. There is no place else in the world that I'd rather live.

The little town of De Pere, a suburb of the city of Green Bay, which is either a bay in its own right or a tributary of Lake Superior (no one knows for sure, and no, it was not discovered by Vince Lombardi), must have appeared as a pristine land of milk and honey to the first Europeans who set eyes upon it. Of course, the Indians who lived here were probably less impressed by that time (1634), as they had already been in the area for about ten thousand years and were likely used to the scenic grandeur. When the Europeans arrived, the area was populated by the Chippewa, Ho-Chunk, Menominee, Sac and Fox tribes.

They say that the very first European to see Wisconsin was a French interpreter named Etienne Brule who, in 1622, explored an area around Lake Superior. He must have been stunned by its size, thinking he had discovered a second Atlantic Ocean. An account of Brule's exploration is included in Samuel de Champlain's map of what he named New France. A second interpreter of Champlain's, Jean Nicolet, who had been charged with finding a water route to the Pacific (1634), found instead that a fortune could be made by fur trapping, provided that the pelts could be delivered to Montreal and then shipped to France.

Nicolet was befriended by Ho-Chunk Indians and built a trading post before returning to Canada the following year. But his vision of riches collapsed after a series of brutal Iroquois attacks, and one hundred years elapsed before French trappers were able to establish a permanent fur trade.

The most famous exploration of Wisconsin during this era of discovery was that of Father Jacques Marquette, a Jesuit priest, and Louis Joliet, in 1673. After hearing apparently fantastic stories about a mighty river flowing into magical lands, Marquette and Joliet, filled with a spirit of adventure, embarked from Michigan on a canoe expedition and entered the Fox River at Green Bay. Reaching the Fox's westernmost point, they portaged to the Wisconsin River and then paddled on down into the Mississippi near today's Prairie du Chien. After a dangerous but exciting journey, they traveled as far south as Arkansas before finally turning northward and making for home. Eight years later, French explorer Renee La Salle entered Rapides Des Peres (Rapids of the Fathers) on his way down the entire length of the Mississippi and discovered the territory he named Louisiana in honor of the King of France.

At the conclusion of the French and Indian War in 1763, the victorious English did practically nothing with the area except continue the lucrative French fur trade, and after the American Revolution ceded the territory to the U.S., which built forts at Prairie du Chien and on the Fox, from which it again fought the British during the War of 1812.

Oddly enough, before Wisconsin became a dairy state in the mid-nineteenth century, it was best known for its lead mining and the creation of wild boomtowns like Mineral Point, Platteville, Shullsberg and New Diggings. Prospectors, carpetbaggers and soldiers of fortune arrived from every part of the country to mine the "gray gold," as lead was often called. Because many miners did not build homes but chose instead to live inside their mines (perhaps to discourage claim jumpers), they unwittingly supplied the state with its nickname: "The Badger State."

The rise of dairy farming roughly coincided with the tapping-out and decline of leadmining in the immediate pre- and post-Civil War period. For one thing, our climate is extremely favorable to dairying. Summers are relatively cool and rainfall adequate; eighty percent of rain falls in the April–October period and covers the growing season. There is much less tendency toward droughts than in nearby farm states like Iowa and Illinois, and conditions here favor grasses, hay and grain crops, which nourish our herds. Nine out of every ten acres of cropped land in Wisconsin are used to raise feed-crops for our cattle. Also, the topography of large parts of the state, rough and glaciated, has conditioned the land surface to utilization for hay and pastures rather than cultivated crops. Cattle, originally imported by drovers from Illinois and Indiana, thrived immediately in the excellent pasture-lands of the state.

Dairy farming has proven profitable because dairying and animal husbandry helps maintain soil fertility and makes it fairly easy to reconstitute land partially worn out by cropping. In the beginning, newly arrived German and Swiss immigrants took naturally to dairying, as they had experience in animal husbandry in Europe. The number of cows and heifers two years old and over kept for milk, rose from 11.8 per thousand acres of land in 1880 to 57.8 by 1930, when the average farm size grew to 113 acres with crops of corn, alfalfa and oats raised for feed. Wisconsin exported large numbers of its finest cows until reaching the high point of 85,000 in 1931. Holsteins are particularly strong in the southeast of the state, where over half of all farm income is derived from milk production and the cow population is dense; Guernseys proliferate in northern and western areas, while Jerseys are far less numerous throughout. Holsteins are predominately the choice of most Wisconsinites.

The Wisconsin Dairyman's Association, organized in 1872 by the pioneer W.D. Hoard, helped focus attention on dairy farm development. Hoard was nearly single-handedly responsible for invigorating and expanding the burgeoning industry and started the magazine *Hoard's Dairyman* in 1885, which has been published to this very day and in which the Allen Farm was prominently featured in a 1964 article. Dairy agriculture and cheese production is vital to the state's economy. They create over 100,000 jobs and infuse 20 billion dollars into the economy.

Like all of us, I was born (1952) in a unique region of the country, into a unique environment. In De Pere, life and society are dominated by two all-encompassing factors: dairy farming and the Green Bay Packers. If you've ever watched even a few minutes of an NFL game televised from the "frozen tundra" of Lambeau Field, you may have noticed that both factors are intertwined: the cheese-colored helmets and pants worn by the Packers (their name derives from Curly Lambeau, who co-founded the team in 1919 and worked for The Indian Packing Plant, which donated uniforms and a playing field) and the yellow, triangular, cheese-shaped hats, called Cheese Heads, worn by many spectators, themselves known as cheeseheads.

The Packers, a once and future sports dynasty—Green Bay is still known to the natives as Titletown, USA—and a slice of Americana, predate the National Football League. Lambeau's team joined the American Professional Football Association in 1921, which shortly afterward morphed into the NFL, and the Packers became one of the league's flagship franchises.

Between 1929 and 1944, the team won six NFL titles under head coach Curly Lambeau. An excellent running back during his playing days, Lambeau is one of only seven NFL head coaches with more than 200 wins. For those interested, the others are Halas, Lombardi, Landry, Shula, Reeves and Walsh. When he retired from the game in 1949, the Packers began a period of decline. But by the 1960s and under head coach Vince Lombardi, an American original and arguably the greatest coach in the history of team sports, Green Bay achieved unimagined heights. Of course, it didn't hurt to have a roster full of star players like quarterback Bart Starr, running backs Jim Taylor and Paul Hornung, wide receiver Boyd Dowler, offensive guard Jerry Kramer and linebacker Ray Nitzschke. In describing Lombardi's coaching style, one player remarked ruefully, "He treats us all the same: like dogs." Under his guidance, the Packers won five NFL championships as well as the first two Super Bowls in 1967 and 1968, defeating the Kansas City Chiefs and Oakland Raiders of the American Football League. Even today, Lombardi is the gold standard against which all coaches and managers are judged.

After a losing period following Lombardi's departure from Green Bay, the Packers again returned to prominence in the mid-'90s under coach Mike Holmgren and quarterback Bret Favre (for whom streets are named in Green Bay, as they are for Lombardi), appearing in two consecutive Super Bowls in 1997 and 1998, winning the first by defeating the New England Patriots and losing the second to the Denver Broncos.

Back when I grew up in the '50s and '60s, Wisconsin was insulated from the march of history. That is, until the Vietnam War intruded. Folks here were concerned with local issues, like the pollution of the Fox River with by firms like NCR, Appleton Paper, Georgia Pacific, Sonoco, Kimberly Clark and Riverside Paper. There was always interest in sports, starting and ending with the Packers, but extending also to the Milwaukee Braves, the Milwaukee Bucks, and a little later in 1970 (the Braves had fled to Atlanta in '66), the Brewers. Amateur boxing and local fight clubs in the state received a good deal of attention, as did bodybuilding, weightlifting and wrestling. I remember hearing about a famous pro wrestler and action-film actor, "Hardboiled" Haggerty, who once played football for the Detroit Lions as well as for the Packers. The '50s and '60s were an era of great building projects. Milwaukee's County Stadium went up in 1953, the Brown County Arena in 1955, and Lambeau Field in '57. Those years also saw the construction of an interstate highway network throughout Wisconsin.

Less praiseworthy was the shameful racial discrimination against blacks and native Americans in those days. Surprisingly, Milwaukee was one of the most segregated cities in the nation. Its schools didn't even pretend to integrate after the 1954 *Brown v. the Board of Education of Topeka* Supreme Court Decision ordering desegregation.

By 1976, the courts had enough of Milwaukee's foot-dragging and ruled that the city's schools were in violation of federal law. But not until 1979—a quarter century after *Brown*—did the school board finally get around to implementing a desegregation plan.

The situation was surprising, given Wisconsin's political past, its liberal, progressive spirit—it was the first state to ratify female suffrage. This spirit was personified in the career of former Wisconsin governor, senator and national leader Robert M. La Follette, whose life's work was protecting ordinary people from abuses by powerful special interests.

By the time of the Vietnam War, Wisconsin's political consciousness had been raised, and both the urban and rural parts of the state came to agree with La Follette's critique of militarism and foreign adventures. At that time, the University of Wisconsin at Madison was a hotbed of antiwar protest and student unrest, which climaxed in the bombing of the Army Mathematics Research Center and a raid on draft board headquarters in Milwaukee and the burning of draft cards.

I point these things out because I realize that my attitudes have been shaped and molded by the social climate I was born in. I've absorbed the mental atmosphere of this place. It couldn't be otherwise. Raised to be suspicious of power, the damage inflicted by the Utility on my animals and business—on my way of life—has only confirmed my worst fears about corporate greed and misconduct.

You might say that I'm *of* the farm, that dairying is in my genes. Both my parents were born on farms in Green Bay, Dad in 1917 and Mom in 1921. One of eleven children, my father, who had three brothers and seven sisters, was uneducated in the formal sense, having been schooled only through the eighth grade, like my mother. But he was a smart, industrious farmer as well as a shrewd businessman and bold entrepreneur.

In this he resembled his own pa, my grandfather, Antone, who taught him dairying. Sometime between 1910 and 1920, Grandpa helped organize a consortium of local farmers who determined that they needed to acquire another milk outlet apart from the neighboring cheese factories.

They bought a piece of property on Cedar Street in Green Bay and built a store there. They collected milk from the area, their haulers delivering it along with groceries, salt blocks and milking supplies, and sold it in bulk to White House Milk Company in Manitowoc. At that time, White House, a subsidiary of A&P, processed the product into canned evaporated milk, a new and interesting commodity.

One of my uncles later drove a milk tanker on a regular run from Green Bay to Manitowoc. Grandpa was an owner, and although I don't know what percentage of company assets belonged to him, I do know that he was CEO and general manager of the consortium from its beginning until they sold it to Lake to Lake in the early 1950s. I know, too, that he acquired some Lake to Lake preferred stock which he later transferred to his kids. So I imagine he held a large share, if not the largest, of the consortium's assets. My father told me that the consortium was one of the only co-ops in Wisconsin that paid an annual dividend from its inception until it was sold after some 40-odd straight years.

My parents met in 1937 at a dance hall, the Bluestone, over in Bellevue. They loved dancing, especially polkas, and were regulars there, a couple of whirling dervishes. They also liked the Rainbow Gardens in New Franken but the Bluestone was their favorite. It's tough for me to picture them that way, but then, what kid can imagine his parents as lovers?

Partly because of their Catholic upbringing, I think, they developed strong work habits, self-discipline and self-reliance at an early age. Alongside her mother, Ma milked fifteen cows per day by hand, as well as cultivated corn crops, utilizing horses. Her father, who they say was straitlaced, strict and puritanical, worked in the fields, in addition to moonlighting at the Northern paper mill in Green Bay.

In 1936, at age eighteen, my father decided for some strange reason to try his hand at another line of work. He never told anyone why. Maybe he figured dairying was just too exhausting. Maybe he'd come down with a case of gypsy-foot, or maybe he just got bored with the farm and longed for some excitement and adventure.

I guess he got them all right, but jumped from the fry-pan into the fire. Disappearing into the north woods, he became a lumberjack, logging for the grand sum of five bucks a day. He was a very strong man but I guess he recovered his senses pretty quickly. He soon returned home, went back to dairying (probably with a new-found commitment, since he now had something even more backbreaking for comparison), met Ma in '38, married her one year later after a whirlwind courtship and bought himself a place of his own, only after going to a few of the neighbors asking a lot of questions to see if it would be a good neighborhood to raise a family. He was twenty and she was sixteen.

Because of their backgrounds, both my parents were knowledgeable about the dairy business. That first farm consisted of one hundred sixty acres, an old farmhouse, and a weather-beaten old barn. Dad set to work at once renovating the place. He added a new milkhouse onto the barn, which he then extended, increased his herd to 32 cows and cleared dozens of acres of woodland. By the time of their sixth child in 1948, they'd built a brand-new, five-bedroom brick house. Dad built it with his own hands, with help from my grandfather, his father-in-law. Restless and energetic as ever, thinking always of new ways to grow the business, in 1954 Dad rented another farm and added twenty-seven more milking cows. Business was brisk and our income was good.

In 1962, with all of the Allen kids having arrived, we built our first new barn, 100 stalls. These days I think frequently about my family and those times, partly because I'm remembering them as they were, but also because of the untimely losses, the pain of which never subsides.

Mom and Dad were old and are gone, she in 2003; he passed in 1993, but so be it, that's the way of things. As much as I miss them, they were elderly and that's the natural order. But not my siblings, not while their parents remained alive. The pain and sorrow my father and mother felt at losing their children was indescribable.

Those deaths were unnatural and unforgivable, ghoulish in the case of my sisters. My brother Les, second oldest, who died of a massive coronary, would be sixty-five today. My younger sisters, Josephine, who would be fifty now, and Geri Lynn, four years her junior, were both suicides.

I know that back in my parents' time, and their parents' time, large families were the norm. Unlike today, when families are much smaller and longevity much greater, the medical reality was that childhood disease claimed many infants and young children; you needed to have ten kids if you wanted six or seven to survive. Coupled with the need for labor and parental support, and with the fact that birth control was shunned by Catholics, well, conceiving lots of children is understandable enough.

Because of the family's size, people sometimes ask if I felt neglected or overlooked by my parents, since there was such a wide disparity in ages: twenty years between oldest and youngest. The answer is no. I loved my childhood and my relationship with my parents and my brothers and sisters. Between schoolwork, farm work and recreation—swimming, fishing and water skiing on the rivers and bays—we were all pretty busy; we spent lots of time together and those times were good. We cared for one another and genuinely enjoyed each other's company.

The barnraising of '62 was a proud and happy moment, one forever stamped on my memory. It was like a holiday. My brother Kevin, a year older, and I stayed home from school to help put stones in the footings so we would not need so much concrete. My brother Cletus and my grandfather laid the cement block. We all helped build the rafters.

We invited neighbors and relatives and about two hundred showed up with hammers and raising poles. Everybody ate (as there was no end to the food) in the house and basement. My grandparents came and worked every day until we were through and Grandma cooked up a storm. John Miller, husband of a cousin, who delivered for Rahr's Beer (you might say his job was to bring cases before the bar), brought us a truckful of brew for the occasion.

I learned dairy farming early on, as my father had. By age four I was working alongside the rest of the family. Dad was in charge of the boys and Mom the girls, unless my sisters happened to be working outside the house, in which case my father was their boss. If Dad happened to be away, then Mom was responsible for disciplining the boys. Her discipline was usually restricted to frantic exclamations of, "Just wait till your father gets home!" When he did get home, it was usually all right. My father was a patient and caring dad and none of us could ever imagine being disrespectful to him. If we misbehaved or acted up, sometimes he would speak to us and sometimes not. A sharp look was often sufficient to put us straight.

We learned responsibility from him at a tender age. When we cut hay, the rider-mower would cut closer to the fence than the automatic rear-rake could reach, so Dad would have us go around the field with a fork and rake two feet of hay away from the fence so the machine could reach it. We didn't need the hay and wouldn't go broke without it, but that was my father's way of teaching us to conserve, and also to keep us busy. Raking two feet of hay from around a fence sounds simple, but picture a twenty-acre field. When we were done, our arms and hands felt like quivering mush. By the time we finished hand raking, driving the tractor felt like a heavenly reprieve.

We put in lots of hours, starting around 5 a.m., regardless of whether we were in school or home for the summer. Kevin and I would milk cows before school started and then again when we got home in the afternoon. We'd change their straw bedding every day, and we particularly enjoyed birthing newborn calves, sometimes helping to pull a baby out if the delivery was difficult. We watched them grow, spent a lot of time with them, and that's why it hurt us deeply when the animals became so sick and we were unable to aid them. We felt about them just as my father did.

"I like cows," he told the reporter from *Hoard's Dairyman* who had come to interview him and write a feature on our new state-of-the-art barn. "This barn is easy to keep clean and gives me a chance to give each cow individual care. This is important to me."

Dad taught us everything—how to properly use the milking machines, how to clean and disinfect udders, how to bottle-feed newborn calves and train them to drink from a pail and how to correctly use all the equipment necessary to operate the farm.

None of us ever complained much about the chores. Life didn't seem unduly hard, but then, we were innocent and had nothing else to compare. Once the first hay crop was in the barn, Dad would usually take us out fishing and swimming. Mom would stuff a picnic basket full of our favorite foods and snacks, and off we'd go to the bay of Green Bay or to a lake and rent a rowboat.

My father was an avid fisherman and taught us, even my sister, Bonnie, to bait a hook and catch perch with our small wooden poles. When we got too hot, we'd leap over the side and take a cool dip, diving deep, our goggles on and looking for fish in the dark blue waters.

Every day when Dad would market our milk, he'd take a couple of kids along and toss us nickels or dimes so we could go into the store and buy some candy. He was always making affectionate little gestures like that. At the end of haying, he'd allow us to stand atop a wagon loaded full of hay bales and grasp the hanging hooks so we could swing madly from one side of the barn to the other. It was important to him too that we had fun and were happy.

As I got older, I came to realize how well respected and well liked Alvin Allen was by the whole community. And why not? He had excellent manners and great courtesy, both because of how he was raised and because it was his nature. He was a man of honesty and integrity and humility. Conservative in his personal habits from having experienced the poverty and squalor of the Depression, he taught us to make do with what we had, to resist complaining, to learn as much as possible about every subject and not to hunger for material goods, which could be lost. He taught us to be polite, even to those we disliked, and to keep silent rather than say something nasty about another person. The daily example of his own manly, forthright conduct was the best lesson of all.

Devout Catholics that my parents were, Sundays were always very important in our home. We had to be finished with breakfast by 7 a.m., or we could not eat until after church at St. Mary's (in Glenmore, a hamlet halfway between Denmark and De Pere as the crow flies), because we fasted in preparation for Communion—one hour without water and three without food was the rule. I had to wear a white shirt and tie, normal for those times, which I didn't mind.

Every week found us in the same pew in the same order: Mom first, then kids, then Dad on the aisle where he could stretch out his legs. Inevitably, my father's head would begin to droop during the sermon until his chin rested against his clavicle, eyes shut, solidly asleep. Mom would always lace into him about it afterwards, in the car on the way to Zellner's grocery store, where she would buy our Sunday dessert, a half gallon of ice cream which we got after dinner, provided we behaved. I used to wonder if my father confessed his little Sunday morning naps to the priest when he made his weekly confession on Saturday nights. Mom sure made it sound like those catnaps were mortal sins, one of the seven deadly. I had to make confession several times per month, whether or not I had anything to confess. Strangely, my penance was always the same regardless of the sin: say one Our Father and three Hail Marys.

On Sunday afternoons neighboring kids would come over or we'd go spend time at their places. At the risk of sounding like an old billy goat, I'll admit that we didn't do the things kids normally do nowadays. Possessing much less in the way of toys and other gizmos, we were more active physically, and very creative. We made rafts and floated down the creek that ran through a nearby woods, the woods where we had field trips in the first and second grade, and the same woods where a good friend and I picked flowers for our mothers in the springtime. We constructed tree houses and slept in them. We built go-carts and raced them. Every year on our farm we boarded horses for the winter. Magnificent animals, we loved them, cared for them and rode them to our hearts' content, hitching them to sleds and toboggans for rides in the snow. On weekends during summer, we went water skiing, taking along grills for barbecuing hamburgers and hot dogs.

Christmas was always my favorite time of year—still is, in fact. I recall that early in December, just after Thanksgiving, Mom and her friends would go out into the woods and start gathering evergreen branches to make wreaths, and sometimes we would get to go with her. The women would make gorgeous wreaths for their homes, for friends and for the church, wreaths decorated with bright candy canes and pine cones and tiny silver bells and red velvet ribbons. Those wreaths signaled the holiday to us, that Christmas was just around the corner.

In school, we would start working on the Christmas Program to be performed in an assembly for our parents a week before the close of school. I remember practicing singing carols and hymns, singing them over and over accompanied by Mrs. Vetter and Mrs. Gussart on the piano. *We gather together to ask the Lord's blessing/He chastens and hastens His will to make known...*That old, weathered, one-room schoolhouse holds many happy memories for me.

At home we enjoyed preparing for Santa's arrival. Above our fireplace was a big mirror on which we glued snow and window-frost and fastened a miniature Santa, sleigh and reindeer. We frosted over most of the windows of the house and wrote different Christmas messages and holiday greetings on each. The spirit of the season was a strong presence in the Allen household, and our anticipation of the 25th seemed to build and build as the December days passed.

On the 24th Dad would select and bring home the Christmas tree. He would sweep through the door in a grand entrance, with the pride of an arborist with his prized seedling. Each year the tree had to be absolutely perfect in height, shape and fullness, fit to appear on a Christmas card. He would place the tree on the stand, switch it into various positions and study it with his expert's eye, making certain there were no bare spots. If he detected even a hint of one, he would drill tiny holes in the trunk and insert branches to thicken the area.

On Christmas Eve we had to be in bed by seven. Mom and Dad always attended Midnight Mass (did Dad keep awake?) and would rouse us at about 2 a.m., with the movie camera ready to roll. As kids we wore the old-style pajamas with the flap over the backside that closed with two buttons.

Our brood must've been some sight, but what a sight we'd awaken to! I always marveled at how Santa could decorate the tree, place the gifts, set up the train and tracks, and repeat that magic in houses throughout the world, all in one night. We'd play with our toys, which weren't very many by today's standards—a few colored pencils, notepads and some miniature cars or dolls—till about 6 a.m. when we fed and milked the cows. After that it was off to church at St. Mary's. The Nativity scene inside the church was breathtaking, with four tall Christmas trees surrounding a huge manger just beside the altar. In the afternoon my grandparents and aunts and uncles would come over for Christmas dinner.

As the years passed, Christmas grew larger in our imaginations and even more meaningful. Just like in most families, it went from Mom and Dad buying gifts for everyone to everyone buying gifts for them. When we were older and business was better, Dad would give us money, a couple of hundred or so each, depending on how good a year he was having.

The bucolic life I've described and the sprawling, happy family I grew up in were things I loved and never rebelled against, as many children do. And so I followed in my Dad's footsteps, determined to earn a living as he had, and his father before him. It seemed as good a vocation as any and better than some. Perhaps my feelings about the farm and about my parents, about my father's teachings and about our way of life, may sound hokey to modern ears, and maybe they are. But that does not make them less true. My parents were rustics, products of a vanished time, a simpler, less cynical, more honest, more loyal and certainly much happier time; a time that lives now only in memory. That is all right. Loss of old ways is the price of progress. And the old ways were not necessarily the best ways, especially compared to the technological advances that have improved the quality of life and removed so much drudgery from work.

But technology cuts two ways. What it improves on one hand it aggravates on another. I've always been a proponent of working smarter, not harder, and have attempted to upgrade and improve the Allen farm operation at every opportunity. And I was able to do so, making us one of Wisconsin's largest and best, until the operation was crippled by a new technology: the WPS reconfiguration of the electrical distribution system. Current ran through our property like a dagger through the heart.

Chapter Two
A CURRENT AFFAIR

Farmers often get a bum rap in books and films. Too often, city slickers stereotype us as rubes, hicks, plowboys, yokels, bumpkins and hayseeds, because we live in rural areas and are closer to the earth and to nature than are our cousins crammed together in the asphalt jungles of big cities. But we're nothing if not practical; we learn from experience, and we learn quickly.

Farmers wear many different hats in the course of our working lives: we're alternately veterinarians, accountants, buyers, equipment operators, breeders, mechanics, electricians, plumbers, carpenters and truckers. We're also skilled problem solvers and shrewd businessmen, since our livelihood depends upon how efficiently we manage and increase productivity while reducing costs. This means keeping our herds and crops in optimal condition. Unlike businesses where managers manipulate data and fudge statistics, our alternatives are simple: produce each day or go under, march or die.

Because we place our products into human bodies—feed people—our business is heavily regulated, as it should be, by a number of local, state and federal regulatory agencies, including the USDA and the FDA. Readers will surely remember *E. Coli* and *Mad Cow Disease* outbreaks of recent years. If not for the knowledge, caring, commitment and rapid reaction of American farmers in isolating and correcting the problem, these diseases might have resulted in epidemics of bacteria-caused deaths, as has too often been the case in other countries.

If any of our products prove harmful to people or animals, we should be held accountable (morally as well as legally) and compelled to pay damages. So, should a public utility like WPS be less responsible if it harms the consumers of *its* products? To ask the question is to answer it. And yet, when WPS' product—electricity, current or juice—sickened, injured and then killed my cows, the Utility simply denied the facts and instead blamed my herd mismanagement and incompetence.

But stray current in the earth, electricity seeping through the ground and into my animals, is what poisoned them. But how WPS got away with murder, and how I finally managed to dig out from under their avalanche of lies, well, therein hangs a chilling tale. Chilling because they can do it to anybody. And they are.

Stray current—actually not stray at all but coming deliberately off the wire—can be defined as the continuous flow of any current over the earth that harms the health of humans or other living things.

In the tranquil, bountiful days of the '60s and early '70s, our farm was one of the two or three largest, most advanced and productive in the entire state, which is saying something, because Wisconsin has been the nation's leader in dairy production for well over one hundred years. Wisconsin milk and cheese is rightfully renowned for its high quality and distinctive flavor.

I remember feeling immense pride in our operation and my family. Thinking back on it, I recall one particularly warm sun-filled August afternoon some thirty years ago. We were outside the new barn, my brothers Bob and Cletus and I, having just finished milking and feeding the herd. Dripping sweat, we leaned against the door of the barn and looked out together across acres of fertile cropland, at the tall, thick, unending rows of corn and alfalfa gently undulating in the summer breeze.

No one spoke, but we all had the same feeling: awe and gratitude. That fleeting, insignificant moment is for some reason frozen in memory. As though it were yesterday, I can still hear the buzz of bumblebees hovering lazily above the snowball bushes outside the house, and the contented lowing of the cattle. I can see the high sapphire sky, cloudless and blue as the deep blue water of Green Bay sparkling in the summer sunshine. The sky that day appeared bluer than ever before. That blue profundity seemed to me like the color it must have been on the first day of Creation.

"Guys," I said. "We sure are blessed with what we have. The possibilities are endless. Look at it, the Allen Dairy Farm. From a tiny place with a couple of cows to operating 1,500 acres and a new 500-cow setup. Pa must be real proud. Me too."

My brothers grinned and clapped me on the back. They felt the same pride of place and purpose. Because at the time, it seemed as if the sapphire sky was the limit for our milk production, that we'd be able to continuously increase our herd and productivity until there was no containing us, until we became the biggest and best dairy farm in the United States.

And then disaster struck. During the same period in which we built a new free-stall barn and milking parlor, Wisconsin Public Service, WPS, reconfigured its distribution system, connecting many miles of distribution system beyond our farm, a seemingly harmless event. This resulted in putting Allen Dairy Farm in the path of their neutral return current on its way back to the substation. Because of their poorly designed distribution system, current leaked from the distribution system into the earth, and went through the earth back to the substation. Before the reconfiguration, there was no distribution system past our farm, so there was no current flowing through Allen Dairy Farm property. Because of the poorly designed distribution system, current flowed down the ground wires—wires running down the distribution poles—that were connected to the neutral wire—wire that took the current back to the substation—driving current into the earth and taking the earth back to the substation. Hell streamed into the soil. Had problems occurred right away, we would've understood the cause. But the issue took much time and anguish to identify.

The Utility was aware, and had been all along, of what the repercussions to the farm would be. WPS wanted to use our property as a conduit for their neutral return current. The reconfiguration took place in 1976; WPS' setup was still the same until last month, despite a judgment against it, part of which mandated repairs. In its colossal arrogance, the Utility refused to correct the problem. It was like the tobacco companies, lacing cigarettes with nicotine and tars, the better to hook new victims, but continued for decades to deny this embarrassing fact.

Now, there are any number of difficulties that dairy farmers face each day. The Utility knows this and uses the concepts of "farm mismanagement" and "multi-factor herd health" problems to confuse and to shift focus away from itself. Some of the difficulties farmers can face in the normal course of events include:

- Declining milk production: Every cow is expected to give a particular amount each day. We plan by looking at past history, and by analyzing figures for similar herds and dairies.
- Elevated somatic cell counts: These are white blood cell counts found in the milk and are produced as antigens by a cow to fight infection. The higher the count, the greater the chance disease may be affecting the whole herd. When mastitis, a bacterial infection of the udder, is present, the somatic cell count skyrockets. When the count climbs beyond a certain number, the milk cannot be sold, and cows with high counts are frequently pulled from the herd. Mastitis can be caused by numerous factors, but high counts are attributable to things like poor milking procedures, bad sanitation, *stray current,*

> long lactation periods, and various types of infection. When our cattle became sick, we always quarantined them, separating them from the healthy ones and starting treatment immediately. The sooner you can start treatment, the quicker you can nurse the cow back to health, providing she doesn't die first. Some cows didn't show signs of being sick other than just getting weaker and weaker and soon, didn't even get up anymore.

To properly prepare a cow for milking, the teats must be first cleaned and then sterilized. The milking device, essentially suction cups with hoses, is also sterilized before placement. Under normal conditions a cow will be calm and docile as its milk is removed.

But because of the effects of electric current on the animal, calmness and docility go by the board and the animal becomes nervous, agitated and frightened. As time passed after the reconfiguration, over several years in fact, we began to perceive increased behavioral problems throughout the herd, even among new cows we imported, and among calves. They would become skittish and balky when entering the milking parlor and continually kick off the milkers, sometimes injuring myself and people that worked for me. Milking times drastically increased, since sterilizations and cleaning procedures constantly had to be revisited. Production decreased, and nothing we tried in the way of medicines and supplements seemed to help.

Other troubling signs were the development of sore hooves. It was hard to get cows pregnant and many would abort several times. And of the pregnant animals, many could not carry to term. No medicines prescribed by our vet, no feed or supplements or vitamins established by our nutritionist, seemed able to alleviate the symptoms.

Living with a deteriorating herd, it was clear that we had a major crisis on our hands. Somatic cell counts in much of our milk were in the range of 500,000 to 800,000, nearing the level that milk plants would no longer accept your milk. At times, we had a 50 percent cull rate; the cows either died or were sold for slaughter before they died. Needless to say, those twin horsemen, sickness and death, were seriously reducing our farm's income. Our family's future was in jeopardy.

Yet we felt we were doing everything in our power to heal the cows. After all, we had several lifetimes' worth of dairy experience on which to draw. We ran the gamut of emotional states. At first, we were merely frustrated; later we became depressed; eventually, we grew angry and spiteful, turning viciously against one another. Mutual recriminations flew around the place like boomerangs.

Early one September morning, I was making breakfast in the house at the main farm, the house my father built in 1948. Like all of the Allens, I love a good meal of almost any type. Food is a great pleasure of ours, and we cook with nearly as much gusto as we eat. In those days, long before the "organic" craze, we grew or raised most of our own meat and vegetables. We raised hogs and chickens, and each night collected basketfuls of large, beautiful eggs. My favorite breakfast is a fluffy cheddar cheese omelet and crisp bacon, newly baked rye bread overspread with churned butter, and a big, steaming pot of fresh-ground coffee lightened with pure Holstein cream.

On that particular morning, Dad and my brother Les were in the kitchen awaiting breakfast. We had already milked and fed the cattle and the conversation began typically enough, the three of us chattering on about small matters. Les, brown-haired, at five feet five inches, is shorter than I am but more heavily built; he happened to be in charge of the milking operation in the new barn. At that moment, we were undergoing a violent outbreak of mastitis.

With all his experience, Dad knew this was abnormal and unnatural, yet had no clue (as none of us did) that electricity was shocking and sickening our cows. The only thing Dad could imagine as being the cause of the problem was improper herd management. The conversation took a harsh turn when Les brought up the subject of mastitis.

I was at the stove frying my bacon and eggs and listening, and was prepared to offer any advice I could that might help. But I quickly sensed that silence would be the wisest choice. Although generally a calm, quiet and thoughtful man, my father was never one to beat around the bush when making a point, and now his point was that Les, as well as others in the partnership, was royally screwing up. Les did not take kindly to Dad's accusation, and got loud and belligerent.

"What do you think?" he huffed, his tone bitter. "That we're deliberately giving the cows an infection? Is that what you believe, that we *want* to make them sick? Oh, sure, we just live to make our cows sick to produce bad milk!"

"And so you are," my father yelled back. "That's all I know. Cows don't give themselves udder infections and sore hooves. You guys better get it together. That's all I've got to say. I can no longer continue to pull a feather out of my hat when I pay the bills. You guys better start thinking about tomorrow, before tomorrow disappears."

Les flew into a rage. Eyes bulging, he began screaming at the top of his lungs. The F-bombs were flying around like I'd never heard before. Then they accused each other of all sorts of sins and offenses. They continued to the point where I became afraid they might start beating up on each another. I was in charge of feeding at the time, and remember racking my brain there in the kitchen to remind myself that nothing was missing from the cows' food ration.

We'd been the first farm in the Green Bay area to implement a herd health program, with the vet checking in every two weeks to test for pregnancy and to maintain vaccination procedures. We were doing everything possible and far more than most dairy farmers. One of the saddest aspects of what the utility did to us was that Dad went to his grave without ever learning the full truth. For that, I blame myself. I wish I could've uncovered it much sooner. It would have restored Dad's faith in us, proven to him that we were good at our jobs.

And so the partnership we'd developed with so much amity and optimism began to disintegrate under the strain of dying cows and the financial pressure of plummeting productivity. The number of sick cows overwhelmed Cletus, whose responsibility was doctoring. He felt a bond with his animals, loved them. He became deeply depressed by their plight and his inability to cure them. Subsequently, I'll relate the morbid details of what became of the broken partnership.

High cell counts clearly indicated the presence of mastitis, a very contagious sort of infection. The standard treatment is to quarantine the animal and administer antibiotics. Of course we did this, to no avail. We even went so far as to replace up to fifty percent of our cows each year.

We first became aware of stray voltage about 1988. Back in 1988, we were like everyone else in America. We did not know that our utilities were using the earth to return their current back to the substation. Naive and gullible, we thought along the lines of how the Utility had propagandized everyone into thinking: believing that stray current was just a smoke screen used by dairy farmers who were poor managers and bad businessmen to evade personal shortcomings and escape responsibility. Though we considered the idea of stray voltage affecting our cows to be ludicrous, we took measures to rule it out, since nothing else we tried seemed to be working.

We got our first taste of dealing with Wisconsin Public Service when we asked them to come out to our farm and check for stray current. They came, they saw, they reported…no problem. Just as we'd expected. We went back to the drawing board to figure out new possible causes and come up with new medical treatments.

For the next few years, we got really creative, devised a whole host of solutions. Up until then we'd fed our stock from what's sometimes known as a "poor man's mixer," weighing all the feed and mixing in needed supplements. We then pumped the mixture into the barn feeders with augers (devices with a helical or spiral shaft used for boring holes in the earth or moving any loose material). In this way, we came very close to the nutritionist's recommendations for proper nourishment, but decided that, given our problems, the system was too imprecise. So in 1991 we bought a new state-of-the-art feed mixer to create a perfectly balanced food mixture for each cow, each day. With this piece of equipment, we could be certain that the animals were receiving correct amounts and proper nutrients one hundred percent of the time. Every cow would receive optimum quantity and quality of feed.

The new feed mixer was a sleek, expensive piece of perfectly engineered equipment. It did zero to correct the problems. Milk production didn't improve and the cows didn't get better. They still balked at entering the parlor and refused to permit milkers to be attached. The death rate was high, and sore hooves and other mysterious maladies continued to torment them.

When WPS reconfigured the system, it had placed our farm smack in the middle of the distribution system rather than on its end. The new wires were made of aluminum, connected to copper weld-wire. We thought the revamped system, if we thought about it at all, was probably a good thing. And we believed what WPS told us: that stray voltage had nothing to do with our herd's illnesses.

But we couldn't cure them. Everyone—veterinarian, nutritionist, other farmers— told us our countermeasures were correct and proper Yet there were years that at least 50 percent of the herd left the farm, whether they died or we sold them before they died. There was only one occasion in over 20 years that we were able to start culling for production. It happened to be when I had WPS' down grounds cut or disconnected. It did not last long. After a year and a half, WPS reconnected their down grounds and my world went insane again. I do remember a time when 75 percent of the herd was limping.

Then an equipment dealer I had spoken with and about the problems suggested we consider remodeling the parlor, going from a double-twelve polygon to a double-twelve parallel, thereby decreasing the number of milk-unit kickoffs and preventing employee injuries. An additional benefit would be that we could improve productivity by needing only one person to operate the parlor instead of several.

We decided to take his advice, at considerable expense: $85,000. We'd leave no stone unturned to solve the problem once and for all. At about this time, the U.S. Department of Agriculture, after conducting a study on stray current (so there was such a thing after all?), decreed that a new requirement for milking parlors was that they had to have what is known as an equipotential plane.

This plane essentially used the steel bars which reinforced the concrete floor and connected it to the metal of the parlor. This was supposed to make everything equal like a bird on a wire. Nice idea, but it didn't work. Supposedly, this would prevent stray electricity, then thought to be caused by power spurts and spikes through milking equipment, from shocking the animals. Never in our wildest dreams did we imagine that WPS was at fault.

For we didn't and couldn't have, conceived that the grounding rod WPS installed outside the barn could conduct stray current into the barn through the equipotential plane and thereby electrocute the cows. We were shocked—no pun intended—when we finally learned that current was flowing like a river through every part of our property, yard, barn, parlor, pasture and house.

The new parlor was a work of art, a thing of true beauty. Fully computerized, it could gauge and measure output for every cow, allowing us to concentrate our attention on animals whose production lagged behind. With all of the new equipment in place and functioning—we had balanced nutrition, productivity measurements and tight physical control of the cows—we felt confident we'd taken every possible step to ensure success and were on the path to pinpointing the problem and overcoming our misfortunes.

We were dead wrong. As with the feed mixer, the new parlor was an expensive bust. Cows continued to sicken, production plunged, fatalities increased. To those of us in the partnership who analyzed things realistically (not all did, because realism was painful), it became apparent that the bottom had fallen out of our once-thriving dairy operation and that the business was doomed. Our farm was going under.

It died a slow, agonizing death. After ten years: the loss of my infant nephew to leukemia; the suicides of two sisters; the heart attacks of two brothers; my own heart attack; the death of another healthy brother, 44, from a massive coronary; the death of a slim, healthy brother-in-law, 50, and the death of a good friend, 43, a neighbor, slim, and working for the county; all from the same causes and the deaths of scores of cows, the dream died hard. All of the above-mentioned family lived in the vicinity of the farm and of the rampaging ground current. It wrecked us all in one way or another.

Throughout those years, all members of the partnership remained at each other's throats. Everyone accused everyone else of laziness, arrogance, incompetence, stupidity and greed. By 1989, the partnership had dissolved. I plugged alone paycheck to paycheck, robbing from Peter to pay Paul. In 1995, I installed a new parlor. By the spring of 1997, we had enough of that parlor. It reached the point where I could no longer stomach the acrimony, and in desperation, I contacted the salesman who'd sold us the new parlor.

"Mike, you sold me a piece of crap!" I shouted at him over the phone one rainy, especially dreary morning, "I trusted you. I thought you knew your business. My sick cows are getting sicker and dropping like flies. It's *your* fucking parlor, *your* fucking product. Get my cows to stand still in that parlor. I knew that if they would stand still, they would produce more milk. Get your ass over here and fix the damned thing!" I slammed the phone in his ear as hard as I could.

Mike was genuinely stumped, because he had placed identical parlors in a number of other farms that were humming along nicely. The only theory he could think to offer, which fit the facts of the case, was stray current. He promised to put me in touch with a man who had assisted other farmers with similar problems.

Still skeptical regarding the idea of stray current but at my wit's end, I decided that I had nothing to lose by revisiting the subject. I asked Mike to set up a meeting with his man and began to read up on the supposed phenomenon.

I found out that the most well known and widely circulated piece of literature on the subject is the Department of Agriculture's so-called Red Book of 1991, titled *Effect of Electrical Voltage/Current on Farm Animals.* (Lefcourt, A.M., *Effect of Electrical Voltage/Current on Farm Animals: How to Detect and Remedy Problems*, USDA Handbook 696). It braced me immediately when I came across this passage:

> The first farms in which stray current problems were identified were suffering severe losses of milk production and income. The producers were generally aware

> they had problems and had spent considerable time and money attempting to improve their feeding program, the milking equipment, their milking procedures and hygiene. But nothing seemed to help.

This description was exactly what we'd been going through. The handbook went on to describe how stray voltage shocked and panicked cows, which were highly susceptible to its effects as their environment was usually wet, aiding electrical conductivity. And much of the equipment surrounding them is made of metal, also an excellent conductor. Yet none of us could feel anything like an electrical shock anywhere on the farm, except for when we would put a unit on the cow. Occasionally we would feel a little tingle on our elbows, when our elbows would touch the stainless steel curb. Therefore it seemed highly unlikely to me that current was to blame. But I did not want to leave the subject before I had an expert come to the farm to assess the situation.

So Mike set up a meeting for me with the owner of the Concept Electric company, master electrician Larry Neubauer. Suspicious because utilities were spreading rumors that there were legions of self-styled stray voltage "experts" roaming the Green Bay area who were nothing but flimflam artists, I became even more leery when I found out that Larry required a $1,200 retainer before he would even agree to set foot on the farm. *Great,* I thought, *more money down the rat hole. But what the hell, what's twelve hundred after the thousands I've already wasted? I'm going broke anyway, might as well go out in a blaze of glory.*

Larry Neubauer looked more like a Packer lineman than a master electrician. Burly and taciturn, a no-nonsense type who spoke little, as though words were jewels to be hoarded, I discovered that the gruff exterior belied his deep sense of morality and feelings about right and wrong. As quick to laughter as to anger, Larry was smart, experienced and conscientious about his work; he didn't suffer fools gladly. Knowing him the way I now do, there's no doubt that Larry would've reamed me out like a Dutch uncle if he'd suspected or seen any evidence of farm mismanagement or maltreatment of animals.

Larry spent a few days investigating throughout the property. I spent time with him when I had any, which was seldom, for the press of problems in the dairy operation consumed all my attention. I didn't know a lot about electricity, but started getting an education in a hurry.

"Hey, Russ, come over here and take a gander at this," Larry called to me in his gravelly voice one afternoon when I passed him in the milking parlor.

He had an electronic meter clipped to an archway. Built of metal pipe, the archway connected to a plate that fastened to a stall in the parlor that held cows for milking.

I went over to him and squinted at the meter, nodding as if I knew what I was looking at, which I didn't.

"Eight hundred milliamps," Larry said and shook his head. "That's 0.8 amperes."

Then he did something startling. He shut down the electricity coming into the farm from the pole and took another reading: "Still eight hundred milliamps!" Although I knew little about electricity, I knew enough to understand that something was wrong. Cutting power incoming from the pole meant the meter should have read zero. But the juice was flowing as forcefully as before. Then Larry did something else that astonished me—he disconnected WPS' neutral wire. Then, and only then, did the meter dip to zero.

It struck me that I was being conned, sold a bill of goods by one of those stray voltage hucksters we'd heard about, particularly after his next statement.

"The problem's fixable. But it'll cost big bucks," Larry said, reinforcing my impression of a con artist. Before making any commitment, I intended to have someone else take a look at the farm and recommend a course of action. I told Larry I'd get back to him.

Larry simply shrugged. A few days later he sent me a report on his findings and offered several ways (all expensive) of correcting the problem. He suggested that I call Wisconsin Public Service out to the farm so their experts could conduct a series of tests.

Again, they came, they tested, they reported…no stray electricity. One of the utility's favorite sayings is "have a nice day." They must have to be able to say that when they are hired, because WPS usually doesn't hire electricians to test for stray voltage, they like to hire a nutritionist or someone like that to go to a farm to test for stray voltage, but really to go to a farm to pick apart the farm's management. They certainly don't know how to find stray current, because they usually never find a problem.

Someone, either Larry Neubauer or WPS, was, at best, wrong, or at worst, flat-out lying. Time would tell which.

In the USDA Handbook 696, the "Red Book" referred to earlier, the following prophetic passage is worth quoting in the following context:

> When these results—current flowing even after power was turned off—were passed on to neighbors, and when

> reports began to appear in the popular press, other farmers suspected that they too might have similar problems and were quick to make demands of their electrical power suppliers, farm electricians and milking equipment dealers. In general, these people had little knowledge of how and what levels of electrical current can affect animals, were unaware of methods of systematically identifying and mitigating stray voltage/current problems, and often reacted to farmers with disdain and contempt. Sometimes, no real problem existed; other times, attempts were made to shift the responsibility of diagnosis and solutions to other persons or organizations. When the latter occurred, farmers felt that no one cared, animosity between two or more individuals developed, and the teamwork required to solve the problem quickly dissolved.

Unless Larry Neubauer's instruments were malfunctioning or completely broken down, something was pumping current into the Allen Farm, after he'd shut the switch. But if Neubauer was right, then the source of the current, unless emanating from the space ships of a Martian invasion fleet, *had* to be WPS.

Chapter Three

OF CLAIRVOYANTS AND SOOTHSAYERS

"Didn't you hear me? Take the wax from your hairy ears. There's no such *thing* as stray current," I told my brother Bob when he first suggested we might want to reconsider the electricity idea.

Bob was always logical, a pragmatic problem solver. He was the "doctor"; his post-partnership responsibility was medical treatment of the herd. He'd been going through hell with them and it showed in the dark circles around his eyes.

"I saw a research article that says so," I continued. "These big universities have agriculture experts who study this kind of stuff for a living. They ought to know."

Bob spit and shrugged his shoulders. "Okay. But if not current, then what? What's ailing these animals? Answer me that."

I had no answer.

As I learned later, most of the methods universities employ for researching and investigating stray current are akin to those of clairvoyants or soothsayers, who diagnose and predict using Tarot decks and Ouija boards, or by studying animal entrails or casting horoscopes. In defense of soothsayers, at least they believe their own mumbo-jumbo.

Stray current researchers play shell games, attacking the problem by studying something unrelated. They make experiments in laboratories, controlled environments that have nothing in common with conditions found on dairy farms. They then trumpet their findings as the last word in science.

If their methods come to seem ridiculous, that's because they are. Farms are not labs. Many factors and influences exist on dairy farms that show the symptoms of a stray current problem, which no experiment can replicate. If it could, then the lab would be a farm. So it's no surprise

Wisconsin State Journal MONEY 8B

Stray Voltage Still Plagues Some Farms

Utilities Have Spent $100 Million On Problem In Last Decade, PSC Says

Tuesday, November 17, 1998
Associated Press

ATHENS—Wisconsin utilities have spent $100 million in the past decade trying to resolve problems farmers have with stray voltage, the state says.

But farmers like Tom Oelke still struggle with the problem. He blames stray voltage for the *loss of 75 percent of his dairy herd* and feels lucky he hasn't lost his farm because of the ordeal.

Near Christmas 1994, Oelke's cows started to *die at the rate of two to three a week.* It wasn't until about a year later that a veterinarian who studied the herd suggested stray voltage as the problem, the farmer said.

"My first reaction was, I didn't believe it," Oelke said. "But I felt I had checked everything else."

Figure 1

that the researchers overwhelmingly conclude through faulty methods that stray current is no problem.

The question that leaps to mind is: why are the researchers not out investigating in the barns and in the fields and pastures where cows are dropping in record numbers? Could it be that the researchers are involved in covering up the issue? Do they have a vested stake in doing so? You bet.

University "experts" contend that their findings demonstrate nothing is wrong, that stray current can't possibly be causing my problems and those of other farmers. The state Public Service Commission, that overseer of utilities and supposed protector of public welfare, has set a "level of concern" stating that nothing below this level need be addressed. "Level of concern" is a brick wall behind which WPS and other utilities hide. The research is fundamentally flawed because it ignores a number of important symptoms.

The first key symptom is a *high mortality rate of cows.* As an experienced farmer, I would never deny that a certain number of animals raised on a dairy farm will die of natural causes like disease or infection. Birth or breeding difficulties, even accidents or mishaps, account for more deaths as well. As much as it hurts to lose even one cow or calf, every farmer realizes that animal deaths are inevitable and a part of the cost of doing business.

In 1996 the U.S. Department of Agriculture surveyed cattle mortality rates in Minnesota and Wisconsin. (*Dairy Herd Health and Production Survey Findings*, Minnesota/Wisconsin Engineering Notes, 1998, available online at http://www.bae.umn.edu/extens/ennotes/enwin98/herdhealth.html.) They reported the annual average at 4.3 percent. The culling rate—how many animals were cut from the herd for sale because of low production—was found to be 20.6 percent. Now, an article appearing in the year 2002 in *Dairy Business Communications* (Harlow, "Where Have All the Heifers Gone?" *DBC* at http://www.dairybusiness.com/northeast/Feb02/Feb02NEDBp16.htm, 2002) whistles a different tune. It stipulates heifer mortality rates at 15 percent but shows the national cull rate to be 40 percent.

So, it would seem that the mortality rate rose drastically between '98 and '02. But the problem here, the problem with research generally in this field, is that competing interests grind competing axes. Article number two is concerned with heifers. Heifers are young cows which have never given birth. Milk only comes from cows who have recently delivered a calf, meaning that heifers are unproductive. Heifers may or

Figure 1

Daily Journal Extra *September 15, 2003*

Dairy Farmers Sue Power Firms Over Cows' Deaths

Complaint Blames Stray Voltage in Ground

By Stefanie Knapp

Mike and Linda Cherniske couldn't understand why their cows seemed to die so much more easily at their new dairy farm in Delta, Utah, than they had back East...

"It's like if you look at them wrong, they're going to lay down and die," Mike Cherniske says.

After much testing, the Cherniskes' claim that stray voltage released into the ground by the Intermountain Power Project is killing their cows...

In the last four years, the Cherniskes lost 500 cows. The deaths reached a peak in December 2001. The family lost a cow a day and had between 50 and 60 cows—or nearly 10 percent of their 600 to 700 milking cows—in the hospital at any given time.

"I was running up vet bills that I couldn't afford," Mike Cherniske says.

In the last four years, the Cherniskes have had Veterinarian bills of $300,000...Their farm, Gunn Hill Dairy Properties, is eight miles from the power plant.

Experts eliminated a multitude of problems that were causing the cows to die or become sick. The Cherniskes were stumped, until one expert suggested that the couple test for stray voltage.

may not have been included in the USDA statistics; therefore the two articles are discussing different propositions, even though they appear to be talking about the same thing.

The point is simply that statistics and research can be twisted to say anything by those who have a vested interested in doing the twisting.

The above article shows that Tom Oelke was losing several cows every week until 75% of his herd was dead. We can't conclude very much about his high mortality rate because we do not know the number of cows he had at the start. Neither do we know the length of time involved in the 75 percent loss. All we know is that Tom's mortality rate far exceeds the 4.3 percent average of the USDA study.

But we can at least establish a reasonable standard. If 4.3 percent can be taken as an average, we'll stipulate that three times this average or, even a bit more, say, 15 percent, would indeed be a high rate. We can then clearly define the first key symptom: a 15 percent or more yearly mortality rate.

As I said earlier, disease, infection and poor nutrition could also be causes for high mortality rates. A veterinarian would immediately be able to analyze and correct the problems with medicine, better food and supplementation. But what happens when stray current is present: treatment will have no affect and the mortality rate of the herd will continue to climb.

The situation described above in Figure 2 ("Dairy Farms Sue Power Firms Over Cows' Deaths," *Daily Journal Extra,* 9/15/03, available at www.howarth-smith.com/News/news-11.htm) illustrates the point. My herd suffered a 40 percent annual rate, and a high rate that is irreducible by any treatment indicates that stray current is the culprit. It is an all too common occurrence, becoming more and more widespread across the dairy belt. As for the Cherniskes, they traveled the same dead-end route as I did. They ruled out every conceivable factor that could have killed their animals until only one conclusion remained: stray current.

The utilities and their paid "experts" dispute this conclusion, though, with all their might. According to them, their studies prove without a shadow of doubt that "stray voltage" (a misnomer, because voltage is simply a pressure driving electric current, not current itself) does not affect cows. But then, their research ignores everything that challenges their foregone conclusions.

Never yet has there been a study that pumped in electricity and electrocuted cows in the same fashion as on the farms, and for good reason. The blowback of protests by PETA and the ASPCA, and the resultant media outcry, would be intolerable. However, a real study would *have to* result in just such high death rates, because if the study didn't kill cows then it wouldn't recreate environmental conditions.

The current studies aren't worth the paper they're printed on. There are no studies, none, that have killed even one cow in pursuit of the truth. The best and most humane way of conducting meaningful research would be to set up experimental apparatus in the field, on the farms experiencing the actual problem.

University experts will argue that the only reason cows did not die during their studies was that they removed sick animals ahead of time. This is a half-truth. We know for certain that it was standard procedure to remove cows whenever they reacted to electricity. This method, or lack of method, says plenty about the universities' desire for grinding axes. Forensic economist Dr. Michael Behr, a universally recognized master in his field, analyzed two studies done by Cornell University and then demolished them in his book, *Stray Voltage Research Fraud.* In a noteworthy passage (Proprietary Publication, Northfield, Minnesota, Third Edition 1997), Dr. Behr states that:

> Because a disproportionate number of cows normally leave the herd upon the completion of a lactation rather than during the lactation, a rate of removal during a lactation projected to a full year will tend to understate the annual experimental "cull" rate. Therefore the lowest possible estimate of the 365 day "cull" rate is 3.6 X 365/394=3.3 cows in all of the ten slots in each sample. This turnover not only invalidates the experiment as controlled research, but it is so far in excess of feasible farm conditions it renders the Cornell Research results irrelevant even if they were valid—a big if. If no cows left the herd during lactation, the turnover would be 0.0 during the 305 day lactation. Some cows will not make it to the end of a lactation so the farm turnover rate will certainly be higher than 0.0. However, an average of 3.3 cows per cow slot is a "cull" rate of about 230 percent compared to the more usual 30 percent, or at about eight times the normal cow turnover.

In other words, Cornell pulled cows from their study eight times faster than would a typical dairy farm, replacing them at a 230 percent rate rather than a 30 percent rate for a farm. And then, surprise of surprises, Cornell found no health problems associated with stray voltage in its experimental cattle. After all, why would they? For they utilized the techniques of clairvoyants and witch doctors to arrive at magical, preordained conclusions.

Cornell University is by no means alone in its faulty methodology. Most university-funded studies reach the same conclusions, and for the same simple reason: their experts are on the utilities' payroll. It is always the servant who takes money. I will detail the symbiotic relationship between the parties in a subsequent chapter.

However, that's not to say that no cow has ever died at the hands of researchers. For years, the "experts" have been preaching to farmers about how to treat cows absorbing stray current they claim cannot kill cattle. In this connection, a highly interesting article appeared in a Wisconsin newspaper, *The Country Today,* which sheds further light on university research methods. Following is a key excerpt:

> Ten cross-bred beef cows destined to be part of a University of Wisconsin at Madison research project died earlier this year because of apparent neglect, university officials reported June 20. Elton Aberle, dean of the College of Agricultural and Life Sciences at UW-Madison, said the animals died at Franbrook Farm, a university-run farm southwest of New Glarus. On June 20, university officials issued a news release to report the cattle deaths. The release said officials discovered in February that the animals were "excessively thin, evidently the result of not having been fed a nutritionally adequate diet." Mr. Aberle said, "It was our diagnosis that they were just not fed enough and not fed adequately and got too thin. They were neglected during the *care and feeding* of the herd."

The very same people who love accusing dairy farmers of incompetence and herd mismanagement perpetrated this animal abuse. And then lied about it in court and under oath. Their hypocrisy would be laughable were it not obscene and disgusting. The question that springs to mind is, how would the study have differed if the abused cows had been included in the findings? What other studies used such bizarre and unscientific, to say nothing of inhumane, techniques? We'll never get the answers.

Despite faked, doctored research by universities, and massive propaganda efforts by utilities to convince the world that stray current is no problem, dairy farmers know full well that manmade electrical current is gushing from utility wires into the earth of their farms and into their livestock. Because help hasn't been forthcoming, some farmers have gone so far as to cut off all electricity from the utility and generate their own power using a generator, thinking that generating their own electricity is going to solve all their problems. They don't realize that they can't fix it because of a poorly designed distribution system. Current will continue to flow through their property from the utility's distribution system, whether they use a generator or not.

Others simply cut the utility's down grounds. In fury, but with great satisfaction, I personally cut eighteen of them, and my cows never had it so good as during the year-and-a-half interim between the cuts and wire restoration by WPS. The animals made a remarkable recovery. Milk production skyrocketed to a record high of 85 pounds per cow per day. Still other farmers drained ponds, pools and manure pits, anything with heavy liquid or moisture that was a conductor of electricity. Some were able to reduce current and get a herd response; other farms were simply destroyed.

The second noteworthy symptom of stray current in cows is a high somatic cell count (SCC). This is the amount of white blood cells recruited by an animal's body to battle various types of infection. The higher the count, the greater the infection. SCCs are used to determine the level of mastitis in a herd. Mastitis, or udder infection, is a part of life on a dairy farm. That is why farmers take great care to clean and disinfect not only the udder but also the milking equipment. In addition to maintaining cleanliness and good sanitation, farmers have an economic motive for minimizing the problem. Buyers, whether small buyers or giant corporations like Land O' Lakes, Inc., to whom the Allen Farm sold its entire product, pay less for milk with high counts. And milk with counts above a certain level is not permitted to be sold at all.

In a paper entitled "Evaluating Herd Milk Quality Using DHI Somatic Cell Counts" presented to the Arizona Dairy Production Conference in 2002, Robert R. Peters, professor with the Animal and Avian Sciences Department at the University of Maryland, gave this succinct description of mastitis and its causes:

> Mastitis is an inflammation of the mammary gland. The source of that inflammation is usually microorganisms

> that gain access to the mammary gland through the orifice of the teat end. Microorganisms that cause disease are called pathogens. In mastitis control programs, these organisms are categorized as either contagious or environmental pathogens. Like any disease caused by bacterial infection, mastitis is a numbers game. Greater bacterial exposure translates to a greater rate of infection and more mastitis. To date, most of the progress in controlling mastitis has resulted from reducing exposure of cow's teats to pathogens, and from the widespread use of antibiotics…Controlling bacterial infection also involves the mammary gland's resistance to infection. Two examples of steps which may be taken to improve resistance to infection are supplementing the cow's diet with the micronutrients or minerals and vaccination against coliform mastitis.

The new feed mixer I purchased assured balanced, high-quality nutrition for every cow. We had them all vaccinated and regularly disinfected, but the high counts did not diminish. During frequent exams, our veterinarian was unable to detect the presence of disease. In his paper quoted above, Professor Peters goes on to make the following critical point:

> There is a long list of factors other than intramammary infection that can be blamed for deteriorating udder health. Studies from the scientific literature support the idea that several factors are associated with changes in herd SCC. Such things as the age of the cow…stress…stray voltage…may all be related to SCC changes…

Stray current—and this is a very important point—does not of itself cause infection or mastitis. Like AIDS, it attacks and weakens an animal's immunity to disease, causing a cow to become susceptible to bacterial and/or viral infections, which an unimpaired immune system would ward off. Therefore, a statement in the USDA handbook, the "Red Book," that electrical current cannot directly cause infections, while technically accurate, is misleading. For example, a human being who we say is "dying of AIDS" is actually dying from a common cold or from a bladder infection or from tonsillitis or from staph; it is not the AIDS virus that will directly cause death but the prior destruction of the person's immune system and ability to fight infection. On dairy farms, stray current functions like an AIDS virus attacking livestock.

Most utility companies still use the USDA Handbook 696, the "Red Book," as their Bible regarding stray current. When WPS tested at my farm, they employed a 500 ohm resistor (named for German physicist Georg Simon Ohm, discoverer of relationships between strength of electric current, electromotive force and circuit resistance—Ohm's law,) with the assumption that this approximates a cow's bodily resistance to current. But how do they know? Have some cows told them this? They "know" because Cornell University professors Aneshansley and Gorewit make this easy, convenient assertion in the "Red Book" (Chapter 5). And this pair is on the payroll, earning large dollars from utilities as professional witnesses in stray voltage court cases. In Chapter 3, they state:

> "While some pathway resistances approach 1,000 ohms or more, worst-case resistances of specific cows on specific farms may be as low as 500 ohms." It is possible that there could be "specific cows" on "specific farms" not evincing resistances under 500 ohms. But where they reside is anybody's guess, since the authors provide no particulars despite their assertion. Even though 500 has become a magic number for the utilities, the truth is that the lowest known, tested resistance in a cow is 244 ohms (Appleman and Gustafson, "Source of Stray Voltage and Effect on Cow Health and Performance," article in *Journal of Dairy Science*, 1985).

The weakest points of a cow's resistance to electric current are mouth to hoof (drinking) and teat to teat (milking). This means that if stray current is present, electricity is permeating cows on a near-constant basis. After prolonged suffering, death by electrocution is the result. Larry Neubauer recently did a resistance test on over 1,000 cows, he averaged 227 ohms of resistance.

Full-blown tests have yet to be performed whereby electricity is transmitted from a milking unit directly to a cow, the teat-to-hoof pathway (except for an extremely limited one by Professor Gorewit in 1988). According to Aneshansley and Gorewit, such a study would not be worthwhile because, in their words, "It is doubtful that the milking machine plays a role...The voltage necessary to override the resistance of the milking path would be in excess of neutral-to-earth voltages."

Fortunately, we now have solid evidence that these two eminent professors are one hundred percent, absolutely...wrong. In 1981, a man named Roger Ackerman, along with Professor Martin H. Graham, of the University of California Berkeley, and inventor of the guidance system for the Patriot Missile, patented a device which Ackerman calls (See Appendix A):

> A central milking system with electrical units energized by an a.c. voltage source having an earth grounded neutral conductor. Stray current inherently produced in the neutral conductor by neutral-to-earth voltage is prevented from being conducted to the teats of the cow being milked via the flow of milk from the cow by creating a highly electric-insulating path between all bodies or films of milk which extend to the cow and all conductive components which are connected to the neutral conductor.

Graham's contribution to the project, in his words, was:

> A voltage compensation circuit...coupled to a conventional milking system so that when a milk stream flows from a cow to a milk pipeline, a voltage compensation circuit makes contact with the milk stream. As a result, due to the voltage compensation circuit's output voltage, a current that might be generated by the electric potential difference between the cow and the milk pipeline is prevented from flowing within the milk stream. Accordingly, electric shocks to the cow are effectively prevented.(Graham, M.H.(2001). **Voltage compensation circuit and method for reducing electric shocks to teats of an animal during its milking.** *Official Gazette of the United States Patent and Trademark Office Patents* 1246(4): Pagination May 22, 2001, Patent Number: US 6234108, ISSN: 00981133. NAL Call No.: T223 A21)

Such were some of the desperate, if brilliant, responses to the utility's saturation of the earth with electricity, necessity once again being the mother of invention. The sad fact is that none of this—not the inventions nor the cutting of down grounds nor changing feed nor buying new milking parlors nor replacing equipment—was necessary at all. The true necessity was for WPS to quit using the earth of my farm, and others, as a return path for their electrical current. And that is exactly what their research has shown to be unnecessary.

Forty-nine of the fifty states employ this exact distribution system, all except California. Unsurprisingly, California has no problems with deadly utility earth current on its dairy farms, unless there is a serious on-farm problem. Not coincidentally, California has overtaken Wisconsin as the number-one dairy-producing state in the nation. These facts alone should prompt immediate intervention by both state and federal agricultural and environmental regulatory and oversight agencies, to say nothing of the USDA; however, the truth is that more than a few politicians owe their elections to hefty bribery contributions from the utilities. Once again I repeat: it is the servant who accepts money. Later on we'll examine that relationship more closely.

Back in 1975, just prior to WPS' reconfiguration of the system, our farm was one of the top dairy producers in the state, and Wisconsin in that year was number one in the country. Following the redesign, for the next twenty-five years or so, I had to sell and ship out to other, safer farms most of my cattle, and today I own not a single cow. I also had to sell off all of my equipment, including auctioning off the beautiful, state-of-the-art milking parlor. When they came and tore it out, it felt like they were tearing out my own innards, heart, liver, lungs. Nor do I or any of my family live on the farm today—the stray current is far too dangerous—although I still own the property, 375 acres, and still raise crops. So that's something.

I mentioned that California is now the number-one milk-producing state in the US, and is prohibited by law from having an electrical distribution system like Wisconsin's. The law was passed in 1994. Although a couple of areas have been grandfathered in, there are no systems like Wisconsin's distribution system in California's dairy areas. The following table from the USDA shows the correlation between the absence of stray current and high milk production in various regions:

Steadily falling production rates in the Midwestern dairy belt are quite noticeable here. Wisconsin and Minnesota, which used to consistently outproduce all the other states, have had the highest amount of stray current court cases over the past thirty years, as well as cases currently on dockets. As Midwestern productivity has declined, it can be seen that the far West has accelerated.

Regional Shares of Milk Production, Selected Years

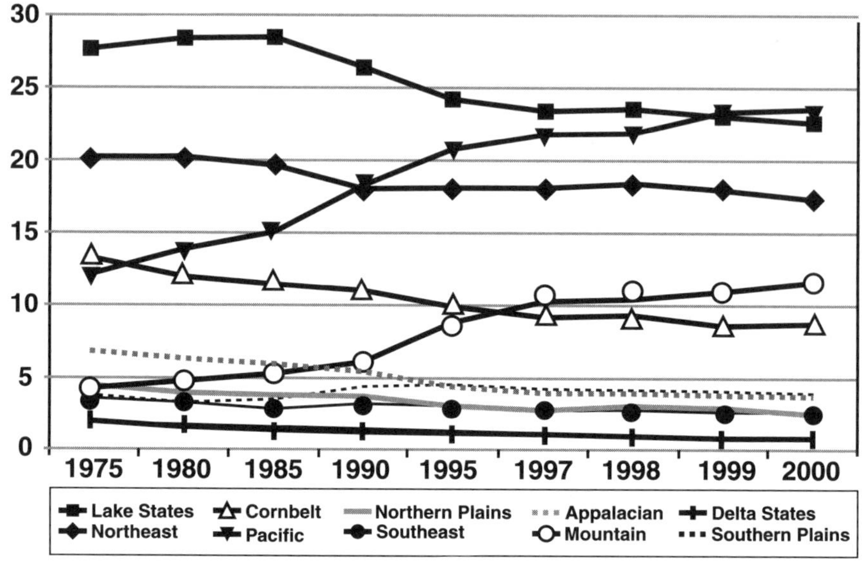

Having detailed symptoms of stray current, they are probably best summarized by my friend, veterinarian Dr. Richard Schulte, one of the most knowledgeable, skilled, upright and honorable men I've ever known. A graduate of the University of Illinois, Dr. Schulte, who is a dairy farmer himself, has practiced veterinary medicine on dairy farms in half a dozen states since 1971. He developed a specialty in embryo transfer and pregnancy work and formed a company named Modified Genetics. He has been a consultant to Brown Packing Company of Chicago developing vaccination protocols. For nine years, he was on the board of directors of the Central Wisconsin State Fair, for whom he did veterinary work, and taught herd health management at the Mid-State Technical Institute. He is a member of the American Veterinary Association, the Wisconsin Veterinary Association and the Association of Bovine Practitioners. He has written extensively on many veterinary topics, and worked with cattle herds ranging in size from six cows to six hundred, in addition to farm animals of every variety, including horses. I've listed his various posts and qualifications because he is an important figure in my story, and one we will hear from at length in several courtroom dramas.

Dr. Schulte categorizes stray current symptoms in the following ways:

SYMPTOMOTOLOGY OF STRAY VOLTAGE

The Journal of Dairy Science published a paper written by R.D. Appleman and R.J. Gustafson in 1985 entitled, "The Source of Stray Voltage and Effect on Cow Health and Performance." In the article it states the "Commonly cited cattle responses include: 1) Intermittent periods of poor production; 2) Unexplained poor production; 3) Increased incidence of mastitis; 4) Elevated somatic cell counts; 5) Increased milking times; 6) Incomplete milk letdown; 7) Extreme nervousness while in the parlor; 8) Reluctance to enter the milking parlor; 9) Rapid exit from the parlor; 10) Reluctance to use water bowls and metallic feeders; 11) Altered consummatory behavior (lapping from the watering device)."

The Texas Agri-Business Electric Council, known as TABEC, distributed a manual entitled "Stray Voltage in Dairies." This manual was produced at the Texas A&M University's Department of Agricultural Engineering and was incorporated into the Idaho Power Company's "Guidelines for the Investigation and Resolution of Stray Voltage Problems." In the TABEC manual it lists the symptoms of stray voltage as 1) Uneven milkout; 2) Cows extremely nervous in the parlor; 3) Cows reluctant to enter the parlor; 4) Increased mastitis (they do not blame stray voltage for the mastitis directly but on the uneven milk-outs. They report a failure of treatments in the udder due to dilution of the medicament by the excess milk in the udder); 5) Reduced feed intake in the parlor; 6) Reluctance to drink water; 7) Increased manure deposition in the parlor; 8) Lowered milk production.

In a mailer to customers in the late 1990s, Wisconsin Public Service Corporation (one of the largest suppliers of electricity in Wisconsin) listed the symptoms of stray voltage as: 1) Nervous cows; 2) Reduced feed and water intakes; 3) Increased incidence of mastitis; 4) Higher somatic cell counts; 5) decreased milk production. The article also states that "Some studies indicate that, under

> certain circumstances, relatively low levels of stray voltage, below those which would ordinarily be felt by people, can have an adverse effect on animals with continued exposure."
>
> To the aforementioned symptomotology we must add: 1) Poor hair coats; 2) Poor reproductive performance; 3) Poor feet and legs; 4) Poor response to treatments for infectious and metabolic diseases; 5) Difficulty in raising calves; 6) Poor weight gains; 7) Abnormal posturing in stanchions and tie stalls; 8) Higher than expected death rates; 9) Higher than normal cull rates; 10) Depressed immune systems; 11) Higher than expected incidence of abortions. These symptoms are the ones commonly reported by veterinarians who work with herds impacted by stray voltage. Where mitigation is achieved, a lot of the symptoms disappear.

Set against soothsayer-professors like Aneshansley, Gorewit and Reinemann, all are university flacks. Cook is from the Public Service Commission of Wisconsin. Then there are the real experts who we will soon encounter: Dick Schulte, along with Scott Lawrence and Barry Hammarback. Finally, there is Don Zipse, who we will also meet; he towers over the university flacks like Gulliver above the Lilliputians.

Making use of part of the "research" by the professors, WPS tried to rob me of my day in court by declaring, first, that stray current did not exist, and then by claiming that the seven-year statute of limitations on stray voltage suits had run out. But if, as they asserted, there is no such thing as stray current, why try to muzzle me with the statute? To properly appreciate these issues, we must peer briefly into the nature of electricity. Which, by the way, was the force behind creation of the Frankenstein monster.

Chapter Four

SECRET OF THE GODS

In writing about electricity and how it destroyed us, I have a mental picture of the god Zeus in a loincloth, bearded and tall as a skyscraper, perched on his mountaintop; hurling down lightning bolts at my farm like flaming javelins.

Had that happened, it couldn't have been worse than what actually occurred. In fact, it might have been better, for we would have seen the fiery bolts raining down and perhaps dodged them. As it stood, we saw and heard nothing and only comprehended the problem through its deadly results. No wonder, then, that the unseen mysteries of electricity, of fire, of magnetism, of lightning, of the aurora, were regarded in ancient times as god-like secrets, obtained by humans through theft (Prometheus) or luck (warrior gods left it lying on a battlefield where it was found by mortals).

Contrary to the idea that an understanding of the nature of electricity is recent, originating with Benjamin Franklin and his famous kite, it's startling to discover that people have been experimenting with electrical power in one way or another since at least a thousand years before the birth of Christ.

According to several legends, in 900 BC, a Greek shepherd named Magnus walked across a field of black stones that pulled the iron nails out of his sandals. The area around the field became known as Magnesia. Much later, in the year 1269, an Italian named Petrus Peregrinus discovered that round lodestones aligned needles with lines of longitude pointing between two pole positions on the stone.

Rene Descartes, the famous French philosopher, in 1638 conceived the theory that light is a pressure wave. In a major experiment, Stephen Gray in 1729 proved that electricity need not be limited to a stationary spot (as by rubbing and friction) but is transferable from point to point by means of conducting wire. The aforementioned Ben Franklin propounded, in the year 1747, the theory of one-fluid electricity, proposing the idea of conservation of charge and positive current flow.

1800—Humphrey Davy invents a battery. 1825—Ampere publishes results of his experiments with magnetism. 1841—Joule proves that energy is conserved in electrical circuits involving current flow and chemical transformations. 1876—Rowland's experiment proves that moving electric charge is identical to current. 1879—Edison re-invents the light bulb using lower current electricity and a small carbonized filament. 1895—Roentgen discovers X-rays. 1905—Einstein publishes the general theory of relativity. 1915 (circa)—concept of quantum mechanics is born. 1970—string theory is comprehensively developed.

I mention these pioneers and inventors as another way of pointing out that the accumulated knowledge in the field of electricity is such that American utilities possess abundant means of safeguarding the transmission of current so it does not harm living things. We get all our electricity, which is a secondary rather than a primary power source, from converting other energy sources such as oil, nuclear power, gas or coal. In the past, towns were frequently built near waterfalls, which drove wheels and provided mechanical energy. Later, as technology improved, hydroelectric power plants went into operation.

An electric generator is a machine that converts mechanical energy into electrical power, based on the relationship between magnetism and electricity. When a wire moves across a magnetic field, an electric current runs through the wire. The giant generators that utilities use contain stationary conductors. A magnet attached to the end of a rotating shaft is positioned inside a stationary conducting ring that is wrapped with a long piece of wire. As the magnet rotates, it creates an electric current (current is simply a continuous spark) in each section of wire as it passes. Every section is itself a separate electrical conductor. All the small currents running through the individual sections make one big current. This big current is then used to generate electrical power.

A utility's power station employs either a turbine engine, water wheel or some other type of device to drive an electric generator that converts mechanical energy into electricity. Steam turbines, wind and water turbines, and gasoline engines are commonly used in the process.

What electricity boils down to is this: it's a form of energy involving the flow of particles called electrons. Electrons (hence, electricity) are subunits of atoms, of which all matter in the universe is composed. An atom has a center, a nucleus, which contains positively charged particles called protons, uncharged particles called neutrons, and negatively charged particles called electrons. There is balance in that the number of electrons is normally equal to the number of protons. If the balance is disturbed, an atom can add or lose an electron. Once electrons are lost, their free, unrestricted movement is what creates an electrical current.

At this point, things get interesting. Because if some outside force, like a dynamo or generator, disturbs the equilibrium and electrons are lost, the loss is only temporary. The lost electrons must return either to the original atom or to another in the same system. This constitutes a closed circuit, which permits current to flow. As an example, if a copper wire were attached to a battery terminal on one end and to a bulb on the other end, no power could be generated because there would be no circuit. Yet, remove the wire from the bulb and attach it to the other battery terminal and the wire will become hot to the touch: a circuit has been completed and electrons are flowing between terminals via the copper conducting wire. Current has returned to the source that created the imbalance.

What's often called "earth current" occurs spontaneously in nature, due to the collision of the solar winds with earth's atmosphere. "Ground current," or neutral wire-to-earth current, is made when a utility's return wire to grounding wires creates a secondary circuit to earth."

Earth current is normal, natural; neutral return current should go back to a utility's substation on a wire. However, utilities instead prefer using the earth as much as possible for one simple reason: it's far cheaper. This return current is what surges through my property to this very day, wounding and killing all creatures great and small, both animal and human. This is why I've gotten rid of my herd, and why I no longer live in the house my father built: the danger is too great. Although where I am living now may not be any less dangerous; Howard, a little town connected to Green Bay, has a lot of bare current carrying conductors buried in the ground. When I turn the power off at peak time, which is usually between six and seven o'clock, morning or evening, I get around one amp of current on the water pipes in the basement.

The Public Service Commission of Wisconsin, that fount of un-wisdom, uses the following definition to assist utilities in fighting off lawsuits by enraged farmers, who I am pleased to call (since I am one, and proud of it) modern-day peasants with pitchforks. In a paper entitled *Stray Voltage History and Current Situation for Wisconsin*, this is how the commission first confuses and then hides the truth:

> Stray voltage is defined as a natural phenomenon that can be found at low levels between two contact points in any animal confinement area where electricity is grounded.

> Electrical systems—including farm systems and utility distribution systems—must be grounded to the earth by code to ensure continuous safety and reliability. Inevitably, some current flows through the earth at each point where the electrical system is grounded and a small voltage develops. This voltage is called neutral-to-earth voltage (NEV). When a portion of this NEV is measured between two objects that may be simultaneously contacted by an animal, it is frequently called stray voltage…

This gibberish purports to show that stray voltage, a "natural phenomenon" found in an animal confinement area, is fine; calling it a natural phenomenon is where I have a problem. It's completely unnatural. What language they always employ! Never do they call things by their correct names. Maybe they hope jargon will obscure the way electricity is grounded.

Chapter Five
OUT FOR JUSTICE

The night before the trial opened, I didn't sleep a wink. I was tossing and turning while dreading the results of defeat: paying court costs including attorney and expert witness fees, which would send me to debtor's prison for a thousand years. I'd lose my home and freedom.

For a few hours, I toyed with the idea of phoning my lawyers and ordering them to drop the case. But the morning arrived and I didn't call. I dressed in a rather handsome suit and tie, gulped down some black coffee and drove grimly over to Green Bay to meet my fate. Never having seen a trial except on *Perry Mason*-type TV shows, I had no clear idea of what to expect. That unknown scared me almost as much as the possibility of losing.

The trial took place in the historic Brown County district courthouse in Green Bay, the Honorable J.D. McKay, Circuit Judge, presiding. Designed by architect Charles E. Bell of Minneapolis, who also designed the state capitols of South Dakota and Montana, the courthouse is constructed of Marquette raindrop stone, with polished terrazzo floors and four large entrances surmounted by two-story ionic columns. Above the roof is a copper-covered dome, clock and bell tower. Portraits and wide, colorful murals adorn the walls, one of which depicts the historic landing of the explorer Jean Nicolet in 1634. The courtroom where the Circuit Court met is one of the smallest of the building's various chambers. Judge McKay's bench, the wood backdrop, witness stand and spectator seating are built of solid oak and are original vintage (1911). Beautifully and brightly lit by gold chandeliers, the Brown County Courthouse is listed in the National Register of Historic Places. Overall, the impression I had was of The Law's grandeur and overwhelming power.

In my suit for damages ($13 million), I had claimed three things: one, that WPS was negligent in providing electric power; two, that WPS created an unnecessary nuisance through that negligence; and three, that the utility acted willfully and recklessly by failing to provide adequate service to my farm. Employing a battery of lawyers headed by their best, the tall, thin, nattily dressed, perpetually nervous-looking Trevor Will of Foley and Lardner, the largest law firm in Wisconsin, WPS had brought in the heavy artillery to pulverize me. They could not tolerate losing for fear of setting off a chain reaction of tobacco-like suits. Because I owned one of the largest farms and was doggedly persistent in pressing my claims, refusing to be put off or discouraged by their stall tactics, the utility was hell-bent on making an example of me as a warning and a lesson to other farmers who might get a notion to mess with WPS. Barry and Scott represented me, along with Dan Murray. Dan would organize the presentation and act as a resource for our team, while Scott and Barry handled the courtroom work and the witnesses. Barry, a lean, handsome six-footer in his mid-fifties with thick, salt-and-pepper hair, was a partner in Hammarback, Murray and Jacobson of River Falls.

Scott Lawrence—beefy, mustached and congenial, a master of the English language with over twenty years' experience in stray voltage cases— had short brown hair and a homey, folksy way about him that put people at ease (and disguised his brilliance). Fifty-five years old, the son of a history professor, he had a degree in math as well as law. He had also studied electrical engineering at the University of Wisconsin at Madison. Scott would prove to be the bane of WPS, garlic to its vampire, and a scourge to their expert witnesses. He was with Lawrence and Des Rocher of St. Nazianz.

I was jittery and tense as the proceedings got under way. Seen objectively, a loss for WPS would be little skin off its nose. Sure, they would receive bad press and lose money. But the negative publicity surrounding the case would quickly dissipate and they would simply raise rates to recoup their payout. In a real sense, I and the other victims of stray current were subsidizing the enemy. WPS would go on without missing a beat. But if I lost, the way of life I inherited from my parents would vanish forever. So I was staking everything on this one throw of the dice. I would not have dared to take on such a powerful entity, though, if I were not confident in my facts, assertions, witnesses, attorneys, and, most important of all, in the justice of my fight.

Still and all, I felt uneasy as I leaned forward in my most uncomfortable of spindle-backed chairs on the first day of trial, toe-tapping against the table's metal foot and concentrating on the Court's instructions. Judge J.D. McKay, sixtyish, tall and husky as a pro wrestler, his jowls clean shaven and the iron gray of his hair matching a stolid demeanor, was asking prospective jurors a series of questions. I nearly laughed out loud when he inquired of one red-faced fellow, "Can you be impartial?"

"No, absolutely not," the man asserted, sliding low in his seat.

"No?" McKay growled, as if he might bite into someone who showed so little regard for the jury system. "And why not?"

"Because my father-in-law lost thirty-five cows to stray voltage."

The judge grunted and nodded. This was good enough even for him.

"All right, you're dismissed."

A little later, I was pleased to hear a woman express prejudice against WPS for shutting off her power while she had a daughter in the house on life support. Thankfully, the daughter had not been harmed, but the episode prompted some understated irony.

"Would that bias or prejudice carry over to this case?" McKay inquired.

"I think it might."

"You're excused," the judge said.

As might be expected in a state like Wisconsin, many in the jury pool had family who were, or had been, or knew people in, the dairy business. McKay dismissed another juror because she had filed a stray voltage lawsuit against WPS, and Scott Lawrence had been her attorney. Several others were dismissed because they had relatives employed by WPS. Then, of course, there were the usual suspects who simply wanted off, especially since the judge had said he expected the trial to last about one month. Some of the excuses were priceless. One man asked out so he could go bowling in a tournament. Another claimed his mind would be muddled if he were forced to serve, since he would worry nonstop about his job. Another asked to be excused to coach his son's Little League team. Still another begged off so she could schedule dental appointments in advance.

For his part, Trevor Will used his challenges to remove jurors who indicated even slightly that they just might possibly perhaps be sympathetic to the idea of stray current. Through clever questioning, Will got jurors to say that they knew people who knew people who knew other people who claimed to have been harmed by electricity. In this way, he obliged McKay to do some of his work for him by dismissing those folks, thereby allowing Will to save precious challenges. Barry employed similar tactics for our side.

Will, though fidgety and constantly fiddling with his pen, was very polished and skillful. After mentioning that stray voltage was really no more than a theory, he shrewdly, almost offhandedly, asked the jurors, "How many of you carry a lucky charm?" More than a few hands went up. Then, "Anybody here believe in UFOs?" The planted idea was that a belief in stray voltage was pure superstition, equivalent to a belief in aliens. I admired the man's skill even as I detested him for it. But by the time the jury was finally seated, I felt a bit calmer. For despite Will's best efforts, the underlying assumption among most of the jurors seemed to be a sense that stray voltage was real enough. This boded well. All things considered, I was content with the final composition of the jury.

Barry's opening statement on that sunny May morning penetrated to the heart of the matter: What our farm was like, the way it had evolved over the years, the deep meaning it held for us, and how the Grim Reaper had come calling one day with his electric death-sickle. I kept wondering what tricks Mr. Trevor Will had up his sleeve, what aspersions he would cast, what tall tales he would tell when it was time to pluck the strings of *his* banjo. I did not have long to wait to find out. But Barry's chords were music to my ears.

"Ladies and gentlemen, good afternoon," he said, looking from juror to juror. "We all know that electricity in the wrong place can injure and kill, and this case is about how electricity from WPS Corporation's utility lines came onto and is coming onto the property and is causing problems to their animals, causing early deaths in their animals and substantially reducing the milk production on that farm."

He pointed to his right at a photograph of our family farm, nestled atop the hill southeast of town, beyond the TV station's antennas on Little Apple Road. The picture could do little to describe the beauty of the farm, its unique smells and sounds, the sense of peace I felt when I arose and started my day in the morning. A peace that the power company shattered, releasing an invisible menace that stalked the property, more lethal than coyotes or wild dogs.

"It's a fairly large farm," Hammarback said. "It's got capacity now of about 400 milking cows and about another hundred dry cows, so about a 500-cow capacity up there. It used to be a big farm. It's not so big anymore, as you'll hear. Dairy farms are getting bigger and bigger and smaller dairies are going away. As time has gone by, the evidence in this case will show that the Allens have done and are doing everything they can to keep up with the modern trends.

"They started out with 32 cows, as I recall, and for a while actually milked them by hand. This was before World War II. In fact, when they first started, I don't think they even had electricity out at their farm, but it wasn't very long before—it was in the, I think, early 1940s electricity came up the hill and came out to the Allen farm and of course once it got there, they started using electricity for the things you do on the farm, for the milking equipment mostly. There are 11 children all together. It started out with Larry, who I think was born nine months and one day after they were married, at least according to us, then Leslie. Larry was born in 1939. Leslie, Jim, Cletus, Bonnie, Bob, Kevin and then came Russ, then Josie, Yvonne and Geri Lynn, the last one, who was born in 1959."

At the mention of my beautiful sisters, Josie and Geri Lynn, who had died by their own hand, I nearly broke down. To me, it was WPS' stray current that killed them. If the cows could only stand so much stray current over time before it took its toll, then my sisters suffered the same, until they had no strength left for life.

No, we never measured how much current was flowing through their bodies. We didn't know it at the time, but we know now that the soil in the neighborhood was heavily electrically polluted. If not for WPS and their willful carelessness, then they'd both still be alive. That was the truth, and nothing would ever change it.

It was up to Hammarback to defend that truth, with his words and photographs. I was helpless.

"Their father Alvin and their mother Alice helped them to learn to milk cows, and how to properly manage and feed animals. Russ actually started when he was about six years old," my lawyer continued. "Now, the way you milk a cow, then and even now, is you clean the cow's teats off. You put the milker unit on. You need to massage it at some point, and you need to get the milk out of the cow, and that's the same today as it was yesterday. Now, of course, there are parlors where the cows come to you instead of you going to the cows out in the stanchions like it used to be, but that's the way they started out. Well, it wasn't very long with the children coming along when the facility they had there got a little bit too small for them. So in 1962 they had one of the last barn raisings here in Brown County. The family got together, and they had big meals spread out in the basement and everybody came over and helped..."

Yes, I still remember that day. People came from all over the state to help us, at least a few hundred friends and relatives. It was one of the last barn-raising bees ever held in Wisconsin. That barn was the featured story in the *Hoard's Dairyman* that Yvonne and Geri Lynn graced the cover when they were girls in '64. We raised the rafters in one day, which were built from lumber cut from our woods. And the food, oh, the food! There were platters and trays galore;every kind of bread—white, black, pumpernickel, rye, hard and soft rolls, sweet rolls, too; and tender roast beef and sliced ham and chicken; every type of fresh Wisconsin cheese; eggs and bacon and sausage;

everything you could think of; and for dessert, apple and pumpkin and pecan and shoo-fly pies; and chocolate cake and cookies and little cream puffs and flaky pastries, all homemade by Ma and my sisters as well as aunts and cousins and dozens of other well-wishers. An unforgettable camaraderie and bond formed between us that was a joy to recall, the kind that forms when people work and sweat together for a common goal and a common good.

I was only ten years old. That barn still stands decades later. But it's empty now, dead silent, without cattle. All that's left in there is the flotsam and jetsam from the ripped-out equipment we had to sell when we couldn't save the animals and continue milk production. Now only the cats wander in and out of the hay loft.

Hammarback paced in short circles, drawing the jurors back to the photograph again and again. "The barn that was put up in 1962 would house a hundred animals to be milked," he said. "So they worked their way up from 32 animals in 1937 to 100 by 1962. As time went on, the operation increased in size. Then they had about 230 cows milking toward the end of the '60s, and in 1972 started thinking that, gee, you know, maybe we really need to expand. Maybe they really wanted to have a state-of-the-art facility. So they went all over the place. They went to extension services. They went to different fairs to see people. They went to WPS Corporation to get information on the proper ventilation to use in the barn, and together the family actually built most of the barn. By 1976, late '75, they built what's now called the east barn. It was a free stall type of barn with a parlor in it. Now, as you probably know, a free-stall barn is where the cows have certain areas that they can roam around in on their own. They're not confined to the little stanchions anymore. They have stalls and their own area, and they can go into those stalls. Eventually, they would be able to handle about 500 cows, milking and drying.

"As the cows are being milked, of course, once they have a calf, they start giving milk. That's how this whole process works. The idea is to have a cow have a calf about every 13 months or so, 12 or 13 months, right in that range. By keeping them having calves in a nice regular manner then the milk production stays up. They have a calf. It keeps the cows healthy, and, of course, they reproduce. As these cows get older and are replaced, there is culling. That is that some of the cows that are sick or don't produce quite as much are taken out of the herd and these new cows come in. It takes about two years for a new heifer, as they're called, to come into the milk stream..." Hammarback went on in some detail and I watched as some of the jurors seemed to grow a little restless. The courtroom felt stifling and hot. I wiped the sweat from my brow with the back of my hand. Was May usually this hot?

I knew my lawyer wished to clearly make the point that we were more than competent farmers: Those Holsteins were healthy, and they were big and solid and clear-eyed and beautiful. The farm was well organized and well run. Pa had taught us how to do that, and we could grow. Still, even though it's been my life, the details of farming are not to everyone's taste. I watched one juror close her eyes, and I worried that she'd fallen asleep. Or could it be that she was just concentrating?

I didn't know for sure, but the panic started to set in again. Were we losing the jury? All Hammerback's words and photographs would never be enough to fully paint the picture of the farm we'd built and that WPS nearly took from us. "...When they moved into that barn, about 1975, what they did was consolidate this other herd from GV, move them up there, so they started with about 230 animals up there in about 1974, '75. Now, at the same time that they did that, and unbeknownst to them of what was going to happen, they had three-phase electricity from WPS Corporation brought up to their farm," Hammarback said with a horrified inflection as if he were telling of an army of fiends filing out the gates of hell and marching in ghastly formation onto the Allen Farm. "If you look at this exhibit, I'll point it out to you, it's 1004 C, but it comes down Little Apple Road here which goes to the southeast and the three-phase stops right at the Allen Dairy, and from the Allen Dairy outward, it's only a single-phase line."

Awash in memory, at this point I started to get really worked up. Barry was leaning on one elbow against the railing of the jury box telling my story, and telling it well, when he mentioned electricity. My jaws clenched and I balled my fists under the table, thinking of how my father had died believing the lie that we had screwed up the operation he built.

Barry continued by telling the jury how "troubling" it was that the down grounds from the three-phase transformer on the pole pumped heavy juice into the steel of our barn, tormenting our animals.

But a nervousness began gnawing at me and I grew agitated, feeling Barry had made a big mistake by understating the problem. I shot a look at the jury to see if the truth of what I knew was registering with them: That the torture and death of cows by electricity, and the executions of the Rendering Plant rifleman was beyond "troubling" to the animals. I felt like vaulting over the table to the jury and shouting those very words at them.

As Barry persisted, striding up and down in front of the jurors and speaking more animatedly, I calmed down a bit.

"Back in 1975 and '76 they didn't know about electrical problems. They were just milking their cows. One of the things Russ will tell you is that they always had a problem with cows kicking, and they wouldn't milk right."

Tell them? I couldn't wait to tell them, take over that witness stand and shout it to the high heavens. Wouldn't milk right? We all had the welts and bruises, the black and blue to prove it. Every lost pound of milk was a strip of flesh off my back, and I wasn't getting any younger. Just wait, jurors, wait till I get hold of you. I'll chew your ears off.

"Now a cow needs to be contented," Barry was saying, "and the evidence in this case will show that the animals in the Allen barn were anything but. They would go into the milking parlor and what would happen is they would jump and kick their milkers off into the manure." He then paused.

"You're going to hear common barn talk," he said, somewhat sheepishly. "I don't mean to offend anybody, but that's just how they talk about it. The cows would get covered with manure and the frustration would be amazing."

Beyond amazing. But the real manure, the real shit in which I was up to my eyeballs, was what came from WPS and its university hirelings, who would soon flock to this courtroom to earn their pay by denying reality and standing truth on its head. This whole place was soon going to be overflowing with shit, until it came out of all our ears. But I was ready, and I knew Scott and Barry were. They'd been shoveling Utility shit for twenty years now, cleaning the stables almost single handedly. I was confident that WPS wouldn't be able to slip anything past them.

Barry now switched over to discussing the partnership and pointed to Exhibit 1010, a milk-production chart, something I wanted the jury to become familiar with for what was to come later. He moved his wood pointer across the easel the way a maestro waves his conducting baton, showing them an average of the past twelve months. Production was lower than it should've been. To my satisfaction, Barry focused on the insidious nature of stray electricity, a theme we needed to pound home. Every time the farm made a small gain, we would quickly lose it and more. One step up and two steps back. They were killing us on an installment plan.

Though the day was bright and beautiful, the May sunlight streaming in the courtroom windows, a sadness overwhelmed me as Barry recounted the destruction of the partnership. I kept thinking about my dad and how he died without knowing the truth. And my dead brother.

"His brothers and his brother-in-law went out—one of his brothers passed away."

The current did him in. Ruined his heart. Age forty-four.

Pausing to let this sink in, Barry shook his gray head somberly.

"Russ Allen was the last one there running the farm. His brother Bob still worked for him, works for him today, and Bob's son Scott, who does the feeding, came on as soon as he was old enough. Cletus, one of his brothers, started a farm down the road, dropped out because they could never figure who is causing these cows to get mastitis and why milk production wasn't what it should be. They would argue about it."

Argue? We almost killed each other, I kept thinking, *for no reason.* We destroyed what had always been loving relationships, for no reason. Because a utility decided to use our land as a pathway for its current. *For no reason.*

"The person that was taking care of the hospital barn would have it full and they'd go to the people doing the milking and say what in the world is going on?"

We practically strangled, shot and knifed each other. That was what was going on.

Barry refocused their attention on the production chart, comparing ours to the Wisconsin average. Then he introduced the villain of the piece, told the jury how I'd gotten WPS out to test the milking parlor for stray current.

"Didn't test anywhere else on the farm. Didn't test the line. Didn't change out any split bolts or fix any of the 1940 series wire, and told Russ and his father there was not a problem. Russ believed them, and down the road they went."

The lying bastards. I looked over at WPS' lawyer, Trevor Will. He showed no reaction. Just kept twirling and playing with his ballpoint pen.

Figuring it was something else, we questioned everything, down to the very foundation of the barn. In 1974 we'd installed a slatted floor to eliminate the labor needed to remove the manure. When the cows went, it fell through the slats into a pit below. The top-ceiling ventilation, recommended by WPS, drew air upward and expelled it. At the time, this system was considered state of the art.

"They try to make changes in the facilities, and by 1992 or 1993 things are just not where they should be."

I remember it all as though it were only a moment ago. We did everything we could in every area to improve operations. Searching desperately for answers, I found none, and then, luckily, ran into Mike Hoerth, the dealer who sold milking equipment.

"Mike told Russ, hey, why don't you build a new parlor? You could run all this with one person. It would pay for itself in labor. Besides, have you looked at Patti?"

Patti, our herdsperson and good friend, was black and blue, nearly had her brains kicked out by shocked, unruly cows as she was milking. But she worked hard and long to got the job done. She was and is an excellent herdsperson and a good friend.

But then Barry brought up a painful subject: The death of my sister, Geri Lynn, who had done the milking before Patti. Geri Lynn couldn't bear to go on any longer after her baby boy Jesse, fifteen months old, died of childhood leukemia. To this day I ask myself: Did stray current play a role in his death? I cannot really imagine that it didn't. Despite having other children to care for, over several years Geri Lynn sank deeper and deeper into the quicksand of depression, until she refused to go on.

"Geri Lynn, his sister, did milking," Barry was saying, " but she passed away—actually, it was a suicide..."

They found Geri Lynn's body in a car, a shotgun in her lap. I shuddered in horror at the image; but then another image, a beautiful one, replaced it, of Geri Lynn in her wedding dress, young and vibrant and beautiful. Her sweet voice was calling to me, reminding me of why I was here, strengthening me for this ordeal. *Russ,* she said. *Remember: The reason you're in this place is not simply to get money for the death of our cows and business. It's to get justice for the death of our family harmony, the partnership, the death of our mental and physical health. Remember me,* she said. *And Les. And Jesse. And Josie. And Pa.*

I can only imagine how terrifying it must have been for her that lonely, dark night of the soul. I couldn't forget her agony the day the doctors at Milwaukee Children's Hospital shocked us with the news about baby Jesse.

For 15 months, Geri brought Jesse to work with her, my mother would watch him while Geri was working. Once Jesse got a little older, Geri would bring him outside with her; we ate breakfast together and played with the little guy. We all got so attached to him. The hurt look on my mother's face when Geri drove up without Jesse is one I will never get over.

We were in the hospital room—Geri Lynn, her husband and myself. Leaning against the metal handrail of the bed Jesse lay in, I watched the overhead monitors to which he was wired, listening to the rhythmic beeps, like a clock ticking down the final hours of his life. The room reeked of alcohol and antiseptic. As nurses in white shuttled in and out of the room, we were all sobbing at once. Tears poured down our cheeks as we clung to one another, praying and cursing at the same time, in utter disbelief.

"How could he be perfectly fine yesterday?" I kept stupidly repeating, over and over. Of course, he wasn't fine. We found out about the leukemia that day; he was gone the next. The sickness must have incubated for a quite a while.

Jesse. I pictured him in his little blue overalls and tiny work shoes that Geri Lynn often dressed him in, a bright, multi-colored bib tied around his neck. He was cute as a button, and as sweet as his mother, with a disposition as bright as that bib.

Who can understand the despair she must have felt as she took up a hacksaw and sawed off the barrel of the shotgun? What were her last thoughts? Were any of them about us? Was she frightened? Irrational and crazy? Or methodical and calm? Maybe by then she was focused solely on the Other Side. "Jesse, honey," she might have whispered. "Everything will finally be all right. Mommy is coming."

The voice of Barry Hammarback abruptly interrupted my reverie and brought me back to the here and now.

"After that point in time, Patti Bowers stepped in. They put in a double-twelve, parallel parlor so they could milk the cows from behind. Between the back legs. There's a kick plate up there so the cows can't kick you."

At least that was the idea. But try telling that to Patti. "Frankly, the cows were worse than ever. They'd gone through milking equipment. Housing. Nutrition. More cows get mastitis. They got more and more frustrated."

We were slap-happy by then, like punch-drunk ex-prizefighters.

"Russ called up Mike Hoerth and said, 'take your blankety-blank machine out of here, look what it's doing to my blankety-blank cows. I can't take this anymore.'"

My language was slightly more colorful. I lobbed a dozen or so F-bombs at Mike that exploded like grenades. A decent man who deserved better, Mike came out right away and tested the equipment, finding everything in good shape and functioning properly.

"You should check on stray voltage," Barry recounted.

And I did. Got Larry Neubauer out and on the job. He measured the current at 800 milliamps.

"Mr. Neubauer, he's a rough and scruff character," Barry told the jurors, in case they were expecting to see a man looking like a white-coated engineer instead of a Packers lineman.

"The testimony will be that the ground fault circuit interruptors in bathrooms, when you push the button they trip at six milliamps. At six milliamps, 6/1000 of an amp, a child cannot let go under some circumstances. That's the strength of that shock.

"The testimony will be that utilities in Wisconsin can't contribute more than one milliamp of current in cow contact areas since about 1989. So what we're looking at then is 800 milliamps or 800 times the contribution. What happened is Mr. Neubauer disconnected the connection to WPS' transformer, called the neutral, and this 800 milliamps went away. He recommended an isolation transformer that keeps the neutral from bringing current onto the farm."

In April of 1997, I called WPS and had them come out. Supposedly they tested extensively. They showed me a bunch of scribbled numbers, some sort of computations, and said I had no problem and have a nice day.

"Mr. Allen," Barry informed everyone, "looked at him and said, 'Well, I sure do have a problem.' One of the people from WPS, Mr. Vern Peterson, said, 'We can offer you a neutral isolator, same thing as this transformer.'"

Strange that they would offer me a solution to a problem they claimed was nonexistent. This Peterson, by the way, an inveterate liar who would figure prominently in the trial, eventually got his just desserts from Scott Lawrence. He probably still sees Scott in his nightmares. It was almost worth suing in court just to watch Scott dissected him, peeling back one lie after another like layers of an onion.

I remember standing in the front yard with my brother, Bob, that sultry, overcast July afternoon, discussing the situation. I could hear the motor of the tractor in the distance, and the deep grumbling of cows coming from the barn. This was in '97. I relayed my conversation with Neubauer to Bob, whom I told emphatically: "Current *has to be* what's affecting the cows."

He nodded slowly and spat. "What in the hell are we waiting for? Let's fix the problem."

So, WPS installed the Dairyland Isolator. The weather forecast for the upcoming weekend called for extreme heat and humidity, which meant that loads on the electrical system would be immense. Sure enough, the cows went wild. Larry came out and put a meter on the pole, demonstrating that the Isolator was closing—shutting down—under the stress of peak usage.

Desperate for a solution, I decided to buy the isolation transformer Larry Neubauer had described. Problem was, no one would lend me $58,000, its cost. Lenders realized that the farm was failing and did not believe the problem could be corrected. I tried seven banks, to no avail. I could understand their concerns. Even Farm Credit, my primary lender, denied my application. The county agricultural agent did his best to help secure a loan, but even he, with his many connections, was unsuccessful.

"You're screwed," he admitted to me finally. "If your primary lender won't provide a loan, no one will."

With nothing to lose, I decided to request a hearing before the appeals board at Farm Credit, hoping that a personal plea might move them to reverse their decision. Though short, it was one of the most humiliating experiences of my life.

Entering a long rectangular conference room with a long table in the center, I was greeted by five individuals sitting together at one side of the table—three men and two women. They motioned me to sit facing them on the opposite side. Courteous but abrupt, they gave the impression of a tribunal that had already made up its mind and did not want to engage in conversation.

Nervous and emotional from the outset, I began by blurting out, "I'm stuck with a high cull rate and low milk production, as you know. But the problem is fixable!" I felt my lip begin to quiver and tears welling up. "I can't watch my pitiful cows suffer and die any longer. I can't stand watching the sick ones shot in the head by the Rendering Plant. God only knows to what extent the stray current is hurting my family and employees."

I told them about Jesse and recounted the failure with the Dairyland Isolator. "I need something that won't open and close," I said. "Once that's in place, my production will increase, I'll make money and pay off the loan. But I need your help. I've got to have the unit." I felt tears on my cheeks and wiped them away with my finger.

No one said anything. They neither offered me a tissue nor a cup of coffee nor a five-minute break. They talked among themselves for a bit, then asked me simple questions about the farm and the problem—too simple, I thought.

Finally, the tall guy with the gray hair and the face of a wooden Indian, who was evidently a boss of some kind, said, "Since we do not see a sufficient amount of collateral to issue the loan you've requested, we encourage you to consider selling part of your property—sections or land parcels—to obtain funding. We also suggest you make other concessions allowing Farm Credit Services to reconsider your application. These could be a combination of the liquidation of existing assets and/or concessions by other creditors to accomplish correction of the stray voltage problem. As the matter currently stands, we have no choice but to uphold the original decision denying the application."

So, that was that. What I said to them hadn't mattered one bit. After listening to some more useless suggestions and explanations, I slunk out of there with my tail between my legs, glad at least that it was over.

By the time I got home, I was angry. I said bitterly to Bob and Scott, "I should be *buying* land for expansion, not selling. If isolation worked, I could. This is asinine."

Then, for the first time in a long while, my luck changed. On a crisp, clear, late September afternoon, I was out in the fields, chopping corn when my cell phone rang. It was Don Adams from M and I Bank. He heard about my predicament and came out to the farm to see for himself.

"Russ," he said, "you've got a terrific place here. I want to see you up and running. We can use the farm equipment for collateral." Let me spell out this man's name in letters a mile high: DON ADAMS. I won't forget his kindness and humanity.

Barry continued my story. "So, in December of 1997, Larry Neubauer finished putting in this isolation transformer." He went over to the easel.

"If you look at the production graph here, Exhibit 1010, what happened—we're looking at these dates—there's December 1997, the cows were producing, on a monthly rate, 15,000 pounds of milk. By the end of 1999, the cows were producing almost 27,000...Cow behavior became much, much better in the parlor. Things have been better, but still production comes and goes."

To a degree, the severity of the current is a function of weather, of hard, dry or frozen ground, which doesn't conduct electricity nearly as well as soft, wet earth. That's why when the spring thaws arrived, resistance to the current was much less; it passed easily through the ground and into the metal structures of the barn. The herd became miserable and far less productive.

Barry moved from the exhibit past the judge and slowly crossed the room back over to the jury box. "The testimony will be that there is *still* current from WPS on that line that they've asked many times to have removed. WPS does nothing. Except tell them it's not a problem.

"Now, in order to understand what was going on electrically, I hope you'll bear with me. I want to talk about electricity."

I leaned forward with my elbows on the table, hands under my chin, and studied the jury. Most seemed attentive, except for a gray-haired, elderly woman who kept fidgeting in her seat. I hoped this was because the hard seat was making her stiff, rather than because of Barry's remarks.

"The basic concept of electricity," he told them, "is that it makes a circuit, a circle. It's got to make a whole circuit to work."

WPS, we knew, was going to try to bury the jury alive under an avalanche of technical detail, impress and trick them with sheer volume of information. So Barry did a smart thing here by launching a preemptive strike, explaining the problem in simple every day terms that lay people would understand.

"So what happens is if I've got one amp here and that goes to two amps and I've got one ohm of resistance here and that's two amps then Ohms law says two amps times one ohm gives us two volts. As the amperage doubles, the voltage doubles also. What happens, for example, if I've got instead of one amp on this line, I've got 100 amps? If that current path is not properly maintained, then I can have 100 volts on that line."

I noticed that Barry was speaking very slowly and carefully, which was a good thing since I wanted the jury to hear the truth from us, before Trevor Will got up there trying to twist the truth into knots.

"So, it's very important that the return path be properly maintained and that wire size be big enough to keep that resistance down. Now, on a grounded type of system that WPS uses out there, the Wisconsin electric code requires grounding except if it's causing objectionable current; then, the grounds can be removed. There are grounds all along here."

He then talked about the Dairyland Isolator I'd installed, really an "on/off" switch which prevented current from flowing, but which would close when heavy loads were placed on the system, like during the summer heat when air conditioners and humidifiers were running full blast, day and night.

"Toward the fall of 1997, Larry Neubauer came out with an oscilloscope, which WPS had not ever used on the farm, and said hey, this isolator is closing. Not working."

WPS then conducted a series of improper tests. When we told them about it, to my amazement they simply packed their gear and the left the farm.

"The next thing that happened is Russ got funding, had Larry Neubauer put in this isolation transformer. An isolation transformer is like the one I've explained to you—the Dairyland Isolator—except there's no step down."

Unlike the Dairyland, this was supposed to restrict the current altogether.

"In the barn at the Allens', they have what's called an equipotential plane. An equipotential plane is an idea that came up some years ago in Wisconsin, that they were measuring voltages in the barn or parlor, and so, if we put everything on the same electrical plane, there won't be any voltage. There won't be any current through the cow. It seemed a good idea, like the bird sitting on a wire who is safe there."

I was starting to get nervous again. This was critical information, setting the stage for what was to come, and I hoped the jury was absorbing it. As I watched them, their faces were turned toward Barry—a good sign.

"In 1987 at the University of Minnesota, they did this study of 116 barns that had neutral isolators on them. What they found is that the major significant factor was that production went up after you installed an isolation transformer."

Electric current is like water flowing from a hose. Voltage is like the pressure pushing the water, so that the water either trickles out or shoots out. Depending on how open the spigot is, the water is under high or low pressure, or perhaps somewhere in the middle.

Now Barry showed them another exhibit, 1067. "Here's our neutral wire coming from the substation." He tapped it with the black rubber head of the pointer. "Here's our two wires running to the transformer on the farm." Tapped again. "Here's the neutral that carries it back, and here's the down ground that goes into the earth. Here's that little inner tie that goes from the primary to the transformer, and you can see that here's the current coming down into the service entrance panel here and then into the grounding block and into the earth but more importantly onto the steel work, onto the milking unit, by, through and on the cow, back to the earth, all to get back to its source, which is now a ground rod at the substation."

There it was. That was the issue in a nutshell. Barry turned and glanced toward me, and I nodded at him, indicating approval.

"Once this electricity's in the earth, it's got to get back to the substation, but how much or how far it goes, they don't have a clue. There's no way to aim or control it. It's uncontrolled electricity, current going into the ground. It's going to flow back." His tone was one of a man who recognized a potent menace threatening the biosphere.

"The dose isn't important. The fact is that there's electrical current in the vicinity of these cows. Studies done by science advisors in Minnesota verify that if current's in the ground, step potentials as low as seven millivolts, 7/1000 of a volt, affect the productive capacity of those herds."

This was well said by Barry, and perfectly true. But WPS, I was convinced, was probably going to counter with something like this: *Yes, yes, there's probably a bit of earth current passing through the farm. But it's harmless, below the threshold our experts consider harmful. Allen is simply a bad farm manager, that is why his cows do not produce to his liking and fall sick. All his claims must be rejected.*

Barry now did another smart thing, talking as if from the cows' point of view. People around here like animals; they've been raised with them in dairy country, and even if they never farmed, are used to seeing cows dotting the landscape.

Barry pursed his lips. "There's electrical charge on the water. She puts her tongue in the water, she gets a shock. Cows have been trained from birth that electric fences are bad, so a cow doesn't drink. Doesn't eat. Doesn't produce milk. Everybody says 'stray voltage,' but it's really stray current."

Pausing for a moment, he looked around the courtroom at the spectators in the wooden benches, then meandered across the carpet and sidled up to the jury.

"Dr. Andrew Johnson will be here to testify that the cows on Russ Allen's farm are definitely affected by the current. The electrical folks will tell you that this happened the minute they—WPS—put that down ground next to the feed room. The current rises and falls depending on soil conditions and loads, but the current's been there ever since."

In December of 1997, when there was a huge reduction in current because of the isolation transformer (and the lack of electrical load, plus the cold hard ground that inhibited conductivity), we increased milk production 30 pounds per cow per day, from 50 to 80 pounds. Nothing else had changed—not me, not my people, not the cows, and Barry made sure the jury understood this clearly.

"The farm has the same barn. It has the same management. It has the same nutritional quality. Nothing changed other than the current on the barn's steel. But there's still current on the farm today. Now, Mr. Allen has done a couple of things to try and stop that."

I cut their hellish down grounds with the righteousness of a surgeon castrating a serial rapist and, at Larry Neubauer's suggestion, even dug a trench round the property and lay down a copper wire ring, what farmers call a Ring of Life, to divert the current away.

"We'll show you pictures of this as the trial goes on."

But what I really needed was the Great Wall of China to protect my farm. Here's how dangerous it is, even today as I write: My girlfriend, Lori, refuses to live in the neighborhood with all that current, especially after she saw all the injured cows. She changed her mind when I showed her the one amp of current on the water pipes in the basement of our Howard home at peak time, with the power off to the home. Howard is a little town connected to the city of Green Bay.

Beginning to get really worked up, Barry began to really lace into the Utility. "Why should WPS be responsible? They're using wire used in WWII. It's still supplying all of these loads. They don't do the maintenance. The only records they showed us were that they tested the poles out there just to make sure they didn't break! We've got an exhibit of the dozens and dozens of split bolt connectors sitting out on Little Apple Road. We've asked that they be removed, and WPS has said no. They won't listen. Russ Allen has sent a letter to Mr. Weyers, the president of WPS."

You couldn't make this up in a million years. Mr. Weyers! Is there a reason for people's names? I picture him with a curled, waxed mustache and wearing a top hat, multicolored electric cords dangling from all his pockets.

"Saying, 'please stop putting current on my ground.' He got a letter back saying, 'We always put current there. We're going to continue to put current in the ground. And by the way, quit cutting our ground rods.'"

WPS was nothing if not haughty, their arrogance a result of unbelievable wealth and unchecked power.

Next, Barry made another pre-emptive strike, cueing the jury in about the bought-and-paid-for experts WPS was going to bring in as heavy artillery. The Utility was probably going to stake the whole outcome of the trial on their testimony.

"The bottom line," Barry said to the jury, his eyes narrowing, "is what do the cows say? Because there will be lots of studies that come up, *funded by utilities*, very suspicious, and we'll dig into some of the details. Lots of college professors are making millions, I'm not kidding, testifying for utilities on studies that we don't think have any merit at all. We submit that WPS knows it because their own expert told them." He said this in a weary, *that's how the powerful get away with murder* tone of voice. It was the old soothsayer syndrome.

One fortunate son, Roger Mellenberger, a real slickster who we will soon encounter, received $140,000 for services related to my case alone. These were charges billed to WPS just for work done before trial. Nice work if you can get it. You need to be a damned good liar and still be able to live with yourself. Mellenberger fits the bill.

Barry turned next to the damages we were asking, $13 million. "How do you calculate what Russ sustained in damages? Dr. Michael Behr, an agricultural economist, has calculated for us. It's not very difficult to do. You figure out where the herd would have been but for the electricity. You calculate the difference between the two. Simply the production that's been reduced. We're saying it should be around 26,000 pounds on a monthly basis, on a rolling herd average, about 82-83 pounds of milk per cow per day, and right now they're at about 70.

"The veterinarians use something called 'differential diagnosis.' That just means what things cause symptoms to occur in a herd. Stray voltage can cause decreased production. You've got nutrition, housing, reproduction, for example, genetics, breeding techniques and so on. The veterinarians, Terry Homan for example, he's going to tell you they looked at all these factors."

Barry walked over to our table, looked at some papers, went back to the jury and launched another pre-emptive strike, pointing out a few flaws in our operation before Trevor Will did. And making the critical point that these flaws were irrelevant.

"Was there a big enough nutrition change to cause 30 pounds of increase in production? No. If anybody could cause 30 pounds of increase by changing nutrition, all the nutritionists would be rich." Several of the jurors chuckled at this.

"The housing that was once the cream of the crop in 1975," he continued, "is just not anymore. Nobody wants to do slatted floors. Ventilation isn't particularly good. It's dark inside. What happened, though, is that the terrible slatted floors, poor ventilation and darkness still existed after the 30-pound production increase. So, they ruled that out.

"They looked at reproduction and said the bulls haven't forgotten what they're doing. So they evaluated the milking equipment. There was no change in the milking equipment. There were no changes in veterinary care, and no diseases that could do it."

As he began drawing near to his peroration, Barry's voice rose and he quit pacing, standing still and looking calmly at the twelve people who would decide my fate.

"Because stray current is a cause of all this, the conclusion was, in fact, it had to be electrical current on their farm. That's what the case is going to be. WPS is simply not maintaining their line. They don't follow their own procedures. The state requires a maintenance program. There's not a shred of evidence that the line has been maintained since the time the second barn was built in 1974. These split bolts on the line are failing. They leak current in the ground. They're not providing adequate service to the Russell Allen Dairy Farm, and that's what the testimony in this case will be."

He fell thoughtfully silent. The courtroom was silent, too; even the spectators stopped whispering with one another. Nervous though I was, I was glad to see he had the full attention of all twelve, plus alternates.

"This is the last time I get to talk to you until the end of the trial, at which point we make closing arguments. I hope you'll find this case interesting. It is complicated. Please bear with us on that and thank you for your kind attention."

Actually, I thought Barry was mistaken on one point: the case did not seem complicated to me. But Barry was preparing the jury for the mountain of data about to descend on them. Hopefully, our experts would negate WPS'. But that was not at all certain. These WPS witnesses had lifetimes of experience giving testimony. They were cunning, and slippery as eels, as Scott Lawrence had forewarned me. They had to be in order to command those huge fees.

Next up was Mr. Trevor Will, and I was under no illusions as to viciousness of the attack he was about to unleash on me. I could feel my heart pounding and the sweat starting to come out of my armpits as Will rose from his seat. He had the look of a devil in his eyes.

Chapter Six

OPENING SALVO

Trevor Will of the high-powered legal firm of Foley and Lardner rose from behind the defense table and positioned himself center stage, halfway between Judge McKay's polished oak bench and the jury box. Will scared me. Tall, thin and elegant, sandy-haired with some distinguishing gray about the temples, he wore an expensive, charcoal pinstripe suit and a pale blue shirt. His navy and gray polka-dotted tie exactly matched the folded handkerchief peering out from the chest pocket of his unbuttoned suit jacket. Black wingtips buffed to a high gloss, he looked as if he were preparing for a spread in *Gentleman's Quarterly*.

But his good looks didn't worry me. What worried me was his reputation as a legal titan, one of the brainiest, winningest attorneys in the region. My palms felt sweaty and I rubbed them against my knees. That Will had not previously argued stray voltage cases in no way reassured me. WPS was not paying him massive fees just to watch him lose.

"Good afternoon," he said to the jurors, and I had to admit that he had a pleasant, friendly, Midwestern voice. "It is my great privilege and responsibility, with Mr. Maassen, Mr. Meckstroth and Mr. Whaley, to represent Wisconsin Public Service—WPS—in this matter. Our job here is to bring you the *whole* story. What you heard isn't the whole story. I'll try to be entertaining enough to keep you awake."

When he said this, I hoped he might get so bogged down in the technical details of stray electricity that all twelve jurors plus two alternates would become lost and fall asleep. And he wasn't going to be discussing how to attract the opposite sex or win money at poker.

"We have some things to look at on the computer screen over here, which is why we have the lights off." Everyone's attention shifted to the monitor as Will quickly sketched his case.

"There are three things I am going to talk about. One is Wisconsin Public Service's conduct, since that's been made an issue. Two, is stray voltage really affecting Mr. Allen's cows? And three, what actually happened at the Allen farm over the years?"

Electrocuted animals. Low milk production. Huge financial losses. Heart attacks. Leukemia, suicides. That is some of what actually happened, though I doubted he would get into *that*.

Next, Will proceeded to try to absolve WPS' ancient electrical distribution system for any part in my problems, which would have been a neat trick even for Thomas Edison to pull off. He pointed with his index finger to a spot on the diagram up on the screen.

"This case is not about transmission wires," he said. "You've all seen these big poles, called H-frames. This is about the distribution system—14,400 volts. That's the voltage of the lines that serve Mr. Allen's farm. There's a picture," he pointed out. "On the left side, there's a three phase. You've heard about the primary neutral. The low wire is the primary neutral. These wires on top that have the insulators, those are hot wires or phase conductors. This is the kind of system, three phase, that goes from the Mystery Hills substation to Mr. Allen's farm. Past Mr. Allen's farm the distribution line changes to a single phase." He paused to give the jury a chance to study the picture.

"There is one hot wire on top. The primary neutral is right here, on which the current flows back to the substation called Mystery Hills."

The real mystery was how WPS was still using a system built during Franklin D. Roosevelt's first term.

Pen in hand, I intended to keep score in my little blue spiral memo pad of any hits Will scored. Like a boxer, he was just feeling out the jury. My nerves had calmed somewhat. The jury appeared interested; no one had started dozing. But it was early. Only Judge McKay seemed a little bored. He kept glancing about the courtroom as if trying to spot something of interest.

"This is a picture of a pole on a single-phase line," Trevor Will said, pointing. "A green arrow shows the grounding wire. It is attached to a long, steel rod driven into the earth. When you heard of Mr. Allen cutting grounding wires, that's the wire."

He began to go technical, and my eyes were glued to the jury to try to gauge its reaction. I kept coming back to an elderly lady in a floral print dress. If he lost her, that could be significant.

Will explained that a multi-grounded system was designed for protection against accidents like lightning striking a pole, or a car slamming into a pole and knocking down a hot phase conductor. It was designed to protect voltage stability so that equipment like computers or motors was not damaged by voltage surges. And he made a key admission, one I hoped the jurors would pick up on and really think about.

"There is current that will flow down from the primary neutral, into the earth, through the earth and back to the substation. The multi-grounded system is designed to protect people, property, and you'll hear why it is standard."

WPS had to destroy the farm in order to save it? How can current leaking down their down grounds into the earth through our home, farms and places of employment—how is this protecting people and property?

"So what is this case about then? The central issue is current being put into the ground-flowing through the earth, up the legs of Mr. Allen's cows, through the cow, and out the other side." His tone was one of deep skepticism.

"This is not a case about how much current is on the ground wires, how much is on the primary neutral, because those things don't affect the cow."

Ah, but they do.

"The evidence is going to be clear you have to have current through the cow. Now, let's talk about my client's conduct and what the evidence is going to show." By evidence, he meant what WPS' parade of professional witnesses would swear to on a stack of Bibles. For big, fat fees, of course. Services rendered.

"It's going to show that the company acted properly. The second thing it's going to show is that we acted reasonably and appropriately under the circumstances."

I wondered what those circumstances were. My circumstances were dire.

Will still had the ballpoint pen in his hand and tapped it against his palm. "WPS is a public utility. That means it is regulated in the state of Wisconsin. It has to comply with laws passed by the legislature."

I felt anger surging upward in my chest and lodging in my throat like an indigestible pit. WPS, a huge, rich corporation, gets laws passed—buys them—to its benefit by contributing cash to the campaigns of Wisconsin legislators and politicians. It's an old story; he who pays the piper calls the tune.

As an example of how things actually work around here, Governor Jim Doyle recently accepted $41,550 in contributions from two utilities after a state panel controlled by Doyle appointees approved the controversial $220 million sale of the utilities' Kewaunee nuclear power plant. WPS and Alliant Energy asked the Public Service Commission to approve the sale. Over half the campaign contributions were accepted by Doyle one week before the sale was approved. These facts are in the public record. Money makes the Wisconsin political world go 'round. Just maneuver your hirelings into positions where they can ram through laws to enrich you.

"It has to follow those laws whether it likes them or not. How does that apply to a multi-grounded neutral distribution system? Well, the Public Service Commission essentially complies, and Mark Cook from the commission will be here. Is he trying to tell us that it is illegal to install a safe system that doesn't leak current in the earth, exposing everyone to it each day?"

Cook was in WPS' pocket, plain and simple.

Barry Hammarback now rose to address the Court. "Your Honor, I object to any statements of law imposed at this point."

Judge McKay suddenly sprang to life, glad, I supposed, to have something to do. "Well, I haven't heard any yet," McKay said judiciously. "I don't know what direction Mr. Will is going in—I certainly understand your objection, but your objection is overruled because I haven't heard any such statements."

Having slipped this in, Barry eased back down in his seat.

Will continued. "There's a specific regulation that applies to multi-grounded systems. You need a ground at every transformer and not less than nine per mile on poles in rural areas such as where Mr. Allen's farm is. Wisconsin Public Service has always complied with this."

What, did he think the jury would give credit to WPS for doing what it was supposed to?

"The second legal requirement," Will said, "is an order of the Public Service Commission called docket number 115. The Public Service Commission commenced a hearing in 1996—the subject was stray voltage and what should be done about it. There were scientists, farmers, utility representatives. Based on evidence offered, the commission issued an order."

Like God issuing stone tablets to the Israelites.

"That order established what's called the *level of concern*. Cow contacts voltages flowing through the cow, which is permitted under PSC order. The utility's contribution cannot be more than one milliamp. You will hear from Mr. Cook that this is not a level of harm."

Of course Cook would say that. But he was aware of the true situation, in fact had made the statement, "You farmers are on the right track regarding stray current, but I will never admit it." He knew which side his bread was buttered on.

The idea that there was no harm at that level upset me. No harm? Those pie-in-the sky, antiseptic, unreal University of Wisconsin studies funded by Utilities always came down that way: The animals could take voltage. In the ivory tower of academia, maybe. But not on Wisconsin dairy farms. Half the world's suffering is caused by grand theories bearing no relation to reality. This Cook was a puppet who would speak words put in his mouth by the Utility ventriloquist. What riled me was the idea that stray electricity was okay up to a point. Hell, there should never be *any* manmade current in the earth, just as there shouldn't be manmade toxins in the water or air. I scribbled a note on my pad: THIS IS RIDICULOUS.

"For purposes of measuring current through cows, the Public Service Commission said assume cows have a resistance of 500 ohms."

That's a huge assumption. Suppose it's 200. Or 100. Or 27. Larry Neubauer recently measured resistance and found out after measuring over 1,000 cows, they averaged 227 ohms of resistance. All I know is that living things are not built to have current flowing through them at any level. That I've seen clearly. It's no accident that electric shock is often used to torture prisoners in wartime, so powerful is the effect. By WPS and PSC logic, prisoners could be shocked up to a certain level without harm. I doubt that many doctors would agree with such an idea.

Will went over and placed both hands against the rail of the jury box. Smiling cheerfully, it looked like he was attempting to bond with the jurors. "So if we know that under the level of concern the utility can contribute one milliamp, that translates into half a volt. When you go out on a farm, you take measurements and check the voltage differential. You divide it by 500 ohms to find out how much current is flowing through the cow. Has WPS taken measurements on the Allen farm? Yes, they have. In 1988, for the first time taking measurements in the parlor, they didn't find anything problematic."

Of course not. That was preordained. I thought about how much suffering and heartbreak could have been avoided if only they'd acted quickly and honorably. The history of my family would have been quite different. Forgetting the dairy business, my brother, baby nephew and two sisters would in all likelihood be alive. What would they have been like today? I felt tears start to well up and fought to choke them back.

Will suddenly drew from his lawyer's arsenal the equivalent of a sawed-off shotgun and let loose a blast.

"Remember the comment by Barry—they only measured in the parlor. And the comment was made that Mr. Neubauer found 800 milliamps in the barn. In fact, when Mr. Neubauer *supposedly* found the 800 milliamps, that was on an arch in the parlor. It doesn't tell you if 800 milliamps was flowing on the metal, doesn't tell you what the voltage differential is."

I watched carefully to see how the jurors were receiving this. One middle-aged guy in a red polo shirt kept sneaking looks in my direction, but whether from sympathy with me or for another other reason, it was impossible to tell. Each time I made eye contact with him he would look away guiltily.

"You need to know what the voltage is on the floor to know whether or not you're going to have current flowing through the cow. Mr. Neubauer didn't do that. We've been out to the farm. We didn't find evidence we were violating that standard. What if we violate that standard? We'll get to that.

"You heard about Docket 106, the first PSC that addressed stray voltage. You heard that it dealt with split bolt connectors. And what you will hear is that Wisconsin Public Service had communication with the Public Service Commission—actually a tight embrace—about split bolts and we have never been ordered to remove all of ours. Moreover, one of the plaintiff's experts, Donald Zipse, has testified that it's appropriate to use split bolt connectors when you are connecting two copper wires. That's the situation on the distribution line on the Allen Farm—copper to copper."

Will now left his position at the far end of the jury box and stationed himself at the near. "You also heard a reference made to: there wasn't enough current returning on the primary neutral, we were violating guidelines. The first thing that needs to be said is that the guidelines do not apply to the situation at the Allen Farm. It is a guideline, not an order or a legal standard."

In other words, WPS won't do anything it isn't legally forced to do.

Will now shifted and assumed an injured rather than accusatory tone. "You heard criticism of the company's maintenance procedure. It's not accurate, as the evidence will show, to say we haven't done maintenance on that line since World War II or before. In 1975, the line was upgraded. It's had pole inspections two times since then. And as problems occur it is maintained. It is inspected the same way as any other line. Mr. Allen is being treated the same way the company deals with all its distribution lines."

Then, another blast from his sawed-off shotgun. "On the topic of how we behave, it was reasonable and appropriate under the circumstances. Every time we were asked to come out to Mr. Allen's, measuring for stray voltage, we did that for no charge."

Was he kidding? Did he want praise from the jury? When a multibillion-dollar corporation causes millions of dollars in damages, it has a moral obligation to come out to alleviate the problem at no charge.

"We were asked out on three different occasions." He walked over to the dry-erase board, picked up a red marker and started writing. "I'm going to write these down because they're important. First was in about 1988. There is a record of the readings taken. Next was in April 1997. Mike Moore, who you'll hear from, was out there. They did an investigation in the parlor, took spot checks in the barn. The spot check numbers were so low that Mr. Moore didn't even write them down."

WPS disbelieved Larry Neubauer's readings, but we were supposed to believe Moore? WPS field reports were notoriously untrustworthy. One of their technicians, Jerry Held, who had tested at my farm, had been fired by the utility when caught falsifying readings at another farm, skewing them in favor of WPS. Maybe he was a sacrificial lamb.

"They didn't actually have a chance to talk with Mr. Allen because he was out somewhere."

Probably, I was attending to sick cows. "They left a report for him," he continued. "They didn't find anything that suggested a stray voltage problem.

"The next time was July of 1997. Two people visited—Vern Peterson and Jerry Held. Mr. Peterson is the engineer. They were there to install the Dairyland Isolator you heard described, the one Mr. Allen asked to have that would separate the primary neutral from the secondary. They installed the isolator, started testing again, and were getting some strange readings. The reason they were getting strange readings was because the reference rod they were using to take readings had been placed, unbeknownst to them, near a pipe that served the manure system from the old barn. Mr. Neubauer came out the next day and figured out what the problem was."

I nearly fell off my bench. I looked up at Barry and he just smiled. Larry Neubauer, my electrician, had to figure it out for them. Will had belittled the "supposed" readings of 800 milliamps that Larry had registered, but now trusted Larry's explanation of these readings over WPS' own expert engineer, Peterson?

But Trevor Will had his story and he was sticking to it. "Mr. Peterson verified the problem. They went to move the reference rod and take new readings. You'll hear that Mr. Allen was fed up with them. Told them to go home. They never completed their testing because he said he had had enough."

This blatant lie was, for me, a red cloth in front of a bull. I could feel the hair stand up on the back of my neck. I whispered, "Take it easy, Russ. Relax. There's a long way to go in this trial."

I took several deep slow breaths. This was only the barest of beginnings. My blood was up because Peterson had outright lied, without even having the tiniest grain of truth behind the lie. I never told them to leave. Said nothing of the kind. WPS had all the notes from the Peterson visit that day, taken by his assistant. And there was no mention of my kicking them out or impeding them in any way, as there would have been had I done so. These were the kind of people I was up against.

"From then on," Will said, pacing slowly in front of the jury, "Mr. Allen was working with Mr. Neubauer. However, that wasn't the end of the story. WPS people still continued to call on Mr. Allen to assist him with, for example, a wiring program that offers money in the event a farmer needs to rewire his system. The company was to come back out and test the Dairyland Isolator in early winter 1997, but Mr. Allen said he wanted it removed."

The damn thing didn't work; it was useless. So, why would I want to keep it?

Will's voice turned sad and grew low as he explained how I foolishly refused to accept WPS' helping hand. "And later on the company continued to talk with him about changing his electric rate and going to a different type of structure. But we were never asked to come back and do any more measurement after July of 1997. We weren't called to come out, you'll hear, and check on what Mr. Neubauer was doing.

"We inspected our distribution line at the Mystery Hills feeder. And we treated Mr. Allen, you'll learn, the same way the company treats all the customers."

Well, then, all the worse for the customers. I am treated the same as any other customer. There is truth to that. You would have to look long and hard to find a farmer they have helped, always below the level of concern, "have a nice day."

"You'll also learn he hasn't had any complaints about adequacy of service—all his motors work fine, all his equipment works fine, the lights come on."

Think of that: the lights came on when we flipped a switch. Now there's something to crow about. And then, quickly, one more blast from his shotgun.

"You'll hear evidence that when the company finds a stray voltage problem as defined by the Public Service Commission, that may mean replacing the distribution line. It may mean replacing the neutral conductor. It may mean adding ground wires or repairing connections. But we bring the system into compliance. What's happened here is *this system has always been in compliance.*"

I bristled at this, looking straight at Judge McKay and then at the jury, thinking: *According to this utility, nothing it ever did was wrong.* I lost my cattle, half my family, and nearly went bankrupt, all to hide my mismanagement of the farm and to blame everything on WPS. Was it plausible that a sane individual would behave this way? To WPS it was.

"Another thing you're going to hear about is the buried ring that Larry Neubauer suggested that Mr. Allen put in—the copper 'Ring of Life,' to divert current. It runs in a big circle. What Mr. Neubauer and Mr. Allen had done is they had created in effect a giant transformer generating underground current through induction from the magnetic field that runs along WPS' distribution line. How do we know that? WPS did an experiment. So when you hear talk about how much current is in these rings, that has nothing to do with current flowing down the ground wires. Nothing to do with current flowing in the cows. It's a sideshow being created by Mr. Allen."

In short, I was electrocuting my own cows and blaming WPS. Will was coming hard after me. I gritted my teeth and awaited his next blast.

"Are we saying he's a bad farmer? No, but he has certain limiting factors that prevent him from getting where he thinks he may be."

And to where I in fact used to be. But that was before they reconfigured the system and used my earth as a conductor.

Trevor Will was on a roll now, animated, nodding vigorously and making eye contact with each juror. He looked authoritative and sounded persuasive. And that concerned me.

"Over time he's had consultants make suggestions he has not implemented. Another factor is the barn, which has a slatted floor. It can cause herd health problems with pneumonia. It's dim. He has not put in brighter light, which has been shown to increase milk production."

Somehow, some way, the cows managed to produce an extra thirty pounds of milk per day per cow in the "dim light" when the current lessened during the dry season and in the winter when the ground was frozen, minimizing conductivity.

"Statistics," he went on, firing away, "show that if you go to artificial insemination instead of using bulls, that produces an increase in milk production. He has not done that either. So to the extent his herd is not performing, it's not due to any electricity but to the decisions he's made himself."

There was a period of silence as Will again slowly paced the length of the jury box, lingering in front of one juror and then another, hands clasped together behind his back. Then he launched into a subject I knew was near and dear to WPS' heart: Monetary damages.

"And that brings up the topic of damages and the $13 million. I need to say a bit about that."

I figured on more than just a bit.

"I'm not going to tell you who Dr. Behr is. I'll let you be the judge of that when he's here and we can talk with him about his qualifications. But you should know what he did. To prepare the damages estimate, he spent less than a day at the Allen Farm, between five to eight hours. Looked mostly at milk checks, receipts that say how much milk Mr. Allen sold to the dairy. He got some cow count numbers, some of which were missing…"

Suddenly, Barry shot up and cut him off. "Your Honor, " Barry said sharply to Judge McKay, "at this point I'm going to object. Argument should be reserved for the end. This should be what the evidence is going to show."

"This is," Trevor Will retorted with irritation, "what the evidence will show."

Judge McKay bent a steely gaze on Will, indicating his intention to ensure proper legal conduct in his courtroom. "Mr. Will, restrict yourself to what the evidence will show and don't argue."

Will nodded deferentially. "I will, Your Honor. The evidence will show he got cow counts, milk weights. The evidence will show he plugged those into a computer program to generate several hundred reports a year. You will notice there was a fifty percent above the state average line on that graph we saw. You might wonder what's the basis for assuming Mr. Allen would have performed at that level. That is the level Dr. Behr just assumed—he did not tour the farm."

According to WPS, everybody on my side was incompetent. Dr. Behr spent as much time on the farm as their witnesses, maybe more. I hoped and believed the jury wouldn't buy it once they heard the compelling testimony of my witnesses. That was probably why Will was trying to demolish Behr before the fact, to taint him in the jury's eyes.

The attempted demolition continued. "He did not inspect the nutrition, did not inspect the electrical records. He simply decided that based on one month's performance, that was the level at which the farm should have been performing forever. You will also hear that he made assumptions about when damages started, all the way back to 1978.

"Mr. Allen's other veterinarian put damages much later than that. Dr. Johnson said three years before 1997. Dr. Schulte said 1990. They can't even agree."

It was a good thing that at this moment Barry rose with an objection. Or else I might have leaped up and started screaming at Will for all his misstatements.

"Your Honor, object to the argument," Barry said forcefully.

McKay again looked daggers at Will. "Apparently, Mr. Will is not going to make that point," he growled.

Chastised, Will said hurriedly, "I will move on." Then he resumed his assault on Dr. Behr.

"He also plugged in feed costs that you will hear are unrealistic. All those things influence the damages number. And we will have Dr. Wayne Knoblauch, who is on the faculty of Cornell University and an agricultural economist, explain the flaws in this methodology and why the damage number is improperly high. So you might wonder what is going on here."

Now *that* was certainly a pertinent question. Just what in hell *was* going on here? We would find out soon enough.

"I will tell you what I think in that regard. I think the evidence is going to show that some time in 1997 or 1998, Mr. Allen and Mr. Neubauer decided that Mr. Allen was going to file a lawsuit."

He was implying a conspiracy. This was so ludicrous that I laughed out loud. Everyone turned and looked at me and Barry shot over a dirty look that said, *keep quiet, this is serious*. Serious, maybe, but definitely ridiculous. Did Will actually expect twelve souls to swallow the notion that I conspired with a local electrician and hatched a plot to defraud WPS of millions? The answer came immediately.

"The evidence is going to show that since that time they have been operating the farm and installing electrical changes, ditch wires, managing the herd in a way intended to show that there is electrical damage."

Clearly our conspiracy required cow carcasses, faked evidence and lots of props.

Then came Will's peroration. Standing two feet away from the jury, his voice rose to such a volume that I was certain people out in the hallway and the offices could hear every syllable.

"When you listen to all the evidence, you're going to conclude that there is no evidence that current has been flowing through his cows at anywhere near the level required to damage them. The explanations for production changes are mundane. Things like nutrition, feed, cow comfort, vet care,temperature, weather. You'll see how those correlate with his cutting the ground wire. And when you're finished hearing all the evidence, you will conclude that WPS is not responsible for any injury to Mr. Allen's cows.

"Thank you very much for your attention and patience. I'll have an opportunity to talk with you again at the conclusion of the case."

So, there it was. The battle lines were drawn, with witnesses, like shock troops, mobilized and ready to enter the fray. The case would put expert versus expert, engineer versus engineer, electrician versus electrician, agronomist versus agronomist. Or at least until it was my turn to take the stand. I was no expert, only a simple farmer trying to make a living and preserve an 80-year-old family business and way of life. Though tense and uncertain of the trial's outcome, I was dead sure of one thing: I was in for what promised to be a hell of a ride.

Chapter Seven

THE TRUTH AND NOTHING BUT...

As court opened the following day, another warm bright flawless May morning, our strategy was to employ the very best witnesses first. Both Barry Hammarback and Scott Lawrence believed that the jury's first impressions would be critical. Not only that, the early witnesses, experts from various fields, would provide an overview of the case and prepare the way for my own testimony. But, as Scott said, expert witnesses on both sides tend to cancel each other out. My attorneys knew that in the end it was my credibility that held the key to victory. Or defeat.

I felt the pressure of the moment as I entered the courtroom flanked by Scott, Barry and another of my attorneys, Dan Murray, a brown-haired, handsome, well-built six-footer in his mid-forties. Dan and I sat in the first row of spectators' benches just behind Barry and Scott's counsel table. Though my resolve to fight WPS was firm, I felt the burden, like a weight, of seeking justice for the dead: relatives, animals, dairy business. Though testifying and getting raked over the coals by WPS figured to be extremely unpleasant, akin to undergoing a root canal, I was eager to take the stand, make my case and get it over with. If the outcome of the trial depended on my words, so be it. I could live with that, regardless of how things wound up. At least I'd be active, instead of just sitting around letting the world make up its mind about what to do with me.

But it was not yet my time. First on our witness list this morning was one William Rasmussen, a River Falls, Wisconsin police officer and an agricultural field investigator for Barry. He had spent considerable time at the farm, taking photos and measuring current at various points. Many of his technical findings, as well as pictures and current measurements at utility poles in the vicinity, appeared as exhibits, which the jury was able to view. Rasmussen provided background logistics.

The first truly crucial witness was Dr. Andrew P. Johnson, a.k.a. the "Udder Doctor," whose testimony, I hoped, would be powerful. The doctor was a renowned veterinarian who had consulted for over twenty years in twenty-six countries and almost every state in the U.S. He was one of the foremost dairy experts in the world and I was honored to have him on my team.

In his mid-fifties, Dr. Andy Johnson was a physically commanding presence. Over six feet tall, with a full head of thick, gray hair and an iron-gray beard to match, the doctor weighed more than three hundred pounds, and every ounce of his bulk was majestic and imposing. But as learned as he was, Dr. Johnson had a knack of making specialized knowledge easily understandable for lay people.

Under Barry's probing, Dr. Johnson related some interesting bits of personal background and established his bona fides for the jury.

"The source of that glut in your briefcase there that you brought up?" Barry, standing about five feet from the witness box, asked with a smile.

"My file on the Russ Allen case," Dr. Johnson said in his deep, clear voice. Bulging with a sheaf of papers thick enough to choke an elephant, the briefcase could be neither completely closed nor latched.

"You don't have a leather one?" Barry said.

"I couldn't find one big enough," the doctor replied, and the jury chuckled.

"How long have you been consulting?"

"From about the mid '80s to current."

"Who do you normally consult for?" Barry asked.

"Farmers form the vast majority of my work."

"Do you consult in litigation for just plaintiffs or also for defendants?"

"There's a mix," Dr. Johnson told him. He went on to say that having been raised on a dairy farm, he was always interested in cows and had been on his high school dairy-judging team at their FFA.

Barry wanted to drive home to the jury that this witness, unlike the university professors who would be testifying for WPS, spent his time in the real world, in the field on farms, and not in controlled, air conditioned laboratories.

"Have you evaluated dairy farms looking for problems with production?" Barry inquired.

"That's what I do. My expertise is milk quality, cow comfort. I also look at nutrition, disease, housing, milk quality, records management. I have the ability to look at most aspects of a dairy operation."

"After you've looked, what do you try to do?"

"Generate a report or have a discussion with the farmer, prioritize items I think are affecting profitability and a game plan of how to address those items." He rubbed the side of his nose with his forefinger.

"Has the work you've done taken you to national organizations?" Barry asked. He wanted the witness to emphasize his world-class credentials.

Dr. Johnson duly obliged, speaking proudly of his qualifications and accomplishments. "I belong to the American Association of Bovine Practitioners, and I'm currently on the board of directors. I'm active in the National Mastitis Council. I'm active in the Wisconsin Veterinary Medical Association, and past president of that association. Dr. Graeme Mein and myself worked on the national standards protocol to evaluate milking equipment—our protocol was adopted by the ASAA."

"What types of awards have you received?"

"I was awarded Wisconsin Veterinarian of the Year, the top honor our state gives. I was awarded the Dairy Preventer Medicine Award by the Bovine Practitioners, the highest honor in North America."

Barry stood silently, letting the jury absorb the information.

"Bovines," Barry said, "for those of us that are veterinary-challenged, are cows, right?" Everyone in the courtroom laughed.

"Yes."

Barry next asked about differential diagnosis, which simply meant looking at all facets of a farm problem, determining root causes and creating solutions. Dr. Johnson became animated as he described his methods his eyes widened and he gestured with both hands.

"What's made me successful is trying to look at the big picture. Even though I might be going to a dairy for a milk quality problem, I notice if housing is bad or there's no feed in the manger. I try to focus on areas of biggest concern."

He went on to say that he was frequently asked to give seminars and lectures on farm problem solving, both in the US and abroad. "I have a 10-day training session for veterinarians in Spain on how to properly evaluate equipment and milk quality issues. I've done training at universities around the globe, had major milk machine manufacturers ask me to train their people."

After this low-key, introductory Q and A, Barry paused for an instant and then moved nearer to the witness box. Like a thunderclap on a clear day, he said loudly and from nowhere to Dr. Johnson, "What is stray voltage?"

Taken aback by the sudden change of subject, the doctor hesitated a moment. Then he slowly shook his head. "Stray voltage is a huge can of worms. There's a lot more we need to learn. Especially about nontraditional stray voltage, which is earth currents. It's an issue where there's some type of electrical phenomenon causing cows discomfort." In other words, exactly the type of current that had wrecked my farm.

"From a veterinarian's point of view," Barry said, "can you explain what stray current does to cows?"

His answer to this question would be one of the most important pieces of information the jury would hear. Leaning forward and resting my elbow on my knee, chin on hand, I studied the jurors' expressions to see if I could catch any hints or clues.

Dr. Johnson nodded. "We get dairies where it may only access cows at certain areas and only affect the water they drink, which affects the feed they eat, which affects the milk production they give. On other dairies it irritates them in the parlor and they don't let their milk down. Don't milk out, and it causes mastitis. On some dairies it causes immune stress and cows are susceptible to diseases. So, there's a wide variation of what it does."

Barry next wanted to know if there was a dispute as to the level at which current injured cows, and Dr. Johnson explained that indeed there was. The "experts," particularly ones from Cornell and the University of Wisconsin who would be testifying for WPS, kept raising that level, saying in essence that cows could take ever-increasing amounts of electricity without harm. Dr. Johnson then made a sensible and valuable statement, the same one I'd been making for a long time.

"I'm one of the people who listens to the cows," he said. "The cows are the ultimate answer as to whether the level is bothering them. My level of concern has never changed."

Music to my ears.

Barry nodded and turned toward the jury. "About a level of concern when it relates to voltage, what do you mean?"

"Well, how much voltage does it take to bother a cow? The level is whatever makes the cow change to an abnormal behavior, from the way she eats, drinks, walks, milks. The level is what bothers the cow.

"I'm not an electrical engineer, I'm a cow veterinarian. The way I look at traditional stray voltage is what we use to measure cow contact points or AC voltage—we would simply take a voltmeter and test between cow contact points, say, hoof to mouth and see if we get a voltage reading." As might be caused by live wires or loose, unshielded cables.

"Now we have something called ground currents, a different issue I'm not sure anybody has complete understanding of. In my opinion it's going to be a much huger issue than traditional stray voltage ever was."

"Why do you say that?"

"It's causing much higher levels of damage. With traditional stray voltage, you went in, you measured it, you found it, you fixed it. Life went on. With the new stray current, man, it takes a long, long time to figure it out. We haven't even been able to fix it."

At that I felt like applauding. It was a ringing affirmation of how I'd spent years pulling my hair out trying to identify the source of our troubles. The jury seemed very attentive, a heartening sign.

"Let's say," Barry went on, beginning to pace slowly, "you go to a farm and measure for traditional-type stray voltage and don't get anything, but you see the cows are nervous and don't milk right. Does your investigation end there?"

The doctor shook his head vigorously. "Oh, absolutely not! On most farms that I deal with, we don't measure voltage but the cows clearly tell me there's something wrong. I will refer the farmer to somebody to measure for ground currents. Because I'm not an electrical engineer."

Barry left the vicinity of Judge McKay's bench and went over to the jury, leaning with his back against the railing of the jury box and facing the witness. "We've heard talk about levels of concern, measured in terms of voltage. Back in the early '80s, was there a level of concern?"

"When we look at the purest research," Dr. Johnson said, and I could tell from the lilt in his voice that the teacher in him was enjoying the moment, "it was originally talking .35 volts cow contact and a volt on secondary neutral. Then it changed to half a volt, then to a volt. What I have stayed with—if it's under half a volt and the cow's not telling me anything, I'm satisfied. If it's between a half and one, I'd call it a gray area and it depends on what the cow's telling me. Over a volt, I want it reduced. *The most important thing is what a cow is telling me.*"

My cows always told me the truth. They died.

Barry began strolling around again as he posed his questions. The movement must've kept him mentally limber. I shifted position several times as my backside was getting a little sore on the hard wood bench.

"The cow tells you these things," Barry reiterated. "They can't speak, or maybe they can, but how do you tell they're suffering?"

The doctor glanced up at Judge McKay, looked at Barry and smiled puckishly. "You know, I'm not Dr. Doolittle, I don't talk to cows directly." Everyone laughed, even stone-faced Judge McKay. "But they talk to me in what—I've never seen a cow lie."

Neither have I.

"They tell you by saying, 'Gee, I've got this wonderful feed in front of me but I won't eat it. I've got this nice milking machine but I won't let my milk down. I have farmers doing a really good job of getting me ready to milk, but I don't want them touching me. Gee, I have a competent veterinarian come out and treat me, but I don't get better.'

"So, I mean, many of the farms are 'cliff' farms—it's easier to drive over a cliff than figure out answers—they do so many things right but the results aren't good."

There was my biography, and Dr. Johnson was recounting it for the world. Several jurors scribbled notes. The point seemed to have hit home.

"Is there any such thing as a level of concern for any particular farm?" Barry wanted to know.

"Every farm is different. So, I can't sit here today and tell people on this jury that I know exactly what that level is because I'm totally befuddled if anybody knows what that is. *I think it's a much lower level than anybody wants to address.*"

Truth was a wondrous thing. I muttered several amens and a couple of hallelujahs.

"What's the typical way you are called to a farm?"

"Usually, the farmer or his veterinarian calls and says, 'Here's a farm that no matter what we do we can't get any positive results.' It could be a cell count issue. It could be cows not responding to therapy. I try to find everything but stray voltage because it's such a tough deal."

There was a silence. The courtroom was still. Barry resumed, "You said that you once consulted for WPS?"

"They had me do staff training, who then went to farms to give them ideas of issues they could look at to help the farmer determine if there's other problems on the dairy."

"When?"

"Early to mid '80s."

Barry then landed a strong pre-emptive punch. "Anybody from WPS come and tell you, 'Get out of here? You don't know what you're talking about?"

The doctor grinned. "Not yet, anyway."

This was a left hook to the solar plexus. It would now be much tougher for WPS to dismiss or repudiate Dr. Johnson's opinions and expertise.

That expertise really revved up under the next round of questions. With Johnson in teaching mode, Barry built a solid foundation for my case, troweling on mortar and laying down brick after brick. Vacating the jury area, he sidled over toward the witness stand again, stopping about five feet away. Facing the doctor, he folded his arms across his chest. I noticed his suit jacket stretch and wrinkle up along his back and flanks.

"When there's a thought that stray current might be causing a problem," Barry said, "what are the differentials, as you say, used in the diagnosis?"

The doctor's brow furrowed. "Housing, cow comfort, milk quality issues. My mastitis triangle would be: The man, the machine, the cow, nutrition, disease, labor and management."

"Management?" Barry asked.

"Labor—I would put those together, employees and management, both. Records, if available."

"Electricity?"

The witness nodded. "Electricity would be one of them. Among all those, there's different avenues. Like with nutrition, you've got dry cows, calves, heifers."

"Reproduction?" Barry asked.

"Reproduction would be another. Feet would be another."

"Feet?"

Just behind me in the second row of spectators (the courtroom was surprisingly full, with maybe twenty-five or thirty in attendance), I heard someone whisper, "This Dr. Doolittle guy is pretty cool."

I smiled and made a mental note to inform Dr. Johnson of the remark.

"Feet."

"Feet?" Barry said again, quizzically.

"Feet," Johnson replied, relishing the word. "Foots. Feet." He was talking about current-caused sore hooves and grinning. "We could come up with others, but those are the major things. Water, you could throw under nutrition."

"Why is water an issue?"

"Because," the veterinarian said, "if they don't drink, they don't eat feed, and don't make milk. It's 97 percent water, you want to make sure the cow has adequate water."

"If the cow doesn't drink water, she doesn't eat feed either?"

Now Kurt Meckstroth, one of Trevor Will's colleagues and another of Foley and Lardner's high-powered hotshots, rose from behind the defense table and launched an objection. Forty-eight years old, with short brown hair and a round paunch, Meckstroth had previously worked for the Judge

Advocate General's, the JAG, office. Perhaps that was where he'd picked up his haughtiness. He'd proven during depositions that he knew very little about dairying; Scott Lawrence, an expert on the subject as it related to stray current, had repeatedly exposed his ignorance.

"Object to the question, Your Honor," Meckstroth said in a nasal twang. "Leading."

Judge McKay nodded, and a small shock of gray hair tumbled over his forehead. "That is," he said.

"I'm sorry?" Barry said directly to Meckstroth, displeased.

"LEADING," Meckstroth retorted, and his voice rose and dripped with arrogance and condescension. I pegged him as a typical smartass, the class know-it-all. Meckstroth looked like conceit on the hoof.

"Mr. Hammarback," Judge McKay said, "try not to lead. I'll sustain the objection."

Barry frowned and said sarcastically, "I'll see if I can." To me it seemed dangerous to risk antagonizing McKay, but I figured Barry knew how much he could get away with.

Still, I was surprised to see Barry bristle over a minor point. But I was glad that he was being strong for me. Obviously, he wanted the proceedings to go the way he wanted them to go, and disliked Meckstroth's interference. As this was only the beginning, I expected some big explosions along the way.

"What happens if a cow doesn't drink?" Barry asked Doctor Johnson.

"Well, she—she'll die. If she doesn't get adequate water, she doesn't eat as much. Water drives dry matter intake. The more she drinks, the more she eats. If cows don't drink what they should, don't eat, they're under stress. And poor performers."

Barry was maneuvering toward the nutrition/feed aspect of what I'd done with my animals to show how well tended they'd been. He knew that attacks on my management methods would play a big role in WPS' defense.

"Some of the terms you're bringing up, I'm going to ask you to explain—you're the first person to talk to the jury about cows. What's dry matter intake?"

The doctor appeared delighted by Barry's questions and shifted eagerly in his seat. Extremely articulate and personable, he seemed eager to impart his knowledge.

"Well," he said, "every dairy cow eats feed, a combination of hays and different silage and grains and proteins and minerals. The total she eats every day is called dry matter intake.

"A lot of dairies feed a total mixed ration—what you feed to the cow may weigh a hundred pounds, but if you take the moisture out it would only be about fifty-two pounds of dry matter. Water can have a big impact on 'dry matter intake.'"

"Okay," Barry said. "Of this list, housing, milk quality, nutrition, water, disease, labor and management, records, reproduction, *feet*, electricity, can you tell us whether these things relate in a major way to the production of that dairy?"

Meckstroth suddenly popped up like a jack-in-the-box. "Objection, Your Honor," he said in his condescending twang. "Leading."

"Sustained," McKay grunted.

Barry ignored this altogether and continued calmly with Dr. Johnson. "Are these things related to the final output of the dairy?" he asked.

"Well, absolutely!" Johnson gushed. "Every one can have a major, minor or additive role. Each can impact the bottom line and final output of the cow."

Barry then asked, "Have you heard the term 'limiting factor?'"

"Sure. Another easier way to refer to it is 'bottleneck.' Where does the bottleneck occur that keeps that cow—that farm—from getting to the next level of production, of success? The limiting factor may be the type of housing they're in or the feed quality or some disease preventing them from reaching goals."

"Can you give me an example of a bottleneck?"

"All those would be examples, obviously bad housing, improperly balanced ration, not breeding cows properly, having sore feet. Electricity could be a bottleneck."

He'd spoken the magic word. Having dispensed with the appetizers, Barry was now preparing for the main course.

For a long moment he was silent, letting the word "electricity" hover over the courtroom. Then he said, "Dr. Johnson, you've done reviews and examinations of research in the area of stray voltage?"

"Yes."

"In this case we've named some treatises or studies on stray voltage. Were you aware of that?"

Johnson said that he was.

"I want to ask you questions about those," Barry told him, preparing the ground for later testimony. "Paper 87-3039, our Exhibit 1103, titled 'Production Record Analysis of Dairy Herd Response to Neutral Isolation,' written by Dr. Appleman, Gustafson and Brennan. Are you familiar with that?"

He was.

"Do you consider Dr. Appleman an authority in the area of stray voltage?"

He did.

"Do you find that paper authoritative?"

He found it so.

"There's a Dr. Polk who wrote an article on bioelectrical magnetics. Exhibit 573, entitled 'Cows, Ground Surface Potentials and Earth Resistivity.' Do you find Dr. Polk to be an authority in the area?"

He believed Polk was.

"There's an article from the 1985 *Journal of Dairy Science* entitled Source of Stray Voltage and Affect on Cow Health and Performance, again, written by Dr. Appleman and seconded by Mr. Gustafson. Do you find Dr. Appleman to be authoritative in this respect?"

The doctor said yes.

"Why?"

"Because" Dr. Johnson explained, "it uses a large population of cattle. About 80 herds with roughly 50 cows per herd average, so it's about a 4,000-cow study, deals with isolation at the secondary neutral and shows the impact of that isolation on milk production."

"All right," Barry said. "There's Exhibit 1107, paper 87-3033 entitled 'Grounding Resistance and Ground Currents in Dairy Facilities' written by Lavern Stetson and Gerald R. Bodman. Do you find Mr. Bodman authoritative in the area?"

"Yes."

"Exhibit 1114, paper 89-3540, 'Stray Voltage Measurements: Correct Techniques Using Shunt Load Resistors,' written by David Winter and William Dick. Is Professor Winter an authority?"

"Yes," Dr. Johnson answered.

"Exhibit 1124, 'Behavioral Studies of Dairy Cattle Sensitivity to Electric Currents,' by R.J. Norell, R.J. Gustafson, R.D. Appleman and J.B. Overmeier. Do you find Dr. Norell and Dr. Appleman to be authoritative in this area?"

Dr. Johnson said he found them authoritative.

Barry continued on in this vein for a good while, striding about and using the Q and A to demonstrate that stray electricity and its effects on dairy farming was a serious subject worthy of deep study and not, as WPS would've loved for the jury to believe, simply an excuse on my part to defraud them of millions.

Apparently, this tactic was not to the liking of Kurt Meckstroth, who now came up with what was, to me, a picayune objection.

"Your Honor," he said peevishly, "Mr. Hammarback identified several authors from some papers, but when we asked Dr. Johnson whether he considered them authoritative, he didn't ask whether he considered *all* the authors authoritative. There's no way to tell what part of the articles were written by which of the authors. Unless Hammarback can establish that Dr. Johnson considers all the authors authoritative for all the articles, I would object."

Judge McKay looked at Barry, who had walked over to the counsel table as though to get closer to Meckstroth. "Mr. Hammarback," the judge said, "do you wish to respond?"

"Yes, Your Honor," Barry said quietly, and rifled through some papers on his table. He looked at Scott. "By signing authorship, certainly that's the acceptance of responsibility for the contents of that paper. Whether we need multiple authority raises questions, but the question really is, is the information in the article reliable?"

McKay turned to the witness. "Do you, Dr. Johnson, consider the authors referred to, authoritative?"

Dr. Johnson stuck out his lower lip. "I have a problem with some. There are several I don't consider authoritative."

McKay nodded. "The Court has a problem in that I don't believe all the authors were even identified."

"True," Barry agreed, quickly dropping the subject by asking the witness, "On Exhibit 1107, 'Grounding Resistance and Ground Currents in Dairy Facilities' written by Lavern Stetson and G.R. Bodman. Do you consider those gentleman authoritative?"

Dr. Johnson hesitated. "What I'm having trouble with is—they are both authorities in the field. That is different than, do I agree with their opinions?"

This was a nice way of saying: Just because some pompous professors had penned articles from the ivory tower, this didn't mean their conclusions bore any relation to reality on the farm. Clearly, Dr. Johnson disagreed with the above-mentioned authors, on whom WPS was going to heavily rely.

Barry resumed his questioning. "R.D. Appleman and R.J. Gustafson?"

The veterinarian shrugged. "I would have the same answer."

"Dr. Johnson," Barry said. "I want to get back to these articles in a bit. But in addition to work you've done on Wisconsin farms, the Public Service Commission held hearings on stray voltage. Did you participate?"

"In the first one I did."

"That was Docket 106?"

"Correct."

Barry asked him to read off a portion of the docket, which he did.

"It says: 'By describing basic screening tests, the commission is not saying further investigation is not warranted if the basic tests do not indicate a problem. There's no substitute for observation of actual behavior in an animal…'"

How true this was. And then came another crucial bit of information from the udder doctor.

"So," Barry said, "the utility in Wisconsin after 1989 goes out and does a stray voltage test, let's assume they don't find stray voltage but the cows are still nervous and jumpy. In your opinion, what's their obligation?"

The veterinarian answered without hesitation. "Their obligation is to go looking in other areas and make sure all the bases are covered."

"As a result of statements in Exhibit 1024 or Docket 106, has there been work done regarding equipotential planes?"

"Yes."

"Why?" Barry asked. "Tell us what happened."

This brought Kurt Meckstroth leaping up out of his chair. "Your Honor, I object! He's not an electrical engineer."

Barry turned and glared daggers at him. "I can lay a foundation," he snapped.

Something told me that Barry's and Meckstroth's animosity went beyond the fact that they were competing attorneys. The two were antagonistic personalities, natural enemies. Like the mongoose and the cobra.

"Lay a foundation, Mr. Hammarback," Judge J.D. McKay ordered. "If you can."

"Sure," Barry said breezily. "Dr. Johnson, do you know what an equipotential plane is? Explain to the jury what that is."

The doctor shifted position in his seat. "What they do is try and bond all areas where the cow's at, so she—like a bird on a wire—where there's no differential between different places a cow exists so everything is on an equal plane."

"And the idea behind that is to do what?"

"Spread out or dissipate electricity, take it away from the cow."

"In practice, did you see equipotential planes go in after utility recommendations?"

"Absolutely," Johnson replied.

"What observations did you make of dairy cattle exposed to the planes?"

"On a lot of farms it seemed to make things worse."

Barry waited. "Did you do some work to try to figure out why on some farms it was making things worse?"

"Well, I asked a lot of questions. When I disconnected the plane on some of my dairies, things got better. When you're on an equipotential plane, your traditional stray voltage tests are meaningless because there's no differential between cow contact points."

"Based on your experience, is it appropriate to test for stray voltage on an equipotential plane?"

"No," Dr. Johnson said forcefully. "Because you're going to get extremely low readings that are not telling me what the cow's experiencing."

Barry nodded and resumed pacing. "Back when you did your training at WPS Corp., tell me what you explained to the WPS personnel about things you looked for in housing when trying to evaluate stray voltage."

I nodded at Barry when he looked toward me. This was an excellent line of questioning, since WPS was going to make a huge issue of farm mismanagement. Barry wanted to show that not only had I conscientiously tried a thousand possible remedies for the problem, but that a world-renowned expert like Dr. Andrew Johnson knew and agreed with what I'd done. And saw the problem in the same way that I did.

"Things like bedding, stall design, ventilation, different types of housing and how to identify potential bottlenecks," the doctor said.

"Bottlenecks caused by electricity?" Barry asked pointedly.

"Right."

"Can you remember whether you discussed symptoms of stray voltage?"

"Definitely."

"Is it important to look at other factors when someone thinks they have a stray voltage problem?" Barry asked him.

"I try to look at all factors," Dr. Johnson replied.

"Mr. Scott Lawrence hired you to look at Russ Allen's operation?"

Scott had hired him and Dr. Johnson came. Barry was now into matters critical to my chances of winning. I could feel my pulse quickening.

"What understanding was given to you of what was going on at Russ Allen's?"

"That he had an unresolved electrical issue. I was to look at the facility and the dairy management and render an opinion as to whether I felt stray voltage caused damage to his dairy."

"What do you charge?" Barry asked.

"My normal rate for farm visits is $150 an hour."

"Uh huh. And to date about how many hours have you gotten in?"

Dr. Johnson's brow wrinkled as he made some fast mental calculations. "Seventy, eighty hours. Somewhere in there."

The doctor's inspections cost me a lot, about $12,000. He was worth every nickel.

"When you went on the farm, tell the jury your observations," Barry said.

I leaned forward, tensing, and felt myself inching up toward the edge of the pew. How would he phrase his answers? Forcefully, I hoped and expected. I needed the jury to understand my situation clearly.

"Initially, I try to look at all the facilities, walk around and look at the cows, talk to people, watch milking, look at milking equipment. Just do a general overview and then start focusing in on different areas."

Barry walked over to the exhibits and pointed to a photo. "Showing you Exhibit 1001, taken about 1974 or '76. Is that the basic farm layout that you saw?"

"Right," Dr. Johnson said. "That's basically what it looks like today." He pointed. "This is the old dairy barn. This is the fresh cows, hospital cows and pre-fresh cows getting ready to have babies. Dry cows were on this end."

The pictures and the memory of what had once been a thriving farm and my home since birth brought a stab of pain to my heart and a well of tears to my eyes.

"They have a calving area here," the doctor went on, "a pre-fresh area, and this is the feeding area.

"This is the free-stall barn. The parlor is in this upper corner. Here. This is where the adult milking cows were housed when I was there. The lot behind the barn, that's where two of the milking cow groups could go outdoors, and they also had lots where some of the dry cows and pre-fresh cows could also go outside."

Barry offered this exhibit, 1001, to the Court, which asked Meckstroth if he assented to its inclusion in the record.

"I would like to do some *voir dire* on the witness before this is admitted, Your Honor," Meckstroth said petulantly.

But Barry, mongoose to Meckstroth's cobra, could not let this pass. "No, Your Honor," he said, perturbed. "If he objects, I'll—"

Do what? I wondered. Barry appeared to dislike this guy, and I couldn't blame him. Meckstroth was more arrogant than Trevor Will and far more obnoxious. The crazy thought crossed my mind that it would take very little provocation for Barry to storm over and bulldog him to the courtroom floor like a cowboy taking down a steer.

Barry stopped himself from whatever he intended to say and said instead, "If he objects, I'll just call Mr. Allen later and lay the rest of the foundation."

But Kurt Meckstroth was not one to easily yield the last word. "I wanted an opportunity to go through the changes between what appears in the picture," he complained, "and what the witness saw on the farm. That's all, Your Honor."

Judge McKay settled the matter. "Let's just hold the picture until Mr. Allen can lay a foundation for it."

Barry relented. He had bigger fish to fry with Dr. Johnson. "Okay," he said amicably. "I'm going to show the jury, by pictures, what types of things you looked at. Photograph 1168?"

"That's his milking parlor."

The jurors were all studying the pictures, I noted. The more they knew about my farm, the better.

"That's the double 12 parallel parlor, where cows are milked." Dr. Johnson then described the process of milking, a comforting and fulfilling sight that I'd always loved. Heck, I'd cut my teeth on it. "Cows are brought in from where they're eating and sleeping into the holding area. They are moved through the parlor and come—12 on the left side and 12 on the right—and as they milk 12 they bring in 12 more and keep repeating that process until all cows have been milked."

"The people that milk the cows stand where?"

"In the center of that stall, called 'the pit.' They stand there and the cows are up so they're working on their udders."

"Can you tell me," Barry said as he meandered over toward the jury, "what the advantages are of a parallel parlor like this?"

Dr. Johnson explained that in the parallel parlor the cow entered and made a 90 degree turn, enabling milking between her rear legs. You could fit more cows in a smaller space, actually a row where they can stand side by side. More cows can be milked in a smaller space than in a herringbone polygon-style parlor. It's also more user friendly in that people aren't forced into as much walking, and it's generally safer—there are less injuries to personnel who have an excellent view of a cow's udder.

"How," Barry wanted to know, "do you get injured in a parlor?"

Dr. Johnson raised his arms and opened his hands, palms upward. "Lots of ways. You can get hands stepped on. You can get kicked. You can get your arm caught up under the cow."

Barry wanted to know about some of the equipment in the photos of the parlor interior.

"Well, that thing, hanging down with silver, is part of a milker unit, called 'the shell.' That holds the inflation and that's what goes on the teats to remove milk. Those four shells of inflation go the claw, which brings the milk together. Milk goes from the to claw to the milk hose, down the pipeline. The pipeline takes it to the receiver, which takes it to the bulk tank."

"And the panels up on top?" Barry said. "What do they do at this steel valve part?"

"That's where controls are. Dairy information pads, where they can put in ID numbers or get pounds of milk."

Barry now went over and pointed to Exhibit 1172, a photo of cows in the parlor. He asked Johnson if it showed a typical milking regimen.

"These are cows coming from the holding area. Cows that are being milked, when released, go east, west, and the next group will come in and fill the stalls. A cow coming into the parlor—here is the ID antenna. There is a transponder on the cow's neck so as she walks through, it reads the transponder and tells who the cow is. It will send information down to the stall she's going to. So you have the number of the cow in that stall and can keep records on each individual cow. If you notice, they have a yoke that keeps them in place. When that cow's done milking, the yoke comes straight up and the cow walks out of the parlor."

Having taught the jury a little about processes, the doctor began recounting the situation he found at my farm. "I did a complete evaluation of the system," he said.

Barry asked for his findings.

"Everything checked out normal on that visit of 12-28-01, except the regulator was dirty and needed service. I told Mr. Allen he needed his dealer to service his regulator."

He was referring to Mike Hoerth. Dr. Johnson mentioned that when troubleshooting on a farm, he usually gathered historical information from both the farmer and the equipment dealer.

Then he made a powerful point in my favor: "Talking to the dealer and looking at records, seeing a regular service program tells me that things were routinely taken care of. Also, everything tested out very well."

We'd scored. Not a touchdown, maybe, but a field goal that split the uprights. More good news followed. Barry asked the doctor to describe the free-stalls we had on the farm, the use of which demonstrated our care and concern for our animals:

"A free-stall," Dr. Johnson explained, "is a little individual home for cows. We call them 'free-stalls' because the cow is free to come and go as she wants, where an old stanchion or tie-stall barn would lock her in position until we release her. The theory behind free-stalls is the cow can, when hungry, get up and eat. If she wants exercise, she can do that. Or she can come and lay down. It's a little individual bed for each cow to call home."

Barry put up Exhibit 1165, a photo of the free-stall barn with its clear rows. Eager to describe and explain everything, Dr. Johnson asked McKay if he could leave the witness stand and go over to the exhibit. McKay allowed him.

"Each of these," he said, pointing and tapping with his index finger, "in between each of these bars is one of those individual cow homes, so to speak. This is the alleyway where she can walk to and from the feed. And then she gets up and lays on a mattress with—I think—this looks like soybean straw or some type of organic bedding. And they lay down and chew their cud."

Dr. Johnson related how the barn, with its slatted floors, had been the "Cadillac" of barns, a state-of-the art structure when installed in 1976. But by today's standards it was technologically backward; for one thing, it was not nearly as well lit and bright as those built in the 21st century. These facts would loom large when WPS began berating my operation.

"At the Allen farm," Barry said, "are there different groups of animals?"

The doctor nodded. "Four. Post-fresh, high-producing, pregnant and breeding, and low-producing." This was a typical division. Barry asked why, and Dr. Johnson explained that it was done to facilitate certain programs for each group.

"This feed bunk out here, is that paved?"

"Cement."

"Are there benefits to cement?"

"Cement isn't a bad idea. We don't want to deal with the risk of mud when it does rain."

"Exhibit 1162 is feed at the outside?" Barry asked.

"Yes, the cows are eating their ration outdoors."

I'd always loved to see that—those beautiful, solid, contented black-and-whites eating on a clear summer morning. How I missed the sight! Hell, I missed everything having to do with my cows. Maybe, just maybe, if we could win here and get a fix from WPS, I could get another herd.

"When you say 'ration,' it's a TMR or total mixed ration. Can you explain to the jury the idea behind it?"

Dr. Johnson patted his ample midsection with both hands. "A dairy cow has four stomachs. They eat hay and corn silage and hay leaves. They are one of the few animals that can turn these forages into milk. Then we add concentrates, corn for energy, and protein mix and then all of the minerals. We want to make sure they're evenly mixed so the cow's getting some of everything. We end up with a kind of smoothie." Several of the jurors chuckled at this, perhaps picturing bovine shakes.

Barry escorted the doctor back to the witness stand. "And so as you're going through the Allen farm and thinking about differential diagnosis, in terms of housing, how would you evaluate what you saw?"

Johnson wrinkled his nose. "Didn't like the housing real well. Old technology."

Though I felt a twinge of embarrassment, I knew this was true. The benefit was that the criticism would further reveal Dr. Johnson's objectivity and enhance his already high credibility.

"Okay," Barry said. "In looking at milk quality issues, what conclusions did you come to?"

"I thought they had a good milking parlor, a good milking routine." He went on to discuss our sanitizing of teats prior to milking by pre-dipping—spraying on the sanitizer—then wiping and cleaning the teat prior to attaching the milker unit. Once that was off, we sanitized once again, or post-dipped.

"Your opinion of the practice?"

"I thought it was fine."

"With the regulator that required cleaning, would that affect production?"

Suddenly, Meckstroth sprang up again. "I object to the form of the question, Your Honor," he said to the judge—a little shakily, I thought. I could see him sneaking sidelong looks at Barry. Maybe he'd gotten the same vibe I had—that Barry might like to start brawling. I hoped he did think that. Let him be intimidated. "There's no foundation, it's leading," he told McKay.

"Just restate the question," McKay said quietly.

Barry did, ignoring Meckstroth totally. "Can you tell me what affect, if any, the regulator that required cleaning would have on production?"

Dr. Johnson shrugged. "I don't think any."

"Did you see anything about nutrition or water that figured into your differential diagnosis?"

"I did not see anything that concerned me," the doctor said, and I looked at Dan, seated beside me, and smiled.

"In terms of labor and management," Barry said, "what observations did you make?"

The doctor answered quickly, and with an intensity of feeling that surprised me. "He had some really *good* people, key people very good at helping run his dairy. He had very good people and the farm was well managed."

There! I thought. *Take that, WPS.*

Barry now switched to a new tack. "You have been called on to evaluate damages in lawsuits resulting from stray voltage, have you not?"

"Yes," the doctor replied. "I have."

"You feel you're qualified to do that?"

He nodded. "I'm qualified to show the actual damages. On cases, I've been able to show a dollar loss. I can stand behind my numbers."

Barry nodded back. "And so in looking at Allen's farm records—let's back up. On that parlor system, parallel 24?"

"Double 12," the doctor corrected.

"Double 12. What did that have in the way of computer outputs?" He meant outputs for capturing production data.

Meckstroth now popped up yet again, a nuisance, and this time he sounded highly annoyed and agitated. "Your Honor," he said, and emphasized the word *honor*, "I object. There's no foundation for that question. Dr. Johnson testified earlier that he didn't look at records."

Barry turned around and cast a baleful glare on Meckstroth. "He said he looked at the daily records that were there," Barry told him in a voice barely beyond a snarl.

"Well," Judge McKay said to Barry, "I understood the answer earlier to be that there weren't many records—minimal, not a system. Incomplete."

Dr. Johnson now addressed the judge directly. "I said the only records I could look at came off his milking report." In fact I'd never had a sophisticated software and hardware system for collecting statistics. But I had enough of a record-keeping system to know exactly what my output was at all times for all cows.

"To that extent," McKay said, "you can answer the questions. But only to that extent."

I could've sworn I saw Barry grin wolfishly at Meckstroth.

"So," Barry said walking up near to the witness stand, "you saw what came off the system?"

The doctor had.

When Barry asked another question regarding data analysis, Meckstroth again objected, and this time Barry let it drop, going on to more fruitful subjects.

"In terms of electricity on the Allen farm, did you look into the history of that?"

"All I did," Johnson replied, "is look at history, because you had electrical experts, and I was there well after isolation."

"Was there anything in the electrical history you found significant?"

"A whole lot of things. Mainly the response they got when they isolated the farm in '97 and responses they got by putting the rings around the dairy. So when you look at the significant increase in milk production, *that was something never achieved except at the time of isolation.*" This was pivotal testimony from an unbiased expert with no axe to grind, strongly bolstering my contention of harm from WPS' electrical distribution system.

I could not tell if the jury grasped the significance of this point. They seemed attentive; only a few looked away from the witness or from Barry. The elderly lady in the floral print dress was one. And the guy who never made eye contact with me was another. She was staring out at the spectators; he was gazing down at the floor. But who could tell what, if anything, that meant?

Apparently, it meant something to Kurt Meckstroth, because he suddenly requested a bench conference.

Barry did not let the interruption break his momentum. We were going good now. "Dr. Johnson," he said, resuming, "are you familiar with Exhibit 1010?" It was a chart made from calculations by Dr. Behr of actual production.

"Yes."

"Okay. For the moment, I'm going to ask you to assume that the chart accurately depicts production on the Allen farm at the times designated on the graph of Exhibit 1010. First, what is this darker, wiggly line?"

"That's the rolling herd averages of the Allen farm," Dr. Johnson told him. "That's taking into account all the cows on the dairy divided by the pounds of milk shipped. It shows the rolling—what their average production per month is."

Barry waited as to let the jury digest the information. "So, that's all the cows on the Allen dairy?"

"Yes."

"Cows don't all produce milk at the same time?"

"Correct."

"When you say 'all the cows,' that takes cows that produce zero milk into this average too?"

"Yes."

"This line across the middle, what's that?"

"The Wisconsin state average, what dairy cows have done over time."

"Okay," Barry said. "If you look at production on the Allen farm over the years, as you have in this case, and asking you to assume that in December of 1997 an isolation transformer was installed at the Allen farm."

"All right."

"Based on the evaluation you did and taking into account your differential diagnosis, do you have an opinion to a reasonable degree of certainty as to whether or not stray voltage had been affecting dairy cows?"

Meckstroth suddenly leaped up, flung out his arms and voiced yet one more of his complaints. I was convinced this was his way of upsetting Barry's rhythm and distracting the jury. "Object to the foundation," he said. "Dr. Johnson was not on the farm in '97."

Barry did not deign to look at Meckstroth and instead addressed the Court. "He's reviewed the history, Your Honor," he said with weary patience, and waited.

Judge McKay did not immediately respond. "Mr. Hammarback, it's a multiple question. I think you're going to have to break the question down. Then re-ask the question if you would, please."

Barry nodded, shoved both hands deep into his pants pockets and refocused his attention on the witness. Again he went to the heart of the matter. "Dr. Johnson, based on the examination, evaluations of historic records and the on-farm evaluation you made, and taking into account the history about changes in the electrical system, do you have an opinion, to a reasonable degree of certainty, as to whether or not current has been affecting the Allen cows?" This was a mouthful of a question, and when he finished Barry was a little out of breath.

Meckstroth seemed desperate to deflect the question. His only weapon, which might postpone but never prevent the answer, was the objection. "Your *Honor*," he moaned, "there's no foundation for Dr. Johnson ever having reviewed any records from 1997. His testimony was that he looked at some records, a little before, a little after. I think he needs more foundation." He sat down with a thump.

McKay sustained him. Again. I was beginning to think the judge was sympathetic to WPS and against us. If that was right, we had big, big trouble on our hands.

Not to be put off, Barry simply took a more roundabout route to his point, nearly rubbing Meckstroth's nose in his questions. They demanded a foundation? He would give them a foundation all right, and of concrete. "Dr. Johnson, since you were on the farm, have you received records concerning production?"

"Yes, I have."

"Have you looked at those records?"

"Yes, I have."

"And have records been calculated on production historically, by Dr. Behr?"

"Yes, they have."

"And have you looked at those?"

"Yes, I have."

"And have records been constructed of historic production on the farm by Mr. Mellenberger, the Defense's witness?"

"Yes."

"And have you looked at them?"

"Yes."

"And have you taken these historic records and considered them in your differential diagnosis?"

"Yes."

"Have you looked at the depositions and transcripts and materials relating to dairy production and the performance of the herd, in the past and up to today?"

"Yes."

"And have those been helpful to you?"

"Yes."

The Q and A was unfolding rapid-fire, like a machine gun, definitively, with all the "ands" at the start of a sentence, the way Barry clearly wanted it paced. I could feel his annoyance at having to go this route, but admired his lawyerly skill in maneuver.

"And have you looked at herd health information, somatic cell count information, information from the records and from Dr. Mellenberger?"

"Yes."

"And have they been of help to you?"

"Yes."

"And are they the types of records you would normally look at coming to a determination of whether electricity had an effect on cows?"

"Yes."

Barry drew a deep breath, took his hands from his pockets and crossed his arms. "Okay," he said. "And based on all you've gathered and the history you've gathered, have you developed enough information to come to an opinion as to whether electricity had an effect on the Allen dairy?"

Dr. Johnson nodded. "I believe I have. Yes."

Barry spoke his next question with exquisite slowness in order that everyone in the courtroom should hear clearly, that not one word should be missed. "And what is that opinion?"

The answer came in like a wrecking ball from the high heavens. *"I believe that voltage did indeed have a negative impact on the Allen herd."*

There it was. He said it. Barry paused for what seemed an hour. All fourteen jurors sat staring at the veterinarian, riveted.

Then Barry said quietly, "And what is that based on, Dr. Johnson?"

The doctor opened both arms and made a sweeping gesture. "All the things we've talked about. When you look over his records, you see that the dairy went in a positive direction. It's had ups and downs, like all dairies. Changed, basically, over a 20-year production period." He rose, with the Court's permission, and went over to the production exhibit for a better look.

"We look at this point in time," he pointed to a line on the graph, "they're at about 10,000 pounds of milk. Let's go up to the end of '96, they're around 17,000 pounds. So over a 20-year period of changing technology, they went up about 7,000 pounds. All of a sudden in one year, when they isolate the dairy, they go from about 13 to 22, up 9,000 pounds a year with the same free-stall, the same crappy lights, the same everything. We took away one thing—the voltage that was bothering these cows. And look at the response."

Again Barry paused for what seemed an eternity. "Your visit in 2002," he said. "Taking into account the housing, milk quality, equipment, milking technique and overall management. Taking into account the nutritional levels you observed Taking into account the disease, labor, records, excluding electricity and reproduction—did you look at feet too?"

"Always," he said, grinning, and the jury laughed. "They were on a regular foot trimming program." He went back and resumed his seat in the witness chair.

Barry sauntered over to the jury box. "So taking into account the condition of the *feet*," he said with a smile, "do you have an opinion as to what the production of that herd would have been but for electrical influence?" He was drawing a bead on monetary damages.

"Oh," the doctor said, "their herd easily could be a 26,000 pound herd or higher. He's already shown us that his dairy could hit those levels if we take the stress away."

Then and there I was ready for the case to end and the jury to award me full damages. I wanted to get up, thank them all very much for their kind attention, and go home.

Barry continued constructing the case like a beaver building a dam. "The records you looked at from Dr. Behr, production records, went back how far?"

"To, I think, like 1976 when they started the dairy." Meaning when the partnership started.

"Are you satisfied you understand every significant management issue that was made over the years that would account for substantial production changes?"

"We tried to identify all of them."

Barry began pacing along the length of the jury box, stopping in front of the floral lady for a moment, then going back the other way. "If I ask you, back to 1975, whether you have enough information concerning the dairy, can you carry an opinion back to that time?"

"Sure," Dr. Johnson said.

Barry nodded with satisfaction. I took out a pack of sugar-free peppermint gum, unwrapped a stick, put the gum in my mouth and the foil wrapper in my pocket. I offered Dan Murray the pack but he waved it away.

"I'm going to ask you to assume that when that transformer pole was put next to the feed and utility room, there was current coming down the down ground onto that farm and earth that would access the animals. Assuming that to be true—we'll back that up in this case—do you have an opinion to a reasonable degree of certainty as to when that current would have started to affect the animals?"

"Yes, I do."

"And when is that?" Barry inquired.

"From the point that pole was installed." Immediately. "The effect would build as the ground saturated from cow urine and would only get worse over time." And speed conductivity.

"Did you have an opinion as to when the condition at the farm may be the worst?"

Dr. Johnson considered this for a moment. "Well, probably from the early '90's until the time of isolation is when it was at its worst. By isolating, we had a tremendous positive impact."

The session was temporarily suspended for a break, and after conferring briefly with Barry and Dan, I went outside to stretch my legs and clear my head. As I strolled about the lush courthouse grounds, I watched squirrels scampering among the tree branches and felt peacefulness wash over me. Maybe it was the beauty of the day; the warmth of the Wisconsin sun soothed the tight muscles along my neck and back. But by the time we returned that afternoon, with Barry beginning a new round of questioning with Dr. Johnson, I felt the tension return. The weight was back on my shoulders.

This new Q and A was a prelude to the university studies that WPS' hired guns would be citing to refute my claims. Barry and Scott wanted to prepare the jury and begin a pre-emptive process by citing some studies which were favorable.

Everybody resumed their previous positions in the courtroom. Dr. Johnson returned to the stand and Barry got down to work, positioning himself to the right of the veterinarian so the jury could have a clear view of Dr. Johnson.

"What seems like a long time ago, we talked about a bunch of studies. One, by the University of Minnesota, involved isolation. Explain the finding, and I'm referring to Exhibit 1062."

"They found a significant increase in production of dairies they isolated: one volt of the secondary neutral," the doctor told him.

"Okay," Barry said. "From your training, one volt of the secondary neutral would have transferred into what level of cow contact?"

"About .25 volts."

Barry asked him to explain the top graph of the exhibit.

"That shows changes in rolling herd average which is for all herds in Minnesota over a period of time. They look at a year before the time of isolation and a year after isolation." He leaned forward and extended his arm and forefinger to indicate points on the chart.

"Anything about this study that was inconsistent with the performance on the Allen farm after isolation?"

"No," Dr. Johnson said.

"This study is consistent, then, with your conclusion that electricity was affecting the animals on the Allen farm?"

"Correct."

"Okay. And you still believe it is today?"

"Yes."

I turned and looked at Dan Murray next to me, who winked at the jurors, who seemed fixed on Dr. Johnson, and then at Kurt Meckstroth. He was scribbling furiously on his yellow legal pad, head down and hunched over the table. I would've liked to have been a fly that could read and landed on his pad.

"Why?" Barry asked.

"Because changes that occur yet on the dairy are unexplainable as other things, being able to still measure current. I don't think the farm has eliminated ground currents. I'm not sure it's going to ever get 100 percent eliminated."

Barry walked over to the jury box. "The other thing about this study you mentioned—that you didn't rely on some of the people or think they were authoritative in some areas?"

"Correct," Dr. Johnson affirmed.

"Have there been studies done after this Minnesota study that are not reliable?"

Dr. Johnson smiled. "A lot of them."

Barry said in a conversational, unlawyer-like tone, "Tell us about some of those."

The doctor cleared his throat and took a sip of water. "Well, the level of concern keeps going up and up and up by these people who have made a career of making big dollars."

What a shot! On target and dead center, a bull's eye. These men with the big-dollar careers as testifiers would be up soon. Let the jury hear the truth about them from an objective observer. Then, another shot dead center.

"If you follow their level of concern and their level of income testifying in stray voltages, they parallel each other."

Touché. Good night and good luck. But there was much more. Dr. Johnson did not hold back. Many of the studies were, of course, geared to produce predetermined results. And funded by WPS.

"But, you know, on studies when you have eight or 10 cows and two have to be removed because they won't drink, and you base all your conclusions on the remaining six or eight, you've eliminated 20-25 percent of the cows.

"Man, I'd be nervous to buy a prescription drug that killed two out of eight people, but they're going to tell me the drug's safe based on the six who lived. I would like to know about the two who died."

Strong stuff. This stunning analogy really grabbed the jury's attention. I could see jurors shaking their heads and a few even started whispering to people beside them. No wonder my attorneys had wanted Dr. Johnson up first. He was a tremendous witness. And he was not through.

"More importantly, the new data doesn't fit with what I see in the real world. If we go back 15 years or so, I had as many requests to be an expert for the power company as I did for the plaintiffs. But I won't change my position on level of concern. I find it unbelievable that we have problems when people were doing the work independently and there was—it hurt cows then, and all of a sudden, miraculously, this didn't matter. Now it's a level of 4, 8, 10, 12, 20 times higher. It doesn't fit with what I see."

Barry waited, letting the room reverberate with the echo of the doctor's remarks. Then Barry asked, "Cows have not gotten a lot tougher when it comes to electricity?"

Dr. Johnson laughed bitterly. First and foremost he was a doctor who loved animals and cared about the welfare of his patients; it was why he'd entered the veterinary profession. Animals were living creatures capable of feeling, made of bone and blood and tissue. They were not machines.

"I don't think so," he said sarcastically. "We put them in stressful situations, we're getting more milk out of these cows and, if anything, they're more susceptible."

"Are you," Barry asked him, "familiar with the USDA Red Book?" Which the WPS experts would continually point to as holy writ.

"Yes."

"And do you find that to be authoritative?"

The doctor shook his head. "I don't, no. Most of the data in that book is irrelevant."

Another bullseye.

"Explain why you believe that, Dr. Johnson."

"Because," he answered, "they include studies that eliminated cows that would not drink. I mean, let's face it." And here he blistered the utility and its hired guns with scalding truth: "There's been a lot of learned treatises out there, and controlled research has given us answers that aren't true. *Because the wrong people are paying for the information.*

"When we took Fen-phen, Sebestral, cigarettes, supposedly they weren't addicting, didn't cause cancer. But their research has been proven wrong."

Barry paused for several long moments before going up near the witness stand and posing his next set of questions. Dr. Johnson now had control of the jury they were braced, and Barry knew it. They were practically hanging on the doctor's every word.

"I wanted to ask some rule-out questions," Barry told him. "There's this increase from 1997 and on up on the rolling herd average that you can see on the chart. If you look at your differential diagnosis, did you look over records and the facility to see if housing changes were in any way a cause of that production increase?"

"Yes," Dr. Johnson responded. "I did."

"And what did you find?"

"The vast majority of the increase had occurred before any changes in housing."

Barry crossed his arms behind his back and ventured over to his counsel table. "Were any changes made to major structures in housing?" he asked. "Still had the slats? Still had the old-style free-stalls?"

"Yes."

"What changes were made to housing?"

"Some rubber mats were added and some large fans for ventilation. In April, May, June of '98."

"Okay," Barry said. "Did you rule those out as being a cause of this production increase?"

"Yes."

"Had you become aware of any changes that could have caused that increase?"

"I can't think of anything I could do on any dairy that could get that kind of change. I'd be really popular, more so than I want to be." He laughed sheepishly.

Barry smiled. "Okay. And have you become aware of any nutritional changes significant enough to cause that large an increase in production?"

He shook his head. "After seeing Dr. Mellenberger's reports, I did look closer at nutrition. And found no basis that nutritional changes contributed to that type of milk production increase."

"Okay, Dr. Johnson," Barry said, and pointed to the computer. "Showing you Exhibit 1301, what is this?"

The doctor studied the screen for a few seconds. Then recognizing the exhibit, said, "This is the report Dr. Mellenberger made, but did not complete the graph."

Foley and Lardner's law office completed the graph based on the data in his report. The testimony of Roger Mellenberger, one of WPS' key expert witnesses, would be important in this trial.

"Tell us," Barry said, "what this graph shows."

"The blue line shows milk production over time. The blue vertical bar is feed purchased per cow per day."

"So if you go to 1997 and compare it to the years after, what observations do you make?"

"Well," the veterinarian said with a toss of his head, "the most interesting is when I read Dr. Mellenberger's deposition. He testifies the reason milk production went up is because they fed a lot more protein. It's pretty obvious why he didn't print his own graph. Because the amount of protein fed per day went *down* from '97 to '98 and absolutely contradicts his own testimony.

"This," Dr. Johnson said derisively, "is a graph he uses in every trial I've been in with him. So, when I look at this, I say this absolutely goes against his own testimony where we actually fed less protein and got more milk in 1998 over 1997."

"Can you rule that out as causing this increase in production after the isolation system?"

"Absolutely," the doctor replied. He giggled softly. "In fact, there's one period of time where the feed quality is the worst they've ever had and they got the most milk ever. In October, November and December of '97, they were milking about 40 pounds. Then isolation occurred and all of a sudden they went to 54, 56, 63 pounds."

"Is there a seasonal swing in production on a pounds-per-day basis?"

"About a six-pound seasonal swing."

"When?" Barry wanted to know.

"In the summer," Dr. Johnson told him. "Which goes against what we normally see. They had the highest production in summer."

"Let's go back to the changes," Barry said. "In terms of the BST effect—BST for Bovine Somatatropin, a synthetic hormone given to cows and used by most dairies to improve milk yield—can you rule it out as a substantial factor or is it part of it?"

"I think using BST gave them a bit more milk, but it sure didn't contribute to that huge increase. We know there's no claim by anybody in the industry that we get those kinds of production increases from BST."

Barry continued to nail the subject down. And refute Mellenberger while he was at it. "The six-pound seasonal swing," he said, "would that contribute to the production increase after isolation in December of '97?"

"No," the doctor answered. "Their swing was spring to summer. The rule is, where there's a swing up, there's a seasonal swing down. They counteract each other."

"And so at the end of the year it should have gone down six pounds?"

"Correct."

"And did it?"

"No."

Barry paused again. "Three times a day milking was another thing brought up as causing this increase. Have you looked into what impact this would have on the herd?"

"Well," Dr. Johnson said slowly, "I looked at Dr. Mellenberger's results. What people have to realize is half the herd was already on three times a day. Mr. Allen noticed that the cows significantly increased production when they isolated. So, he put one more group—he has four groups of cows—he put one more group on three times. If you take the difference they had times the number of cows in the group, it figures out to a contribution of 1.9 pounds more milk. That's a long way from the 9,000-pound increase."

Barry crossed the carpet and moved nearer to the witness. "If we look at the effects of increase—again, this is Dr. Mellenberger's chart, Exhibit 1299, can you tell us whether the increase had started in Group 3 prior to three times a day milking?"

At this point Dr. Johnson seemed a bit exasperated. Perhaps he was getting tired. "It went from 38 to 59 pounds before they ever went to three times a day milking!"

Barry kept at him about this. "If we go into the fourth group, had they already increased too?"

Dr. Johnson licked his lips. "If we go back to where isolation occurs, they went from 21 to 45 pounds a day, doubled production, without any other changes."

Sensing that the doctor was anxious to press forward, Barry said, "Okay, you've looked at the veterinary care and the dollars for care over the years at the Allen farm?"

He had.

"Do you know Dr. Foust?" Barry asked. "And is he a good vet?"

"Yes."

"He was there before Dr. Homan?" Terry Homan was our regular vet.

"Right."

Barry wanted to know if Dr. Johnson saw any deviation in the level of care from one vet to the other. The doctor replied that they each offered the same type of care but possessed an individual style.

Then came the point: "Can you rule out that change in veterinary care caused the production increase after the isolation system was installed?"

The doctor nodded. "Dr. Homan would be thrilled to death to think he'd made that much impact in that short time, but I don't think the doctor change was significant. I think it was done not because Dr. Foust was incompetent—just all the frustration."

That was exactly right. I just felt so frustrated and helpless at the time, and figured we needed to change things, and that hopefully our luck would change too. Anything was worth a shot.

"In terms of behavior in the milking parlor, have you taken a look at the milker information sheets?"

"Yes."

"Is that a useful system?"

Dr. Johnson shook his head vigorously. "It's a good system. You gotta realize we're milking 24 hours a day. It's impossible for one person, so we have to have communication between people. And it's an excellent way to look at what happened the previous shift. Or if somebody wants information relayed to the next shift."

Barry wanted to further clarify the process for the jury. "They're just written sheets that the milkers write things down on?"

That was precisely what they were. We had a white dry-erase board and colored markers.

"You've looked at some?" Barry asked.

"I looked at a lot of them. There was a lot of frustration. Like, cows didn't want to come in. More mastitis. One went so far as to say he had to beat a cow."

This had not happened. The author wrote it for me to see as a frustrated, boiling point protest, to get my attention over the massive discomfort and unruliness of the cows that had occurred during his shift. None of my people, ever, had harmed an animal, as Dr. Johnson confirmed.

"I've not been able to verify that on the Allen farm," the doctor said. "People get real frustrated, you know. You can't tell a cow what to do and they don't always listen. If they're not cooperative and things are not going well, you get pretty frustrated."

This was a vast understatement. Words like confused, angry, outraged, depressed and heartsick leapt to mind.

"In terms of a management practice to keep track of what's going on," Barry said, "how did you find it?"

"Excellent," the doctor told him, and my eyes went straight to the jury. Hopefully, they would recall this word when Trevor Will or Kurt Meckstroth began their trashing of my operation. "Many of my dairies have implemented similar sheets. They may note that milking unit 22 doesn't detach, or cow 336 had a sore foot, make sure you check her. That's really an excellent way to have communication among different people on a dairy."

Barry next wanted to know, "Did you look at the affects of cutting the grounds on this farm?"

"I did," the doctor replied. There was really nothing of importance that he had not seen.

"What specific times did you look at?"

Dr. Johnson thought back a moment. "I think it was in December versus January 2002 or 2001. And then an 18-day period in June and July."

"Did you find any general trends?"

The doctor then made a ringing statement, one I hoped would resonate with every juror: *"Cutting the grounds was a positive thing,"* he said succinctly. Because production had jumped drastically afterward.

"The other one you looked at, I think, is when WPS came out and reconnected the grounds?"

"Correct."

"Looking to see if production would go down?"

"Correct."

"What was the trend?" Barry asked him.

"The trend was down. But not as significant as I had first calculated."

With statements like this, his objectivity and commitment to truth could not be easily challenged. At least I hoped the jury would still see it that way after WPS sank its fangs into the good doctor during cross-examination.

"Do you believe, though, that that still supports your opinion?"

Dr. Johnson nodded rapidly several times. "Oh, absolutely," he said, with fervor. He fell silent for a moment, then continued, "If I could legally give him advice for the best interest of his dairy farm, I'd tell him to go home tonight and cut his down grounds."

Nothing could've been plainer, more honest, more straightforward or more telling. I was highly curious as to the kind of tactics Trevor Will planned to use on this man to discredit him. Whatever they were going to attempt, they had their work cut out.

As Barry came down the homestretch of his questioning, he asked for a reiteration of all the doctor's key points and a summary of his main themes. Once again Barry leaned against the railing of the jury box on one elbow and faced the witness.

"In terms of the ability of the Allen herd to produce 26,000 pounds or more in their current facility, do you find housing to be a limiting factor?"

"I think so," the veterinarian said. "They could get to 26,000 with what they've got, but housing is going to be limiting."

"So, at some point they need to redo their facility?"

Dr. Johnson crossed his legs. "Like they did in 1974 or '75, they went to leading-edge technology. It's probably time to think about that again."

Indeed it was, and I had. But with cows dropping like flies and milk production plummeting, I simply could not scrape together the funds required for major improvements. Hell, that was why I was here sitting on this plaintiff's bench.

"Is milking equipment a limiting factor in reaching 26,000 pounds?"

"No. Their equipment can take them to a much higher level."

"In terms of rations I think Steve Root—the nutritionist—is providing, are they a limiting factor in any way?"

"Steve Root is a very capable nutritionist," Dr. Johnson declared. "I don't think nutrition is going to be a limiting factor."

"In terms of disease," Barry said, "have you found any disease organism on the farm that's a limiting factor at the 26,000 pound level?"

"Outside normal incidents of disease," Dr. Johnson answered, "like we see on all dairies, I think with the veterinarian they have, I do not see disease as a limiting factor."

Barry had not moved a mite from where he stood leaning on the rail. "Labor and management," he said. "Is that a limiting factor at the 26,000 pound level?"

Dr. Johnson smiled. "I'd be happy to take Russ' people to other dairies. They're very capable. Russ has done a very good job of picking the right people to help him operate his dairy."

Barry turned and faced the jury. "The record-keeping system, is that a limiting factor?"

"No," Dr. Johnson replied. "It would be nice to have more records, but I think they can manage."

And then came the key question of the trial, the sixty-four-thousand-dollar question. Or better, the 13-million-dollar question. The jury's answer to it would determine my future: "In terms of electricity on that farm, in your opinion, is that a limiting factor today?"

The doctor replied without a moment's hesitation. "Absolutely. Until we eliminate all current, this herd will be constantly fighting that battle to get where they need to go."

Barry said nothing. He did not move. Dr. Johnson sat still, quietly awaiting the next question. You could hear a pin drop in the courtroom. There was no sound, not from the jury, lawyers, bailiff, stenographer or spectators. Finally, Barry broke the silence with an anticlimactic question.

"The reproduction program on the farm today, and the feed, are they limiting in any way?"

Again Dr. Johnson shook his head. "No," he said softly, and nothing else.

Barry was done. Dr. Johnson was done, at least for now. There was nothing left for Barry to ask him. As it was late afternoon, Judge J.D. McKay adjourned for the day. I had the remainder of the night to imagine the ambush Trevor Will and Kurt Meckstroth were plotting for Dr. Johnson. With imaginary fragments of a vicious Q and A swirling around in my brain, I pictured Meckstroth forcing Dr. Johnson to recant everything and blame all the problems on me. That was the emotional, scared half of me at work. The rational, calm half was encouraged. From my layman's perspective, Barry and Scott had made a golden decision in putting Johnson on first; he was a great witness for me, had given powerful testimony. As I lay down in bed that night, I could only pray that the rest of my witnesses would testify even half as well as Dr. Johnson. And that included me.

Chapter Eight

AT CROSS PURPOSES

It turned out to be Kurt Meckstroth. This guy would do the deed with Dr. Johnson—cross examine him—and I expected fireworks bigger than the Fourth of July. For Meckstroth had been the one who had originally deposed the udder doctor, and Dr. Johnson had become so enraged that day by Meckstroth's arrogance and belligerence that they'd nearly gotten in a fistfight. I dearly hoped it might happen again so that the jury would turn against Foley and Lardner and, by extension, WPS.

Meckstroth, obnoxious during last month's deposition, badgering Dr. Johnson with idiotic questions he could not possibly answer, had actually made a threatening physical move toward him. Sitting across the table from one another, at times the two were nose o nose and eyeball to eyeball.

"Can you," Meckstroth had asked for the fifth time, triggering a heated exchange, "narrow your visit to a certain farmer who had cut his grounds down any closer than in the last ten years?"

Dr. Johnson had already told him he could not remember, since he'd visited more than a thousand farms during that period. "Kurt," the doctor finally replied, jaws clenched and shifting his 300 pounds uncomfortably in the chair. He sighed loudly, blowing air out his nose like a bellows. "If I could narrow it down to the exact time and date, I could remember the damn farmer's name! Why do you keep asking me stuff that I can't—if I could tell you it was one and a half years ago in June of 2001, I would tell you that. I go to so many farms, more than probably any other dairy consultant you will ever run into. I can't remember. I'm just telling you what I can recall. If you would like—I thought it was unethical for a lawyer to answer my questions, but if you want to go there, from now on I will ask you, Kurt, what you would like for an answer. I've done the best I can. I am not trying to hold anything back. I just flat-ass don't know the exact date!"

Meckstroth, snotty as a child denied dessert, shot back, "I didn't ask you the exact date, Dr. Johnson!"

Now for some crazy reason, Kurt Meckstroth—a better name would've been Quirk Meckstroth—sprang up out of his chair, head low and arms outstretched, looking for all the world like he intended to dive across the table into the udder doctor and ram him like a tackling dummy.

Dr. Johnson clearly read the movement that way. Bug-eyed, prepared to defend himself, the doctor vaulted up. I thought the two would collide like Sumo wrestlers, right in front of the stunned court reporter.

"Come on, mister!" Dr. Johnson roared, goading him on. "Don't you treat me unprofessionally. Don't you dare get up like you were going to attack me! Or, I'll flatten your ass!"

The readiness to rumble shown by the normally genial doctor stopped Meckstroth dead in his tracks. He whined, "I move to strike. I did not get up!"

This was laughable. "You did, too!" Dr. Johnson yelled, assuming the stance of a fighting cock with his legs apart and his chest thrust forward.

Barry Hammarback now intervened to restore a semblance of decorum. "Record should reflect he started to get up," Barry said to the stenographers, unable to suppress a smirk.

"Yes, he did!" Dr. Johnson hollered.

"I take exception to that," Meckstroth said prissily. "I did not start to get up out of my chair!"

"You did too, you threatened me!" Dr. Johnson bellowed.

"I did not!" Meckstroth said. At that point, I wouldn't have been a bit surprised if they'd started sticking their tongues out and making faces at each another.

Still refusing either to sit or to let it go, Dr. Johnson retorted, "You did, too!" Then turning to Barry, "Could I report him to the ethics board?"

Barry leaned back in his chair with hands clasped behind his head. He was enjoying the show, grinning like the Cheshire cat. "You can do whatever you want," he said.

This calmed Dr. Johnson. He resumed his seat. "Thank you," he told Barry. "That is unethical."

Then the doctor issued a stern warning to Meckstroth and shook a chubby finger at him. "You don't like the answers, don't you dare threaten me again. If you come out of that chair next time—you have been warned. I have the court reporter as a witness, three other people here. I will deck you if you attack me again."

"I move to strike Dr. Johnson's entire response as being non-responsive," Meckstroth griped.

"You have gotten an answer," the doctor snapped.

"You can't narrow the time span down?"

"No, I can't." The doctor nearly spit the words out. "Would you like to put in a number? Are you that unethical? Pick a number between one and ten."

Everybody laughed. Except the attorney.

And so it went that day, Meckstroth jabbing and Johnson parrying. And that was merely the deposition, small potatoes compared to the trial. As the principals entered the courtroom and assumed their positions of yesterday, I wondered if history was about to repeat itself.

Meckstroth, in a dark-blue suit, light-blue shirt and paisley, blue-and-white tie, began the questioning by asking Dr. Johnson to restate his opinion on when the current first began to affect cows on my farm. They covered much of the same ground Barry had gone over yesterday, and Meckstroth read into the record portions of the doctor's deposition. No more than 15 minutes had elapsed when their first clash occurred.

"Do you," Meckstroth asked, "have an opinion as to when Russ Allen's herd first became adversely affected by electricity?"

"An estimate based on the records, it was several years prior to isolation. But I can't give you a start date."

"Did I ask you that question on the day of your deposition April 17, 2003?"

"Yes."

"And did you tell the truth on April 17, 2003?"

"Based on the material I had at that point in time, absolutely."

Meckstroth nodded. "Now, Dr. Johnson, you've had an opportunity to meet with Mr. Hammarback since April 17, 2003, haven't you?"

I supposed he was implying that somehow Barry had put the doctor up to giving testimony favorable to me, never mind whether or not it was true.

"Yes," Dr. Johnson said.

"And you knew that unless you changed your testimony it was likely that the Court was not going to permit Mr. Allen to recover any damages for any time period before 1997, the time the isolator was installed. Didn't you?"

Dr. Johnson glowered. "No, I did not."

"Well," Meckstroth said, "you testified that you relied heavily on data compiled by Dr. Behr. And you knew, didn't you, that last week Dr. Behr prepared a new report calculating damages only from August 15, 1991 to the present. Didn't you?"

His tone was that of a grand inquisitor demanding a confession from a heretic.

"No, I did not," the doctor grunted. Perhaps their last go-round was still fresh in his mind and he was anticipating another. "I've never seen that report."

Apparently, Meckstroth felt there was nothing for him in this line of questioning, so he turned his attention to the milker communication sheet on which a frustrated worker of mine had written that he had beaten balky cows. Which never happened. But it seemed that Meckstroth was going to blame cow hysteria not on current, but on animal abuse by my employees.

"The comment made," Meckstroth said, looking at the communication sheet and then reading from it (Exhibit 1743), "was this: 'We had to milk—we had to milk again. No f---ing notice. So the f---ing cows got a good beating because they pissed us off.' Correct?" he asked, showing the sheet to the witness.

"That's what it says," Dr. Johnson answered.

"And the milkers identified themselves as 'f--- this' and 'f---k that,' correct?"

"At least they're not initials," Dr. Johnson said, and the entire courtroom burst out laughing.

"And," Meckstroth continued, not cracking a smile, "they added the comment, 'Next time we want notice.'"

"I guess they hated to milk twice in a row." He meant working a double shift.

"And, Dr. Johnson, you don't know if they got adequate notice about the next time they had to milk, do you?"

"I've never met either one of these milkers, 'f--- this,' or 'f--- that,' so I wouldn't be able to answer," the doctor said dryly.

The courtroom roared.

"So," Meckstroth went on, his faced pinched like he needed a laxative, "you don't know if the milkers beat the cows next time because they were upset for not getting adequate notice, do you?"

Barry objected to the question as being argumentative and McKay sustained him. So Meckstroth tried a different tack.

"Dr. Johnson, you consider the design of the free-stalls in Mr. Allen's barn, to use the words you used in your deposition, to be pretty shitty. Don't you?"

The doctor grinned. "You sure seem to be doing a lot of swearing," he told the lawyer.

"I apologize," Meckstroth said, red-faced, and the jury burst out in laughter again. "All my swearing has been quotes, though." Then he quickly changed the subject.

"Dr. Johnson, yesterday you testified about differential diagnosis. And the idea was to take every limiting factor on a farm and rule them out to see what was holding the herd back, right?"

"Tried to, yes."

"Milk quality is another one of your factors besides housing and cow comfort isn't it? And a key component is the milking practices followed by the milker, right?"

"Yes, sir."

"Another item is nutrition, right? You didn't see any rations from before the fall of 1997, did you?"

Dr. Johnson shook his head. "I only looked at rations prior to and after isolation."

"You think nutrition is a matter of art rather than science?" Meckstroth ambled over to the jury box, looked at the jurors, and then turned back toward the witness.

"A lot of it, sure."

"You don't know Steve Root, do you?"

"Not personally. I don't think he's the nutritionist for any of the people that I work for."

For a moment, Meckstroth stared silently at the doctor. I got the strange feeling something was going on under the surface but had no idea what it might be. I turned to Dan Murray and whispered, "What's up?" Dan simply shrugged.

"You testified yesterday," Meckstroth said, "that you were impressed with the quality of the people that worked for Mr. Allen."

"When I was there, yes."

There was a pause. "Did Russ Allen tell you that within the last month his head milker pled guilty to reckless homicide?"

This was not fair game and shouldn't have been introduced. But Meckstroth, sharp as he was, had slipped it in. Barry immediately jumped up and requested a bench conference. The spectators were buzzing and it seemed bedlam might erupt in the courtroom. McKay ordered the jury out.

When they were gone Barry demanded that the question be stricken from the record. He also demanded a mistrial, which the judge refused to grant.

What had happened was this: My head milker, Sergio, a Mexican who'd been my employee for about two years, had, along with two others, beaten a man to death. When I learned of it, I was dumbfounded, as Sergio had always been polite and quiet and was a conscientious worker. Evidently, there was bad blood, some sort of vendetta or family feud going on. After a night of heavy drinking the three had offered the victim, also a Mexican and a drunk, a ride home from a bar. They drove to a desolate spot, dragged him from the car and beat in his head with baseball bats.

By mentioning this, Meckstroth was looking to refute Dr. Johnson's claim that my workers were excellent. He wanted to portray them as brute savages responsible for abusing the herd, causing cow misbehavior and lack of milk production. His contention, as he told McKay during the bench conference, was that a man capable of murder with a baseball bat was quite capable of animal cruelty. The ploy worried me because, while untrue, it had a certain plausibility. I hoped McKay would disallow the whole subject.

"If this were a vehicular homicide," Meckstroth argued to the judge, "it may not be pertinent. But given the nature of this homicide, it is pertinent. A person capable of doing that to a human being would certainly be capable of hurting cows. That's an issue, whether these cows were mistreated in 1997 or any other time affecting the performance and health of the herd. It is relevant, and whatever prejudice Mr. Hammarback may feel it's going to present, it does not override the materiality of the testimony."

The judge opened his hands and invited Barry to respond.

Barry said to him very deliberately, "I've been practicing law quite a while in Wisconsin, and I consider even the thought of this type of testimony as being outrageous. Furthermore, in light of the Court's pretrial order that such types of conduct be screened before presentation to the jury, it goes directly against the order in this specific case."

Barry was not finished. He was earning his fee in spades; his lawyering here was rock solid and very much on point.

"Wisconsin Statute 906.08, which talks about evidence, character and conduct of a witness, particularly Subsection 2, says you may not prove a person's actions by extrinsic evidence. The only thing you can do is use other acts for truthfulness on cross. So, the defense is trying to prove an extrinsic act about a character that nobody even placed in this trial. The plea of guilty took place a month or two ago, *after* the increase in production. So the argument that somehow the beating of cows had anything to do with a decrease in production is preposterous. Most important, this is incredibly inflammatory. Its object is to incense and inflame the jury about a tangential issue not germane. I move for a mistrial based on misconduct of counsel in violation of the Court's order."

These were powerful and true arguments. I noticed Dan Murray beside me nodding appreciatively as Barry finished.

Meckstroth then made a counterargument that the issue for him was Dr. Johnson's assertion about the exemplary character of my employees, and the question of what type of information (or misinformation) I was feeding the doctor. This sent a nervous shiver down my back. I wondered if the entire case might hinge on this point.

Judge McKay hedged his bets and split the difference. Rejecting Barry's motion for a mistrial, he agreed with Meckstroth that Dr. Johnson's reliance on labor/management issues as part of his differential diagnosis left him open to questions about the character of my staff. He then decided to strike the inflammatory question and would instruct the jury to disregard it.

"I think," the judge told Meckstroth, "there is an area of inquiry which is appropriate. I don't think it's necessary for dramatics. You can make your point, I'm sure you know how, without referring to the homicide. A homicide of a human being and the manner in which cows are treated are far apart."

Meckstroth protested, but it was too late. Just before McKay ordered the jury brought back in, there was a short break. When I returned to the courtroom, I caught Trevor Will and Meckstroth laughing that they got the damning evidence in. They knew it would be hard to erase from the jurors' minds. The judge instructed the jury to disregard the question about the homicide. I was getting an education on courtroom deportment, the law and legal tactics, more than I ever wanted to know. Strange, the twists and turns life takes. Never in a million years would I have ever imagined becoming involved with electricity, much less losing beloved family members to it.

Dr. Johnson resumed his place on the witness stand. Meckstroth began revisiting areas of the doctor's differential diagnosis, arguing briefly with him about small points from his direct questioning by Barry. It seemed to me that Meckstroth's idea was to challenge and try to chip away at many of the doctor's conclusions about the electrical problem on my farm, to create numerous small points of doubt in the jury's mind as to the actual source of the problem, all of which would add up to one big doubt. It wasn't a bad strategy for a defense attorney who did not need to prove anything; all he had to do was create enough uncertainty to prevent them from finding in my favor.

"Dr. Johnson, there's something called 'on-farm stray voltage,' right? That's caused by the electric system that the farmer's responsible for, right? And then there's off-farm stray voltage, the kind that comes from the distribution line, right?"

"Yes, sir."

"You don't know whether that entire measurement Larry Neubauer made was caused by on-farm stray voltage, do you?"

The doctor admitted he did not know, and I squirmed on my bench. Meckstroth was scoring.

"Dr. Johnson, the only conduct of cows on the Allen farm that you attribute to stray voltage was that you thought the cows weren't comfortable going in, or while in the parlor while you were there, right?"

"Yes."

"You never saw the cows lap water, right?"

"No, sir."

"And you don't recall them being steppy when you saw them in December of 2001, do you?"

"I do not recall that," Dr. Johnson said.

I glanced over at the jury. My floral lady was eyeing up the judge, and my I-won't-look-at-you guy was studying the spectators; one girl was fiddling with her earrings, while another just stared at the court stenographer. I wished I had Jo Ellen Demetrius, the famous jury consultant, sitting here to tell me what these folks were really thinking.

"So the bottom line is," Meckstroth was saying to the witness, "you've never seen the cows appear steppy on his farm, have you?"

Dr. Johnson pushed out his lower lip. "All experts have that problem."

"So the answer to my question is, no, you've never seen them act steppy?"

"That is correct."

Meckstroth wanted to know whether the doctor thought a cow could feel less than one milliamp of current, and did he always use a 500 ohm resistor to replicate the animal when testing at farms. Dr. Johnson explained that the literature on the subject suggested this. Meckstroth next tried to show that the udder doctor was behind the times when it came to contemporary thinking on the subject, as demonstrated in the Public Service Commission's raising of the level of concern.

"You consider yourself one of the few people who refuse to change your view of what the appropriate level of concern is, don't you?" Meckstroth said.

Dr. Johnson smiled. "Yes, sir," he said. "Until the cows show me something different, that's exactly where I'm going to stay."

Then, a body blow from Meckstroth. I awaited the fireworks display.

"Well, by not changing your level of concern, you've been able to preserve your expert witness cash flow, haven't you?"

Meckstroth certainly knew all about witness cash flows, since his own witnesses were all wealthy men grown rich racking up gigantic fees testifying for utilities. Psychologists call it projection, when you accuse others of your own faults.

"Yeah," the doctor said with sarcasm. "If you consider my cash flow significant. My yearly income is a tenth—less than 20 percent of what your expert charged just in one case. So that's ridiculous. My percentage—is minimal."

Meckstroth had gotten under his skin and I was glad of it. Maybe now he'd show the jury his true colors.

"I don't enjoy these cases," Dr. Johnson said. "I turn down more than I take, and the bottom line is, we have a problem in the dairy industry that you people spend millions—"

"Your Honor—" Meckstroth interrupted, unhappy about the accusation the good doctor was about to hurl at WPS.

"—trying to fight," Dr. Johnson finished.

"Your Honor, I move to strike the answer as nonresponsive!"

Barry now chimed in. "I think it's responsive."

Behind me, I could hear a number of spectators chattering. They too seemed to be hoping for fireworks.

"I think," Dr. Johnson added, before Meckstroth could get in a word, "it's *very* responsive to your question." This comment made me laugh out loud along with several of the jurors. I liked that the doctor was spunky. In addition to his expertise, he had brass.

"Next question," Judge McKay said, indicating agreement.

Meckstroth's face reddened. "I do move to strike the last portion of his answer!"

McKay peered down from the Olympian heights of his bench. "I'll strike the editorial comment, not the answer itself."

This seemed to satisfy Meckstroth and he quickly moved on. "Earth currents aren't new, are they Dr. Johnson?"

"Been there a long time, but we're learning more about them as time goes forward."

"Wisconsin has had a multi-grounded distribution system for over 50 years, hasn't it?"

"Unfortunately, yes."

"The only thing new is that vets like you who have refused to change your level of concern have encouraged farmers to bring lawsuits, right?"

Another low blow from the master. As soon as Barry heard this impudent question he hurled an objection. "Completely argumentative, Your Honor."

"Am I allowed to answer?" Dr. Johnson asked McKay, and I knew from his tone that, if allowed to respond, he was going to tear into Meckstroth. Meckstroth may have sensed it too because he left the vicinity of the witness stand and scooted over to the jury box. Maybe he figured there was safety in numbers.

"Not yet," the judge told Dr. Johnson.

"I'll withdraw the question," Meckstroth said. He returned to a subject near and dear to my heart— expert witness fees. Talk about people living in glass houses; I was amazed that he dared to raise this topic again, given the names on his witness list. I crossed my legs, folded my hands behind my head, and waited to see how Dr. Johnson would handle him. I already had a pretty good idea.

"You're charging $200 per hour for your testimony here today, aren't you, Dr. Johnson?"

The doctor looked him in the eye. "Yes, sir."

"Now, you don't know how much Russ Allen paid Larry Neubauer for things Mr. Neubauer has done that's purportedly limiting adverse electric currents on his farm, do you?"

How I loved that word, *purportedly*. *Purportedly,* some cows had died at my farm. *Purportedly,* I was out of the milk business. *Purportedly*, my future was on the line in this trial.

"I don't really care," the doctor said.

Meckstroth next went on a Q and A tangent about the doctor's *purported* failure to examine vet records and cull rates on my farm, trying to show that Dr. Johnson had not delved deeply enough into conditions to be able to form worthwhile opinions. Meckstroth won a small point by reminding the doctor of a mistake he'd made in calculating my milk production after I had cut the down grounds.

"Once the mistakes in those calculations are corrected, those production increases are within normal herd fluctuations, right?"

"Correct."

"So that information isn't critical anymore, is it?"

"I'm not sure I agree. If you guys, the utility, weren't worried about the results, you'd probably allow them to be cut."

Frowning at the witness, Meckstroth moved to strike. McKay agreed.

Meckstroth suddenly left the jury and moved over toward the witness, and I half-expected a replay of the firestorm at the deposition. But, oddly, nothing happened.

"So," Meckstroth said, "you've never considered that if you treat 1997 and 1998 as aberrations, the production increase on the farm would have been consistent? That's not something you've looked at."

"No."

Meckstroth was attempting to show the jury that the production increases after my cutting of the down grounds were not caused by the cutting but were the cumulative effect of changes I'd made by remodeling the barn, in ventilation, culling of poor-producing sick cows and BST supplementation; that the lack of current had nothing to do with it.

"You've not made any efforts to quantify the impact of the remodel?" Meckstroth said. "If the remodel that occurred to the old barn in 1996 improved cow comfort and if there was some limiting factor in 1997 that caused the drop in milk in 1997, and if the nutrition problem was overcome by 1998, you would expect to see production increases compared to, say, 1995, resulting from the benefits of remodeling. Correct?"

"Sure."

Now he launched into a Q and A on the deficiencies of my barn, stalls, lighting, equipment, designed to demonstrate that those things were responsible for production decreases and not electrical current.

When he had beaten that subject to a pulp, he went into milking procedures and the ways and means by which cows received teat and udder infections. Dr. Johnson gave him a tutorial, but I became restless and wondered where this was all heading. For certain Meckstroth was not looking to shore up gaps in his own dairy knowledge. I waited for the other shoe to drop.

But what fell was only a small bedroom slipper. Dr. Bodman, another vet who had observed my farm, had made notes which Meckstroth construed as critical of my milking methods and as causing tissue damage to the animals. The notes were open to interpretation, as the udder doctor deftly proved, but Meckstroth kept trying to maneuver Dr. Johnson into conceding that my "faulty" methods were responsible for production shortfalls. Then he again began to harp on equipment problems.

"Now," Meckstroth said, "you testified yesterday that one of the things you did was look at records pertaining to the maintenance of milking equipment prior to your visit to the Allen farm, right?"

"I don't know if I said 'prior,'" Dr. Johnson answered. "But I did look at records. And I also had a conversation with the dealer."

"One of the records you looked at was a printout of an inspection of pulsators done in 1999, right?" These were part of the equipment for milking.

"Yes."

"And that printout reflected that 15 of the 24 units in Russ Allen's parlor in the east barn were set at a ratio of 65 to 35, right?"

Dr. Johnson looked down and thought for a moment. Raising his head, he said, "I'll assume for this that that is correct."

"Well, 15 were set one way and nine the other, right?" Meckstroth said.

"Yes."

"At the time I took your deposition, you had never seen that happen before in a—"

Dr. Johnson nodded several times. "I'm glad you brought this up because it can't happen in a DeLaval system. I went to DeLaval and talked to their engineers. It is impossible. If I can explain how the pulsators work, this is not an abnormality. It was a change made when the dealer was at the farm because the system is incapable of functioning as those graphs were made."

The explanation riled Meckstroth. "Your Honor, I move to strike. My question was did he ever see it prior to that day. I think the answer is no. But I haven't heard him say that."

McKay glanced over at the witness. "I'll strike the answer as nonresponsive and ask you to respond to the question."

The doctor grinned and shrugged his shoulders. "I had never seen that on a DeLaval system ever before. Correct."

"At the time I took your deposition, you didn't even think that was possible, did you?"

"As I explained," Dr. Johnson said patiently. "It is not possible."

Meckstroth paused. "When you conducted this review of the milking equipment records, you didn't detect abnormality in the pulsation ratios in these records?"

Dr. Johnson shook his head dejectedly, as though trying and failing to get through to a hopelessly backward pupil. "The problem is, you're misleading everybody. There was no abnormality. It was normal pulsation, but the ratios changed. A normal change within the pulsators. Something we do all the time on dairies, changing pulsation ratios all the time."

"But at the time I took your deposition, April of 2003—"

"I did not know an explanation for that change, correct, but it was not described as an abnormality for me."

Meckstroth's eyes narrowed. He was struggling to damage the doctor's credibility on this subject, groping for some magic bullet. I doubted he would find one.

"And you weren't even aware there were nine pulsators set at one ratio and fifteen set at another," Meckstroth said.

"I wasn't aware," the doctor replied dryly, "that a Martian flew over that night either. It's irrelevant."

Apparently, recognizing that he could not enter the building through these areas, Meckstroth tried breaking in through a window. He walked over to the defense table and whispered back and forth with Trevor Will for several moments, nodded and then embarked on the next set of questions.

"One of the things you recommend to clients is that they employ a milking order, a process by which the farmer milks cows based on whether they're infected with mastitis, right?"

The doctor nodded in agreement. "Not necessarily group them together, but know who the cows are that need to be dealt with."

"The idea is to milk them last so cows that aren't infected don't catch infection when milking units are attached?"

"That's not really right," Dr. Johnson told him. "Most of my dairies don't have the luxury of milking them last. The key is to sanitize the unit before another cow is milked."

"Well, Mr. Allen never used a milking order, grouped cows with infections so they could be milked separately, did he?"

"Correct," Dr. Johnson said.

"Now, doctor, you never discussed with anybody on the Allen farm whether Bob, Russ Allen's brother, was supposed to be injecting his cows with muse but wasn't doing it, right? Muse is a substance that contains vitamin E and selenium, two substances important in promoting bovine health, right?"

"Sure."

Meckstroth folded his hands together and turned toward the jury. "If the Allen farm was relying on Bob Allen giving muse, and if, unbeknownst to the vet and others, he wasn't doing it, would you expect that would contribute to incidents of retained placentas?"

"Sure could," the doctor told him. "Sure."

"There's a number of things that can affect an artificial insemination-conception rate on a farm, correct?"

"Sure."

"These things include the abilities of the person doing the breeding, right?"

"Absolutely."

"Handling of the semen, right?"

"Sure," Dr. Johnson told him.

"Nutrition?" Meckstroth asked.

"Sure."

"Housing?"

"Sure."

"Weather?"

"Sure."

"It would also include heat detection, wouldn't it?"

"Oh, absolutely," the doctor said.

"That means somebody has to notice when a cow is in heat, right?"

"Sure."

Meckstroth was going in circles. Round and round he went and where he would stop, nobody knew.

"And the way someone can detect a cow in heat is that one cow will mount another, simulating the breeding process, right?"

Not simulating the process, but actually trying to engage in bovine sex.

"That's one way, sure," Dr. Johnson said, and the courtroom laughed.

"They're both female animals," Meckstroth said, "but one cow will actually mount another and the one on the bottom is the one usually in heat, right?"

"That's possible," Dr. Johnson said, grinning.

Meckstroth returned to the jurors. "Housing is a factor in the conception rate because if cows have bad footing they tend to be afraid to mount other cows, right?"

Dr. Johnson shrugged. "I guess that's a possibility."

"And that makes it difficult for the farmer to detect when the cow's in heat, right? So he doesn't know when to artificially inseminate her, right?"

"It's always a possibility, yes."

"You've never calculated conception rates for the Allen farm, have you?"

The doctor drew a long, deep breath of impatience. "Since we've covered that multiple times, I said I did not. That answer would stay the same."

After some more inconclusive and ill-tempered jousting with the doctor, Meckstroth tried to bracket his main contentions in the last bit of his Q and A, leaving the jury thinking about his main themes.

"Dr. Johnson," he said, standing beside the jury-box railing, "it takes a long time to review all those vet records, doesn't it?"

"Yes, it does."

"Takes a long time to do analysis of vet records, doesn't it, Dr. Johnson?"

"Yes, it does."

"And it takes a long time to analyze milk production records, doesn't it?"

The doctor clucked his tongue. "You already asked that two questions ago."

"No," Meckstroth said. "My question now is, it takes a long time to analyze—"

"That's exactly what you asked two questions ago," Dr. Johnson reiterated.

The judge now intervened. "Let's not argue about it," he growled. "If your answer is the same just say your answer is the same."

"You left that work," Meckstroth said accusingly, "to be done by Dr. Mellenberger, didn't you?"

"No," Dr. Johnson shot back, "I left that to other people. As I said in my deposition, other people were looking at that. I did not see a need to duplicate efforts." The doctor must've wondered if the attorney was trying to put him on trial.

"Dr. Johnson, BVD is not caused by earth currents, is it?"

"Not that I'm aware of."

"And Johne's disease is not caused by earth currents, is it?"

"Not that I'm aware of."

"And milk fever is not caused by earth currents, is it?"

"That's correct."

"Dr. Johnson," Meckstroth intoned, smiling wanly, "you're not aware of any scientific research that associates acidosis with electric current, are you?"

"Not that I'm aware of."

"And you're not aware of any scientific literature that associates laminitis with electric current, are you?"

Dr. Johnson shifted position and crossed his right leg over his left. "That's correct."

"And you're not aware of any scientific literature that associates electric current with retained placentas, are you?"

"That's correct."

"And you're not aware of any scientific literature that associates mastitis with electric currents, are you?"

"That's correct."

"And you're not aware of any scientific literature that associates pneumonia with electric currents, are you?"

"That's correct."

All these diseases he mentioned, by the way, were common at various times and in various amounts on every dairy farm in the world.

"And you're not aware," Meckstroth droned, "of any scientific literature that makes a correlation between stray voltage and the suppression of a cow's immune system, are you?"

"That's correct."

The attorney paused, stepped away from the jurors and went over near Judge McKay. "Your Honor," he said, "I have no further questions."

McKay now allowed Barry to question the doctor on redirect, and Barry made the most of the opportunity. He rose from behind the table and moved close to the witness stand.

"Dr. Johnson, let's review a couple of things. All of these things, BVD, Johnes, acidosis, retained placentas,— can those be caused secondarily by stray current?"

"If the cow's stressed, it can be more susceptible to diseases, yes," Dr. Johnson said.

"From an applied science point of view and from field observations, do you have an opinion to a reasonable degree of certainty in veterinary medicine as to whether there's a relationship between stray voltage and stress in dairy animals?"

"Yes, I do."

"What is that opinion?"

Dr. Johnson said forcefully, "The common denominator on dairies with stray voltage problems—these issues are present and made worse when the cow's under stress of current."

"The things Mr. Meckstroth brought up," Barry said to him, "diseases, illnesses, nervousness in the parlor, teat-end injury and so forth, can all those things also be caused by electricity?"

"Absolutely."

Barry waited so the jury could absorb the finality of the statement.

Then, Barry said, "Showing you Exhibit 1032 indicating .657 volts RMS—above one milliamp—measured cow contact with the resistor, does that exceed what you think is acceptable for dairy farms?"

"Yes."

"Asking you to assume this measurement is made with the isolation system bypass and neutrals connected back, and asking you to assume that was the condition prior to December of 1997, do you have an opinion to a reasonable degree of certainty that the Allen herd was affected by stray voltage?"

The doctor nodded once. "I have an opinion."

"And what is that?"

"It was as if cow behavior—showing problems, and finding voltage over 0.5 volts, I would consider that a problem."

"Okay," Barry said. "And the voltages in excess of half a volt, the recommendations that higher levels were safe came out of what kind of publication?"

Dr. Johnson chuckled. "Learned treatises, that I don't believe in. The Red Book and others that go against what I see."

"Dr. Johnson, referring to Exhibit 1113, would you read to the jury paragraph one and paragraph three, please?"

The doctor looked the document over and began to read aloud in a slow, enunciated voice.

"'Number 1: The report is based primarily upon the result of just two laboratory-based research studies. These studies were conducted with relatively few numbers of cows, six to eight, and under well-controlled laboratory conditions, i.e., stresses that fall on the cow from other sources were kept to a minimum.'

"'Number 2: Consequently there is serious question whether the results would be similar had the animals been exposed to the full range of stresses, handling, housing, grouping, weather, etc., commonly encountered on a dairy farm where everything is not ideal and they do not get individual attention on a routine basis.'

"'Number 3: Each of the two stages, according to the USDA publication, began with at least two additional cows. In each instance, at least two animals reacted so violently during the initial phases of the project that they were removed from the actual study soon after it had begun.'

"'This was done because there was fear that continuing to utilize these animals would result in their permanent injury or even death. In one instance it was reported personal conversation with one of the researchers that a videotape taken of these animals showed such violent reactions that the research institution's legal counsel advised that the videotape be destroyed. It was.'

"'This was done for fear someone from outside the research institution would see the video and problems could result because of animal cruelty and welfare concerns. Consequently, the lower range of extremes of a typical distribution of a bell curve were deleted, shifting the results towards higher voltage and current requirements to achieve the same results.'

"'As I have told several of the researchers, I cannot argue with the data they presented. However, when these studies are taken in total, one can only conclude that this is not the proper way to do scientific research and then report the results as though they are meaningful. The report results are badly biased and their worth must be seriously questioned.'"

The doctor finished reading, and Barry again paused.

"Okay," he said. "That was written by Mr. Bodman. Do you agree with all that?"

"I sure do."

Barry began pacing back and forth in front of McKay. "Showing you Exhibit 1311 from the Red Book, is that chart in any way a reliable predictor of the effect of electricity on dairy animals?"

"No."

"Does the chart show what you see in the field? And on a perception level, do they see what you see in the field?"

"No—many cows perceive at a much lower level than that."

Meckstroth asked McKay for a short re-cross, which he granted. I was less nervous and more confident after Barry's redirect; I thought he'd done well in focusing the jury on the key points of the doctor's testimony. Whatever Meckstroth had up his sleeve could not, I figured, be earth shaking. Or he'd have gone there much earlier.

He rose from his table after Barry had taken a seat, and went over to the jury.

"Dr. Johnson. Earlier we looked at answers you gave about the start dates and your view of damages. Mr. Hammarback read you a question he asked at the end of your deposition and the answer you gave, right?"

"I didn't realize he asked it or you. You guys are so good at asking long questions," the doctor grinned, "I'm not sure I remember who asked what. So, I don't think that's worth the trouble. I'll assume it was Mr. Hammarback so we can keep moving."

Meckstroth did not smile. Ever. Maybe he was afraid his face might crack. "There's a fundamental difference between the answer you gave Mr. Hammarback and the answers you gave to my questions, specifically about your opinion of when electricity started affecting Mr. Allen's cows, right?" His tone was snippy.

"Semantics," the doctor said, matching snip with snip.

"Nothing you provided addressed the start date of damages, did it?" Meckstroth looked at the jurors.

"I think we're talking the same thing, but evidently, there's a disagreement," Dr. Johnson said. "I guess you'll have to read the question again."

Meckstroth re-read the question: Whether or not Dr. Behr's analysis of where my milk production would have been but for stray current was accurate. Dr. Johnson believed it was.

"Were you asked that question?" Meckstroth said. "And did you give that answer?"

The doctor shook his head several times as if to say, *What kind of stupid question is that?* "Evidently," he told the lawyer. "You just read it!"

"Nowhere in that answer did you talk about the start date of damages, did you?" Meckstroth intoned, playing a little game of "gotcha."

"All semantics," the doctor shot back. "But I guess the word 'start date' was not in that question."

There was a silence. Then Meckstroth said, "The senior editor of what you refer to as the 'Red Book' is Dr. Leftcourt, correct?"

"Yes."

"The correct title is *USDA Handbook Number 696*, isn't it?"

"Yes."

I had no idea where this man was trying to go. I wondered if he did.

"The Red Book is a federal publication, isn't it?"

"I guess so. It contains papers that are inaccurate," the doctor said.

"Dr. Leftcorte was an employee of the US Department of Agriculture at the time he served as senior editor of *USDA Handbook 696*, right?"

"I'm not sure. I'm surprised he'd allow inaccurate papers to be put in," the veterinarian said.

"Dr. Johnson," Meckstroth said leaving the jury area and coming up closer to the witness box, "the Red Book was published in 1991. So that means there's been 12 years for you to publish an article that proves the Red Book wrong, hasn't there?"

This was dirty pool in my mind. Why was it Dr. Johnson's job to write articles refuting federal studies?

The doctor clearly felt as I did about this and in response, he gaveMeckstroth an earful, dripping sarcasm.

"Let's see, there's been 12 years for a practicing vet to do the work of all these researchers funded by—yeah, that's real good. I'd have to feed my family and do research on my own. Power companies won't give money like they give those researchers because maybe they wouldn't like what I'll find. That's a great answer. That's my job as an independent veterinarian to do the research of all these universities. Great question."

This answer, the only possible answer, infuriated Meckstroth. "Your Honor, I move to strike!" he complained. "Nonresponsive!"

"I," Dr. Johnson said, standing his ground and again ready to rumble, "think it's *very* responsive."

"It was certainly descriptive," Judge McKay intervened. "But it wasn't responsive. I'm going to strike simply because it isn't responsive. You said you don't have time, energy or effort to do the research? Or the money, I guess. Ask the question again, Mr. Meckstroth."

"Let me," Meckstroth said quietly, "start over. Every veterinary school in the world has had 12 years to do research to disprove the Red Book, right?"

Dr. Johnson shrugged. "I guess," he said sullenly.

"And to your knowledge nobody has ever done research disproving the Red Book, right?"

By this point, Dr. Johnson was through playing nice with Meckstroth. Let Meckstroth moan and bitch, who cared? The truth was the truth. "To my knowledge no one has offered funds to do research to dispute the Red Book. That is correct."

Again, Meckstroth flared up. "Your Honor, I move to strike!"

This was his answer to everything he disliked.

Even McKay had had enough of this. "It's his explanation of the answer," the judge grumbled. "I'm going to let it stand. That's his belief."

Meckstroth nodded in deference to Judge McKay. "Now, Dr. Johnson," he resumed. "A few minutes ago Mr. Hammarback asked you to read from a publication, something written by Dr. Bodman who had been retained, like you, as a consultant. That article has not been peer reviewed, has it?"

"I don't know," Dr. Johnson said. "I don't care."

"The peer review process is a process by which people who share expertise in a field review publications and comment on whether they ought to be published, right?"

This was a real can of worms. Since university research was so heavily funded by utilities, there were no voices ever heard dissenting from a pro-utility line. Controlled by professors who were utility surrogates, peer review was essentially a propaganda arm, as Dr. Johnson pointed out.

"That's what it's supposed to do. But we allowed articles that *eliminated* information to be published. Don't forget Fen phen was peer reviewed; tobacco articles were peer reviewed. They haven't been accurate." His unspoken answer was: Neither are articles on stray current affecting dairy farms.

Meckstroth needed to quickly rehabilitate peer review in the jury's eyes since he was going to present witnesses who were involved with it.

"There's been a lot of peer reviewed articles over the last 12 years reaching conclusions consistent with conclusions in the Red Book, right?"

Dr. Johnson smiled. "There always will be. Almost every article that's been peer reviewed over this period of time has been sponsored—written by people that testify against farmers on stray voltage. Or the money has come from the power company. So we're still splitting the same straw."

Everyone was silent. Then Meckstroth finally, mercifully, said, "I have no further questions."

One last question came from Barry. "Dr. Johnson, does peer review matter? Or does proper application in the field matter more? From a practitioner's point of view, can you tell me whether peer review is an advantage in the field?"

"There are times it is, times it isn't. I have to deal with what I see on a farm and what the cows are telling me. I've not changed my position on that."

And so Dr. Johnson ended pretty much where he'd begun: By reminding the jury that when it came to cows reacting to current, the proof of the pudding was in its taste, not in the recipe ingredients. In my opinion, as well as in Barry's and Scott's, Dr. Johnson had testified effectively and built a very strong foundation for me. It didn't seem that Meckstroth had laid a glove on him in the cross. If that was the best Meckstroth could do, I felt a growing confidence that we were in good shape. But then again, what I felt didn't matter; all that mattered was what those 12 folks in the jury box felt, and that I would not know until well after the fact.

Next up was Dr. Michael Behr, the man who had calculated the damages figure for my suit. Meckstroth, the hit man, was gunning for him; he'd already fired shots at him in the opening segment with Dr. Johnson. If he could shake or break Behr, he could break my case. I hoped and believed that would not happen. But who knew what they would try, especially if they began to feel the case slipping away from them. I would not put anything past this utility. Its hirelings would lie about anything. They had proven it time and again in the field. And in the courtroom.

Chapter Nine
THE BIG GUNS FIRE: ONE

By the time the trial was almost two weeks old, I still hadn't been able to shake my dread of losing. I knew my fear was irrational, because the case had been developing smoothly, but today, the 12th day of trial, was a big one, possibly *the* critical day. I had three of my best witnesses in line to testify, and the jury's perception of them could make or break me.

Mike Behr was an agricultural economist and the man who had calculated my damages; Don Zipse, an electrical engineer renowned for his expertise, would refute WPS' batch of experts; and Doug Sutter was a former Brown County, Wisconsin, AG Agent. An objective observer whose job it had been to clear up misinformation about stray current stemming from university research, Doug's mission was to aid farmers by applying research practically on farms. What he learned had stunned and shaken him.

Dr. Behr was sixtyish and very fit, a handsome six-footer with short, sandy-gray hair and a deep, melodious voice. He was going to be raked over the coals by the defense, but for now Barry had him on direct. The doctor began by explaining that he was a forensic economist, one who determines a likely course of economic events under varying circumstances, such as what my production would have been but for stray current. He had two master's degrees in economics and a PhD in agriculture. He had grown up on a farm in Minnesota, taught college economics in Duluth and later emigrated to Wisconsin, where he taught business and economics at the university in Superior from 1969 to 1983. After that, he had opened his practice as a full-time forensic economist.

Barry began by asking him about the types of matters that fell under his domain.

"The bread and butter of this business," he said in a twangy, Minnesota accent, "has been personal injury and wrongful death cases where persons have to take lesser employment...determine what the loss is there. Beyond that, most of the work is where small businesses have been disrupted in one way or another. Most of that has been farm business, such as here."

He mentioned that he'd evaluated some 1,400 separate agricultural losses of one kind or another, and had received a great deal of insight into farm problems as a result. He reeled off an impressive list of organizations to which he belonged: American Agricultural Economics Association, American Law and Economics Association, International Atlantic Economic Association, American Statistical Association, American College of Forensic Examiners. From each of these bodies, he had received prestigious fellowships. Barry introduced his credentials as Exhibit 229.

Dr. Behr explained why he had been summoned by Barry Hammarback to my farm. "To determine the amount of damage, economic loss, associated with the presence of stray voltage," he told the courtroom.

Barry rose from his counsel and slowly made his way toward the jury. "How do you go about figuring damages on a dairy farm?" he asked.

"The manner is the same," Dr. Behr said, "whether it's a dairy farm or someone injured in an auto accident. It is to compare the resources in monetary terms that were preserved after the injury or event and compare that value with the probable value that the person or business would have produced without the injury or damage. The difference between the two—the actual and the normal—is the amount of damage; There's no other way to do it."

Barry nodded. "Dr. Behr, what information did you gather to perform your analysis of the Allen farm?"

"First, information about what actually happened, the presence of damage; Second, information about the economic environment of the farm—that is, what similar farms are doing, by which we evaluate what the Allen farm could have expected to accomplish without stray voltage. The sources of information are twofold: Farm budgeting—data compiled by universities in agricultural parts of the country where farms send in their records for analysis, whether it's hogs or cattle, and what can be expected by way of profits. The secondary category is statistics—price information on products used in farming, which is from the US Department of Agriculture. The third category is people of various types. A farmer would provide some information, and others associated with operation over the years such as veterinarians, electricians and dairy scientists. That pretty well sums up the types of information that go into the mix."

I was glad to have this on the record so the jury would realize we had not simply picked damage numbers out of a hat but had used scientific logic to arrive at the figure.

Barry moved over to the exhibits. "We've got Exhibit 1010 here," he said. "Explain what that is."

"That's a chart," Dr. Behr responded, "constructed from data in my report. The yellow area is a measure of milk production per cows, called 'rolling herd average.' The darker black line also measures milk production per cow, but month to month as opposed to year to year. The straight line is a trend of milk production per cow by year for the state of Wisconsin. The upper line is the normal trend, 'normal' defined as probable conditions on the farm without damage. So I'm saying the upper line is my opinion about the probable milk production on that farm without stray voltage."

Barry paused, allowing the jurors to digest the information on the graph. Then he said, "Dr. Behr, taking into account your experience, training and expertise in agricultural losses, and taking into account the information you worked with on the Allen farm—assuming stray electricity affected animals on the Allen farm from 1976 to the present—do you have an opinion to a reasonable degree of certainty as to whether or not the Allen farm sustained economic damage?"

"I do."

"What is that opinion?"

"It is that it did."

"And based on your evaluation, do you have an opinion to a reasonable degree of certainty as to the total amount of those damages, leading to the present any effects that inflation may have had?"

The doctor looked straight at Barry. "Yes. I do."

"And what is that opinion?"

"That amount is $11,043, 387."

As Barry again paused, I laughed to myself, imagining how all the suits at WPS were cringing over this number. I uncrossed my legs and leaned forward with my elbows on my thighs.

"And," Barry resumed, "have you also done an evaluation of the effects of inflation?"

"Yes, I have."

"And if you include your calculation for effects of inflation over the years, do you have an opinion to a reasonable degree of certainty as to the total amount of damages sustained by the Allen farm?"

"Yes. I do."

"In today's dollars?" Barry asked.

"Yes."

"And what is that opinion?"

"That amount," the doctor said to him, "is $13,870, 967. And I trust that's clear that that's not in addition to the $11 million, but that the $13 million includes the 11 million plus inflation."

I began wondering what Trevor Will and Kurt Meckstroth had up their sleeves for their cross. It was already clear that they were out to demolish this man. Whether they would succeed was something else.

"Thank you for that clarification," Barry said with a smile. Then, to further educate the jury, he said, "What types of things does a dairy farm produce, Dr. Behr?"

Behr folded his hands together in front of his chest. "My approach is to view a dairy farm—a dairy enterprise—as producing three things. Milk, of course. It also produces beef in the form of milk cows that get turned into hamburger when they no longer give enough milk. And the third thing the farm produces is other dairy animals: Heifers who replace the culled cows when they're sold, and usually in numbers that exceed the number of cull cows so there will be extra heifers to either increase the herd or to sell to other dairies."

"When you did your analysis, were there losses in these categories?"

"Yes."

Barry nodded, and then went to the nitty-gritty. "Would you take us through the method used to calculate these losses? For example, you figured a milk loss for the total herd?"

This was critical; I was convinced that this was the area Will or Meckstroth intended to attack.

"Yes."

"Explain to the jury how you calculated the milk loss."

"Okay."

"The cows you have listed on Exhibit 1376—Dr. Behr's report/analysis—this would be milking and dry cows?"

"Right. The next number is the total pounds of milk that these 325 cows caused to go out the driveway in that month—178,897 pounds of milk. Let's go down to December of 1978, you'll see the value of the rolling herd average is 312.75 cows, the average number of cows for January through December of that year. It's a rolling average because every month when we move down to the next month, we subtract the thirteenth month back, add the next month forward and recalculate. You'll notice the average went down slightly. The reason is we added 278 cows in January of '89 but took off 325 cows in January of '78. So we added fewer than we took off."

I watched the jury carefully to see if they were getting it. Surprisingly, they were all attentive to the doctor's presentation. It was a good omen.

"The next column," Barry said, "is rolling herd RHA milk. What's that?"

"Rolling herd average amount of milk, calculated again on the same 12-month period. The next column is annual rate, milk produced month by month."

"And the next column?" Barry asked.

"Sales, dollars," Dr. Behr responded. "The amount of money received for the amount of milk shipped in that month. It's from the milk check stub but you have to subtract hauling and government program taxes. I calculate for each year a milk price, which is here $10.37 a hundred weight. I'm going to use this price to value the milk not produced but which should have been. I do not want the cost of hauling and taxes in because milk not produced does not have to have those amounts paid on it. So I take them out up front before I calculate this price."

Barry went back over to the jury, probably to help them maintain focus in the face of so much information.

"Would you explain to us," he said to the witness, "what a trend line is?"

Dr. Behr shifted in his seat and crossed his left leg over his right. I noticed that his black wing-tip shoes were polished to a high gloss. "A trend line in the case of the Wisconsin state average is calculated by a procedure called regression, which computers do easily, and takes values year by year and calculates a general relationship between them giving an overall picture."

Barry leaned back against the front rail of the jury box. "In your opinion as an economist, is it fair to compare a trend line with actual production?"

"Sure."

"Okay. Back on your report, once you've gathered all the information about what the farm has produced historically, what is the next step in milk loss analysis?"

"To determine what the production would have been without the damage, and this goes to the normal trend. My understanding that the events that relate to the onset of the issues here go back to 1976, when there were electrical installations. I did not start the damages directly in 1976."

"And why not?" Barry asked him.

The doctor shrugged. "I give the cows a couple of years to adjust to the new situation. So I waited until 1978 where the cows dropped, came back up, could not keep going, dropped some more. At the second drop I started calculating damages. The point is that the normal trend line says production is going up at 100 pounds per month per cow, 1200 pounds per year per cow. The reason I picked 100 pounds is that although it varies from farm to farm, about a hundred pounds per cow is a good average.

"Now, a hundred pounds per month is a lot faster rate of increase than the increase in the Wisconsin state average. Cows don't produce that much milk forever, so I flattened it out at 50 percent above the Wisconsin state average trend. Forces in the industry work on the Allen farm as on other farms. They're all subject to certain production increases over time. I set it at 50 percent based on the performance of the farm at a particular time, the year 2000. The herd approached 24,000 pounds per year and was within 1,500 pounds of 50 percent above the state average. The herd did that even though it was coming off a damage situation. There were changes made in the electrical situation in late '97, early '98, isolation occurred and the herd took off with the previously damaged cows. So, it's warranted in my view to conclude that wherever this normal trend line is, it's somewhere about whatever percentage you have here by the state average represented by the peak in 2000."

There was some verbal fencing between Barry and Trevor Will at this point because Will objected to Dr. Behr's mentioning that Dr. Johnson and Dr. Terry Homan, my vet, were of like mind about production numbers and loss calculation. But it amounted to nothing, and McKay let Barry continue the Q and A.

"And so once that had been established, then what's left to calculate?"

Dr. Behr waved his finger. "Simply the loss in milk per cow for that month. And then to multiply it by the number of cows that should have been in the herd without the damage, and we come up with a difference for which I have another page in the report."

"Show us how you got that difference," Barry said.

Dr. Behr nodded vigorously. "We'll start with 1978, the first year. Page 40 on my report. The loss is calculated month by month. The loss for January of 1978 is $9,908. Then I add the months across the page and we get a total for the year of $86,391. Notice two things. The number of cows for January starts out close to the actual number of cows, then it goes up about two cows per month and the progression continues until we get to 500 cows, plant capacity. Which is to say what the farm is geared for and probably would have been operating at without the damage. Because of the damage, however, the actual herd sizes for this entire period were less than that.

"Secondly, we calculated the milk per cow at the annualized rate for 1978, and then we multiply the number of cows by the annual milk per cow, divide by twelve months because we want a monthly figure, and we get 268, 467 pounds of milk that should have been produced or would have been without the damage in that month. Then I multiply that figure by the price of the milk for the year, use that price for every month, and then we get the normal milk sales and subtract the actual milk sales from the normal and get the loss for that month. Do that for every month or every year. Looking at Exhibit 10, those figures get larger in later years because the milk loss per cow is greater. Secondly, as the herd size increases toward the normal 500, the shortage of cows will also contribute to the higher gross milk loss each month or year."

Barry walked back toward the witness stand. "And so in going through your analysis, what are the categories you actually looked at?"

The doctor took a sip of water. "The first three are types of production I mentioned: milk and young stock loss, and the cull cow loss or beef loss, and then the next item is cost saving. That recognizes that you lost milk, but also lost costs. You don't have to feed the lost animals or call the vet for them so we subtract the cost savings of the reduced herd, which comes something over 7.5 million dollars for this analysis.

"The next item is the cost of remedies of electrical damage, installations and capital loss, the purchase of cows that would not have been purchased but for the damage, and the reduced value of the herd in the present. The analysis is consistent with a herd building to 500 cows to the present moment, with the normal amount of heifers and young stock . The loss of inventory or reduction of inventory is an element of damage and is included in the capital loss section."

"And so when you total up all the pluses and minuses that you calculated, what do you end up with?" Barry asked.

The doctor pointed to the exhibit. "Here, at the bottom of the page, is the figure I mentioned, $11,043,387, category by category."

Barry nodded. "Dr. Behr, for reasons that have come up in court here, I've asked you to split the damages between two periods, have I not?"

"Yes."

"Could you tell me which periods you split them between?"

"Before and after mid-August of 1991."

"And could you give your opinion to a reasonable degree of certainty as to damages for mid-August of 1991 including any adjustment for inflation?"

"Through August 14, 1991, the amount is $7,041,394."

"And for the second period?"

"That is from August 15, 1991, through April 30 of 2003. That amount is $6,828,702."

"All right, thank you," Barry said to him. "Do you know Dr. Mellenberger?"

"Yes," Dr. Behr replied. "I know who he is, certainly."

"Have you compared the rolling herd averages that Dr. Mellenberger has done?"

"Yes."

"Are his the same as yours?"

This was pre-emptive. Since Mellenberger was a WPS stooge, they were surely going to try to present his statistics as the final word.

"Well, initially, they were quite a bit different for some months, and since that time, for about a year, I've agreed with him. For about four years he's pretty much agreed with me. At this point we're very close together, but we've done that in our own ways."

"And if you had used his cow numbers instead of yours, would that have an effect on the total amount of damages?" Barry asked.

"Small effect, yes."

"What is that effect? In dollars?"

He looked down at his notes. "Without the inflation adjustment, if we go with his numbers, the damages are $120,633 higher than mine. And with the inflation adjustment, he would be $149,984 higher than mine."

Barry was finished with Behr. Foley and Lardner would now have their crack at the good doctor, and I could feel my pulse quicken as Barry returned to the counsel table and took his seat. I inched forward in my pew, waiting to see who would step up.

Judge McKay nodded toward the defense table. "Cross-examine, Mr. Will?"

"Certainly, Your Honor," the attorney said. He rose and walked slowly toward the witness box, tall and elegant in his navy-blue suit and Gucci loafers. I could not decide who was worse—Will, with the exquisite courtesy he always showed while twisting the knife into you, or Kurt Meckstroth, with his haughtiness and pit bull attack style.

Will began by attempting to take the doctor down a peg, going negative in an oh-so-polite way, making Dr. Behr state everything that he was *not*.

"I want to ask a few questions about your background because that's where the examination started," the lawyer began. "First of all, you have one PhD, not two?"

Will knew perfectly well he had two master's degrees, not two PhDs.

"Right."

"And you're not a veterinarian, right?"

The doctor said he was not.

"Not a biologist?" Will said.

He was not that.

"Not an electrical engineer?"

No, not an electrical engineer, the witness told him.

"You're not trained in dairy housing?" Will asked.

He was not.

"Or animal behavior?"

"No," the doctor confessed.

"Or an animal reproductive expert?"

He was no expert in the field, he admitted.

"And you don't know how to analyze feed or rations, correct?"

Dr. Behr stated he did not know how to do those things.

Listening to this, a newly arrived Martian might've thought Will was talking to a dunce who had accomplished exactly zero in his career.

"You've described what you've done in this case as acting as a forensic economist, is that right?"

The doctor agreed that he had.

Trevor Will now fired a shot across the bow. "And you've defined that as someone who makes a living testifying in litigation. Fair?"

The doctor smiled. "I don't think I included, quote, 'making a living' doing it. But some people do, and some do it on a lesser basis."

"Since you're still here body and soul," Trevor Will told him, stone-faced, "I guess you're making a living at it. Fair?"

"I guess so," Dr. Behr conceded.

Will was good, no doubt. I had to give the devil his due.

"Now, you don't have a degree in forensic economics, correct? You're not aware of any school that offers a degree in forensic economics, are you?"

Dr. Behr shook his head. "I'm not sure there is, although some courses are offered in this the last few years."

"In fact," Will said, "anybody who's an economist can call themselves a forensic economist, correct?"

"I suppose," the doctor said.

Suddenly, I had a mental picture of Dr. Behr barefoot in a log cabin studying books by firelight, educating himself, like Lincoln, on matters of interest.

"There isn't any test you have to pass? And no state licensing or regulations?"

"No," the doctor said.

"In fact," Will said pointedly, sliding over toward the jury, "there isn't even a continuing education requirement, is there?"

In other words, any nincompoop could call himself a forensic economist if the spirit moved him.

"No."

Will had his back to the witness. He was facing the jury and asking questions over his shoulder. He reminded me of a soldier lobbing hand grenades.

"You said you're the only full-time self-employed forensic economist that you know of. You draw a distinction between academic economists and forensic economists, correct? Isn't there a difference?"

Dr. Behr seemed a bit fidgety to me, not a good sign. I hoped it was just discomfort in a hard chair.

"It's different, sure."

"And what you define as an academic economist is somebody who typically has an appointment at a university or other institution, teaches or does research. And a forensic economist's work is also done by academic economists, in their spare time, outside normal hours. Fair?"

"That's the way I got started, right," Dr. Behr agreed.

Will finally turned around and looked at the witness, although he did not leave the jury box area.

"You mentioned you did more than 1,400 cases, correct? And isn't it true that more than 95 percent of the cases are for the plaintiffs or the claimant?"

"For the claimant, yes. Without regard to the legal configuration, yes."

Will smiled at the jurors. "Basically, the person is trying to get money?"

"You could say that."

"And in the stray voltage area, you've been hired to work on over 350 cases, haven't you?"

"Yes," the doctor said.

"All but one was brought by the farmer, right?"

"Yes."

Will nodded several times. Then he turned again and faced the jurors.

"You've been hired for over 100 cases by Mr. Hammarback and his law firm, correct?"

As if WPS and Foley and Lardner didn't have their own stable of house experts, headed up by million-dollar Mellenberger.

"That's correct."

Barry broke in and quipped, "I'm not getting any younger."

This lightheartedness annoyed Trevor Will. "Maybe I would ask Mr. Hammarback to discontinue editorial comments because I didn't do that during your examination," he said sourly.

Grinning, Barry apologized.

"You estimate," Will said to the witness, "that about two-thirds of your income is from testifying in stray voltage cases. And you typically spend about 1,750 hours a year providing professional services?"

"You're probably reading off my deposition, and that's fine. It sounds reasonable."

"That would mean," Will said, "that you're spending around 1,200 hours a year on stray voltage, fair? You went out to the Allen farm in September of 2000, and at that time there was not a lawsuit filed, correct?"

That was true.

"You were at the farm about five hours. And the only person you remembered talking to, for anything substantive, is Russ Allen, right?"

"Right," Dr. Behr affirmed.

I was eager at this moment to go up there and switch places with the doctor. I would've loved giving Trevor Will a piece of my mind...I reminded myself that I would have my opportunity before much longer.

"And you didn't tour the facility?" he asked, going negative again.

"Correct."

"Didn't take any feed samples?"

"Correct."

"Didn't observe milking?"

"Correct."

"Didn't see any cell count testing or data?"

"I probably did, but did not make any particular note of it."

"Didn't inspect cropping operations or observe field hands, right?"

"That's right."

"Didn't review any electrical measurements?"

Dr. Behr shook his head. "I don't believe so."

Trevor Will turned to face him. "The place where you were for most of these five hours was at the kitchen table in the home, correct?"

"Yes."

Sure. We sat around the whole time sipping espresso, playing cards, arguing about the Green Bay Packers and reading *Playboy*. What the hell was Will insinuating?

"And you were—you were basically going through records that Mr. Allen had assembled for you to review?" he said in a skeptical tone.

"Yes."

"Now, you are of the opinion that the three most common causes of dairy farm damages in Wisconsin are Smith harvesters, milking equipment and stray voltage. Correct?"

"That's correct."

Trevor Will spread his arms theatrically. "And yet when you went to the Allen farm, you didn't even ask to inspect the milking equipment?"

"That's correct."

Trevor Will shook his head slowly and somberly, as if in disappointment. "And you didn't even realize there was a Smith harvester on the premises, did you?"

"I don't know," the doctor replied. "But if that's what I said in my deposition, I guess that's true. I know they're not a big issue in this case."

Again, Will shook his head incredulously. "You didn't even remember that Mr. Allen had one. Is that fair?"

Now, *that* was a damned good question. It seemed that nothing about having been forced to come to this courthouse was fair, at least for me.

"That's very possible," the doctor told him.

"All right," Will said. "When I asked you at your deposition, you weren't even able to describe his barn, correct?"

The doctor smiled. "Well, I was not in it."

"And," Will said accusingly, "you also told me that your conclusions in this case would have been no different even if you had never gone to the farm! Correct?"

"Certainly, at that point. Probably even as I sit here today," Dr. Behr said forthrightly.

Will, at last, left the jury area and moved close to the judge's bench.

"After you went to the farm, you prepared an initial report, what you described as a 'quick and dirty' report. This is something you sent to attorneys and client, Mr. Allen and to Mr. Scott Lawrence, didn't you?"

Dr. Behr did not recall.

"At that point," Will said, "you had already identified 1978 as the time at which you would start damage calculations, correct?"

"At least tentatively so, yes."

Once more Will struck a pose of disappointment. "Even though you hadn't spoken to any other experts?"

"That's correct," Dr. Behr replied.

Will paused for several moments, then said, "Doctor, in the stray voltage work you do, often there is an electrical event involved such as isolation, correct?"

"That's often the case."

"And then what you do is look at production data that precedes the date of isolation?"

"Among a lot of other things, that's correct."

"What you look at is the production graph, for an area in time where either production fell or stopped rising at the rate it had been. Is that fair?"

"Those kinds of things. Yes."

"And then you start damage calculations from that point?"

The doctor shook his head. "Not always, but the point is the cows talk to me."

"And so we're clear on the record," Will said, "you don't speak Holstein?"

This caused a ripple of laughter.

"No. But I speak milk production."

Will next tried to show that Dr. Behr's selection of the year 1978 for the beginning of my damages claim was arbitrary as well as unscientific. "It's your view, isn't it," he said, "that starting in the late '70s there was a widespread stray voltage problem in Wisconsin and the Midwest?"

"Again," the doctor said, "there is a strong tendency of what the cows are saying. These problems began in the very late '70s, absolutely without question."

Will stared at him. "At your deposition you said you could have started it in 1980 or 1976. You picked 1978, in the middle; you thought that was fair?"

"Based on what I just said in the late 1970s, probably so." The doctor seemed to be making a strong effort to remain calm and patient.

"But that was done," Trevor Will said, "without going out and doing any kind of analysis of what was going on at the farm in the 1980s or 1970s. Not at vet records, cow comfort, milking routines, et cetera. Correct?"

"In general, correct," Dr. Behr answered.

"Well," Will said, "if we simply 'listened to the cows' as you put it, one could look at the production graph and say, well, maybe it was here they were telling—in 1985, they were telling me stray voltage cropped up, correct?"

"You say 'maybe,'" the doctor told him. "And that's the question. I would still not select that point for reasons I have not had the opportunity to explain."

Will glanced back at Meckstroth, sitting at the defense table. "I'll ask a different question," he said, sidestepping the doctor's comment. "Now, in your analysis of the Allen materials, you were primarily interested in the milk check records, correct?"

"I would say so. Yes."

"There were periods of time for which Mr. Allen did not have records that told how many cows were on the farm?" Will asked.

"That's correct."

"And that's part of where you and Dr. Mellenberger had a disagreement about cow numbers. But the problem is, if you don't have actual cow numbers, you have to estimate or approximate them. And you went about it one way and he in a different way?"

Mellenberger's way, first, last and always, was to earn his big fee by making everything look rosy with WPS and nasty with the rest of the world, making sure his golden goose grew fatter and happier.

"Correct," Dr. Behr said.

"You went about it by saying," and here Will again turned accusatory, "if I know the beginning and ending number of cows, you would approximate the intermediate numbers?"

The doctor replied defiantly. "That's true. I might make some adjustment beyond a linear progression based on large changes in milk sales at a given point. But generally, you are correct."

Silent for a moment, Trevor Will slid over toward the jury box. "So if we have a 10-month gap, and at the beginning you had 10 more cows than at the end, you'd likely just subtract one every month?"

"Probably."

"Isn't it true," he said, "that those are really the records that drive your damage analysis—milk checks, cow numbers?"

"Well, they drive the milk loss section of the analysis completely."

"And they also drive the calf and cull-cow loss?"

"The herd-size figures, yes."

"Now, it's true, isn't it, Dr. Behr, that because you don't delve into some other farm records, you have missed important data?"

Dr. Behr opened both hands, palms facing upward. He shrugged. "That's possible."

"In fact," Will said with the harshness of an inquisitor discovering a new heresy, "there was one case where you missed—"

This brought Barry Hammarback up out of his chair, and not a moment too soon. "I object," he said, "to what may have happened in other cases."

Judge McKay called for an off-the-record bench conference, from which Trevor Will returned duly warned. He remained standing near the witness box but seemed more slow and careful in framing his questions.

"Dr. Behr," he said, "you have missed data that indicated how much milk was being produced in a particular instance. Fair, in one case?"

The doctor looked at him with raised eyebrows. "Well, I don't know. I don't know what you're talking about."

"Do you remember a case where the son of the farmer was—"

This sounded like the beginning of one of those old 'farmer's daughter' jokes, but I wasn't laughing, and neither was Barry.

"Your Honor," Barry interjected without rising.

"Right," McKay said to him. Then he instructed Trevor Will, "Without going into specifics—and if the doctor doesn't remember, just go on to another question along the lines."

Trevor Will's brow furrowed. "Perhaps I have a way of refreshing his recollection, because we discussed this in deposition. If I could show him the deposition it might refresh his recollection."

"As long," Judge McKay cautioned him, "as you don't read it."

This whole exchange really bothered Barry. I was just anxious to be done with this witness, and the hell with all the legal mumbo-jumbo. But I knew enough to see that Will was looking to twist the rules. And that could only hurt me. Therefore, I was glad that my attorney was again being forceful and strong.

"I guess I object," Barry told McKay, and this time stood up to do it. He sounded impatient and annoyed. "I don't see how this is possibly relevant. He's admitted it's possible for him to miss something in a case, and where else are we going with this?"

"The problem," Judge McKay told Barry, "is that the witness hasn't specifically indicated that he misses things. If you're willing to stipulate there are times he does, then Mr. Will can go in a different direction."

"Absolutely," Trevor Will said.

I really couldn't see the importance of this; Dr. Behr was human, and like everybody else, subject to mistakes or oversights. But the eagerness with which Will wanted this stipulation—he was like a dog hoping to sneak a steak bone from the table—made me want him not to get it. Was he trying to portray the economist as an incompetent? If so, it was ridiculous and would likely end up hurting his case. I was for that one hundred percent.

"I think," Barry told the judge, "Dr. Behr is willing to admit it's possible that he could."

Dr. Behr now decided to speak up for himself. "Nobody's perfect," he said. "Including me. I don't dispute that. I'm sure I've missed something here or there, one time or another along the way."

Will, the hungry mastiff, now had snatched his bone from the dinner table and was ready to run away and bury it. "Would you be willing to agree," he asked Behr, "that you've even missed how much milk was actually being produced on that unnamed farm?"

Dr. Behr chuckled. "Could be. I thought I knew when you started this what you were probably talking about. There's a good deal that can be said about that, but I'm not sure that's the right case."

I looked over at the jury to see how they were taking all of this, but they gave me nothing. Some absently watched the witness, some had their eyes on Trevor Will and some just stared off into space or out at the spectators. I noticed a newspaper reporter in the gallery. He had a big ID tag hanging round his neck and was busy writing in a small tablet.

Judge McKay told Will he was going to have to live with the doctor's answers, and to move on rather rapidly. I wondered what the fuss had been about and made a mental note to ask Barry and Scott about it later.

"Dr. Behr," Will said now, "in the Allen case, did you ever do any analysis of the difference between the amount of milk produced shown in the Allen computers, and the amount shown as being sold on the basis of the milk checks?"

Dr. Behr shook his head. "Not really. I've seen those numbers, and didn't do an analysis. It's the milk sold that counts. That's what I focused on."

"But," Will said, "if there was a discrepancy between those two, might that raise a caution flag in your mind?"

"No, not necessarily."

"Doctor, is it fair to say that when you started on this project, your working hypothesis was that there was a stray voltage problem on the farm?"

Will, I suspected, was trying to lure the doctor into admitting a bias in my favor, but Behr would not take the bait. "I'm going there with an open mind as to whether that is correct or not."

"You haven't tried to assess whether it's cow contact or earth currents or something else?" Will asked.

"No. Other people are better qualified."

"And you're not able to apportion—if you assume there was an on-farm electrical problem, you're not able to apportion any of these damages to that as opposed to some other electrical problem?"

This brought Barry up out of his seat with another objection. "That's assuming facts not in evidence, Your Honor. There's *no* evidence of any electrical on-farm problems."

"Well, Your Honor," Trevor Will said, "I disagree. Since—"

Judge McKay abruptly cut him off. "He's answered the question. The question and answer stand." McKay asked the court reporter if he'd gotten it straight; the reporter replied that he had. I felt McKay should've been doing a lot more of that kind of scotching of Will and Meckstroth all along.

"Would you agree," Will continued, "that in preparing the damage claim, the economist must exercise judgment?"

What kind of answer did he expect? Did he think Behr might say, "no, judgment isn't required. I just read lottery numbers in the newspapers and write down whichever ones tickle my fancy"? I'm no lawyer, but it looked to me like Will was biting on granite, couldn't make any headway with this man and was just throwing out questions to see where they would land.

And then the obvious answer: "Sure."

"And the way in which that judgment is exercised can affect the damages result?"

"Sure."

"You mentioned that you received some information about Dr. Andrew Johnson's opinion in this matter. Assume that in his deposition he said that damage from electricity started at most several years before 1997. Assuming that to be true, where on the graph would you start your damage calculation?"

Suddenly, it dawned on me that Trevor Will might be contemplating losing this case and was trying to limit WPS' financial liability. I felt a stab of excitement. Could that be true? Or was this just wishful thinking?

"On that subject," Dr. Behr said, "I would substitute my experience and opinion for his."

Eyes wide, Trevor Will stared at the witness. "You'd disregard *his* opinion?"

Behr said calmly, "I wouldn't disregard it and I would take it under consideration. But I would disagree with it."

Funny, but when Meckstroth cross-examined the udder doctor, his attitude was that Dr. Johnson did not know what the hell he was talking about. Now, in trying to debunk Behr, Johnson suddenly became the fount of all wisdom.

"Assume for the purposes of this question that he's right. Are you able to pick a date? You can have four or five years before '97."

"Well," Dr. Behr said hesitantly, unsure, it seemed, of exactly where Will intended to sandbag him, "if you give me six, I know for sure where I would start it: at the beginning of '91."

Trevor Will watched him for a few seconds. "The next assumption you've made," and his stress on "assumption" was the same as he would've used for "lie," "is the idea that the herd would have had 500 cows, correct? And you have included in your damage calculation 'pretend' cows that don't exist, right?"

This angered Dr. Behr. "I cannot answer your question as long as you characterize them as 'pretend' cows."

Will laughed. "Well, they aren't real, are they?"

McKay broke in here and instructed Trevor Will to rephrase his question, which he did.

"It's been referred to," Will said, "as 'phantom cows' sometimes. You're happy with the term 'phantom'?"

Now it was Dr. Behr's turn to laugh. "No, I'm not. I can't answer based on this word game. I'm not going for it."

Now Dr. Behr fully understood Trevor Will.

"Well," Will said, "*you* called them 'normal' cows. Don't you believe that's a word game?"

"I characterize them as *missing* cows," Dr. Behr said. These were simply cows and the calves that would have been on my farm had it not been for WPS' stray voltage maiming and killing them.

"Did you ever talk with Mr. Allen specifically about what the partnership intended in terms of how fast they would grow their herd?"

"No. Well, not in terms of *x* number of cows per year. But in general terms of the plan, which was clear to me. It was evident what would have been done."

What would have been done is that happy, healthy lives would have been lived on a beautiful family farm by man and beast alike. And I would never have seen the inside of this courtroom. And my brothers and sisters and nephew would be alive. And my heart would be strong and healthy, instead of weak from two heart attacks.

"If you assume," Trevor Will said, "that the family had no particular plan to grow the herd at any particular rate, would that affect your damage analysis?"

Dr. Behr shook his head. "No. In fact, it would support exactly what I did. Which is to go with the normal internal rate of expansion in that herd."

"Let's talk about that," Will said. "The internal rate of expansion you used was 10 percent. Isn't it true there are statistics in the agricultural literature that indicate the normal growth rate of a herd is between 2.1 and 2.5 percent?"

"I don't know," Dr. Behr said sullenly.

"Do you agree that the growth rate is a matter of judgment for the economist?" Hands thrust deep in his trouser pockets, Will eased back over toward the jury.

"Within fairly narrow constraints, yes."

"Would you agree that if you used a 2.5 percent internal growth rate, it's going to change this damage calculation significantly?" Now, with long strides, he walked quickly over to the exhibits.

"You're pointing at the chart," Dr. Behr told him. "It would not change the chart but it would change the total amount of damage, that's correct."

"It would take the herd much longer," Trevor Will said, "to get to 500 cows at a 2.3 internal growth rate. I want to ask you about—the normal trend should be 50 percent above the Wisconsin state average, right?"

"Right."

"That was based in significant part upon the production that the Allen herd achieved in 2000. Is that correct?"

"That's a major factor," Dr. Behr agreed.

"If we look in your report—I'm going to zero in on October, November and December of 1999. Those are the three highest monthly productions at the annual rate that Mr. Allen ever achieved, correct?"

"I'd have to inspect the rest of the list," the doctor said, "but that's probably true."

"He was able to get up over 24,000 pounds three times, in those three months, right? And it was that information you used to support setting your trend at 50 percent above the state average?"

"No," Dr. Behr said sharply. "Absolutely not. I relied on the rolling herd average for that."

"If we look at the rolling herd average, it peaks two, three months later?"

"It peaks in May 2000."

Trevor Will was silent. "Do you know," he asked, "who Brett Favre is?"

Dr. Behr looked at him uncomprehendingly. "Do I *what?*"

"You know who Brett Favre is?"

"Brett Favre?"

"Yeah. Brett Favre."

"Never heard of him," the doctor said, and the entire courtroom, judge and jurors included, laughed. Behr was probably the only person in the state of Wisconsin who had never heard of him. Brett Favre was the star quarterback of the Packers, three times the MVP of the National Football League. Everyone around here knew that. Everyone except the good doctor.

"The reason I ask," Will said, "is this. At least people in Green Bay know that three times in his career he's thrown for five touchdowns in a game. Accept that as true, all right?"

Dr. Behr shrugged absently. He was clearly not a fan of pro football. "Okay."

"You don't expect him to throw five touchdowns every game, do you?"

For now, Dr. Behr was willing to play along with the lawyer. "Of course not," he said on cue.

"And just because something has hit a peak, you don't assume that's the norm, do you?"

"Well, either 'yes' or 'no' is misleading."

Will paused momentarily. "One other thing before we get to that. You indicated that this increasing line here is at the rate of a hundred pounds per month. The data you collected came from farms you had information about from litigation work you did. You calculated an average for all of the farms, right?"

"Yes."

"And on the basis of that average you adopted a hundred pounds as your trend line. And the actual average you calculated is 97 pounds, isn't it?"

The doctor seemed a bit confused. "I remember it was just a pound or two."

"Would you agree," Will asked, "that you have to feed a cow more to get it to produce 25,000 pounds of milk than 20,000?"

"If no damage is past or present, yes, I would agree."

"And it's also true, isn't it, that if you did add something in for additional feed costs it would increase the credit and decrease the damages?"

The doctor agreed.

"One of the things you also talked about in your report was that you included $80,000 for 'mitigation' expenses, correct? And you did not do any assessment of what those dollars had been spent for, did you?" Will said.

"No. That wouldn't be material to my decision."

"And you didn't look at whether the money was spent for repairing defective wiring in the farmhouse? And you didn't assess whether the money you included in the mitigation column was a reasonable expenditure?"

This was absurd. It was like Scott told me they were going to throw shit at the wall and see what sticks. Will was just flinging out words to see where they would land.

"No, sir," the doctor told him, sullen again.

Will walked over to his defense table, bent over, rifled through a few papers, and said with his back to the witness, "Dr. Behr, I want to go back to the 500 cows."

Again? I thought. I was sick of hearing about this. Behr must've been, too.

"You indicated that if someone were to have a 500-cow milking herd, they would have to have more animals on the farm than that. More bovine animals, true?"

Behr shrugged. "Probably."

"You estimated it would take 900 to 1,000 animals in total to maintain a 500-cow milking herd, correct?"

Behr nodded. "What you say is correct."

"And you know," Will said, "that the Allen farm could not accommodate 1,000 animals?"

More absurdity. I could've had a million animals if I'd wanted; just rented enough farmland. There was plenty of it in Wisconsin

"That is correct."

"And if they had to obtain other facilities, that would involve a capital expense, wouldn't it?"

"Right," the doctor replied.

Will went up near Judge McKay. "You have not adjusted your damage calculation for any capital asset purchase to house these additional animals. Fair?"

The doctor looked bemused. "It would be inappropriate to do so because there is no loss. It's not an issue."

Will drew a deep breath. "Let me try again," he said, becoming exasperated. He was not receiving the answers he wanted, which made me very happy. "They didn't build the barn themselves but sent out animals to be raised somewhere else. Wouldn't that increase the cost? Increase the credit to the extent that we're not talking about real animals?"

This idea of imaginary cows really seemed to throw Trevor Will for a loop. Maybe he should've become an imaginary lawyer.

"I can cut through this," Dr. Behr said with a smile. "The cost of buying that replacement heifer or raising it is not going to be greatly different. I included the cost saving associated with raising those heifers irrespective of whether they raise them. You either have the cost of raising or buying them, and I gave credit for that."

Will pursed his lips and looked down at the floor. "Doctor, when I took your deposition, you indicated you wanted to go back and look at the question of whether adjustments should be made to this damage report based on cow capacity of the Allen farm. Remember?"

"I may have said that."

"Did you go back and do that?"

Dr. Behr sighed. "I gave it more thought, but decided we were okay where we were."

"Did you look at the fact of whether Mr. Allen ever used the east barn to house dry cows and new calves? And if, in fact, he was, wouldn't that reduce the plant capacity of the milking herd?"

Dr. Behr tilted his head and stared at Trevor Will as though the lawyer had just arrived from planet Mars. "If that was the permanent arrangement, that's true."

Before the day's court session ended, Trevor Will requested that the jury as well as the witness be dismissed from the courtroom. Will hoped to be allowed to question Behr about a letter the doctor had written in 1985—at the time he and his wife divorced—in which he referred to himself in his capacity as a damages specialist as a "maximizer"—someone going for the biggest possible judgments and settlements.

Apparently, at that time, Dr. Behr had been trying, during a financially difficult period, to build up his forensics business; he charged very high fees.

Will was anxious to get this in front of the jury. That damages number, $13 million, was stuck in his craw like a chicken bone. He hoped to show that Behr's "maximized" calculations were the result of his greed and avarice. Behr's remaining testimony promised to be of the highest interest.

And so next morning, Trevor Will started him off with the topic nearest and dearest to his lawyer's heart: dollar figures.

"Doctor," he said, "you talked yesterday about an inflation adjustment in your damages calculation. And that was $2.8 million, wasn't it? Isn't it fair to say that most of that adjustment is attributable to years before 1991?"

Dr. Behr concurred.

"Okay," the attorney said. "And I think only about $771,000 of inflation was included for the last 12 years. If we look at the years from 1996 to the present, there would be even a smaller inflation component, correct, smaller than $700,000?"

"Yes."

Will nodded. "All together, you've included roughly $1.5 million in your damage assessment for purchased cattle, correct? And you indicate that represents money Mr. Allen spent on what you refer to as 'excess cattle,' right?"

"That's correct."

"By 'excess cattle' you mean cattle he had to purchase to add to the herd because it was not growing internally, correct?"

Actually, it was dying internally.

"Substantially," Dr. Behr said.

"Okay. And you assume that the cattle represented at page 26 of your report for these years were purchased because cows were dying or otherwise unable to produce the milk because of damage from the electricity. Dr. Behr—the dollars shown on page 26 represent cattle Mr. Allen purchased in those years, correct?"

"Correct."

"You assume he purchased to replace animals that were culled or died?"

Dr. Behr shifted in his seat and adjusted the microphone in front of him. "With regard to heifers, there's not a big difference between the cost of whether he raises or buys them."

Will then launched into a lengthy Q and A about cull rates, animal purchases and plant capacity, designed to somehow show that the doctor's damage numbers were greatly inflated. I thought he failed miserably, succeeding only in confusing and boring the jury. They sure looked disinterested; their eyes were everywhere except on Will and the witness. A very good sign.

Apparently realizing after a while that there was nothing in this for him, Will tried another subject. "It's your view, isn't it," he asked, "that most farms in the state of Wisconsin are affected by stray voltage?"

Behr nodded once. He had already told him this. "It's a very, very major problem. In some counties, the percentage affected is very high."

"Didn't you tell me that it was your view that starting in the late '70s there became a major stray voltage problem in Wisconsin? But you have admitted, haven't you, that the Wisconsin trend line does not support your hypothesis that there was a widespread problem that began in the late '70s?"

Behr looked down in his lap, at his notes. "The key word in your question is 'began.' I've never said that in 1979, suddenly there was a major jump upward. It was a gradual onset, so I would not expect to see any jiggle in the state average. Although based on what we see in other states, I'm convinced that the increase in the state average has been slowed by stray voltage."

Will then embarked on a series of questions regarding reports of Behr's findings at my farm and his calculation of damages based upon those reports. He tried to demonstrate that the reports were contradictory, but the doctor defended himself logically and persuasively. Unable to gain traction in any area, Will finally resorted to dredging up the doctor's past personal issues.

"You told us yesterday that you started as a full-time forensic economist in 1983, correct? You had been doing that for a little before that?"

"Yes," Dr. Behr said. "Beginning in 1978."

"And you were offering your services to lawyers at that point?"

"Yes."

Here it came, and Dr. Behr knew it was coming. Barry had alerted him the night before.

"And in 1984 to 1985, you were in the process of building up your practice?"

"You could say that," the doctor told him. "Although prior to that time it was pretty well up to full speed during the last month I was at the university. But what you say is generally correct."

Will again strolled over to the jurors and leaned against the center of the jury box, facing the witness. "There was an event—I don't want to get into the details—but an event in 1984 that caused a dip in your practice, correct?"

He did not dare mention Dr. Behr's divorce, because Judge McKay had forbidden it as irrelevant.

"You got that right," the doctor said with bitterness. Whether his bitterness stemmed from his memory of the divorce, Will's brass in bringing it up or both, I couldn't tell.

"And isn't it true," Trevor Will said to him, "that at that point you described yourself as a 'maximizer'?"

"Yes, I did," Dr. Behr said defiantly.

"And that was in a letter you wrote to an attorney with whom you were hoping to work?"

Dr. Behr shook his head vehemently. "No, not correct! That was in response to him having been on the opposite side of a case from me."

"So he was representing the opposition? And it was in that letter to him you described yourself as a maximizer?"

"That is correct."

"And isn't it also true, doctor, that the reference line in that letter is to your lottery ticket?"

Dr. Behr wrinkled his nose and snorted. "I've forgotten that. But I think that's correct."

Will had nothing further; he had shot his bolt. I felt satisfied that he had not made even a dent in Dr. Behr. And I was very encouraged by the course of the trial so far. For the first time, I began to feel I had a realistic shot at winning the case. Dr. Johnson and Dr. Behr made a powerful tandem; their professionalism, detailed analyses of my farm problems and confidence in their own findings had shone throughout their testimony.

Because Will's cross-examination ended with a whimper instead of a bang, I was surprised when Barry asked for redirect. Will had not laid a glove on Behr, so what need was there to have to rehabilitate him? But Barry knew his business, and so I waited to see in what direction he wished to go.

"Dr. Behr," Barry said, getting up from his chair and buttoning the middle button of his suit jacket. "You indicated you did about 95 percent plaintiff's cases? Why is that?" He went up close to the witness and stood before him with his arms folded.

"Well," the doctor said, "that's where the work is. It's claimant's responsibility to put the case together, and that means initial research.

So that's where most of the work is." He paused, then said, "I get calls from the defense to be available. It's just a matter of sending the plaintiff's report for evaluation. And that's the end of it. There's just not that much work on the other side of these issues."

"How much time do you have in Mr. Allen's case?"

"About 150 hours."

"And your charges in this case so far would be?"

"Not more than $15,000," Dr. Behr replied

"You charge what per hour?"

"$100 an hour. It's the same for everybody."

"Why do you go to the farm when you do one of these cases?"

"It's not absolutely necessary in my opinion. But I go for a couple of reasons, often at the outset: one, to see whether this looks like a serious operation, good or bad, or what's the overall impression. Two, there's often a look at conditions based on milk production history. And if you say to the farmer, 'send me your records,' who knows how long it will take? If you say, 'Let's go the attic together and find them,' it's easier. For those reasons, I usually end up going to the farm."

"Why did you use the term 'working hypothesis'?"

"Well, it's a statement of what the case is about. In this example, stray voltage. My working hypothesis would be that stray voltage was the cause of the damage. It's what I proceed under, qualified only because it's subject to modification by discoveries that there is no damage or that damage is due to something else."

Barry then asked, "Why is it that several reports were made by you?"

Now it seemed to me that Barry's purpose in redirect was to streamline and summarize the main points of the doctor's testimony, bracing the jury. I had to admit I was a much better farmer than I was a lawyer.

"The updating of time," Dr. Behr said. "Although nothing major in this case was brought out in Mr. Will's questioning."

I laughed when I thought how Trevor Will must've hated that remark.

"Some new facts or evidence," Behr continued, "may cause changes or may simply support the conditions of the earlier report. And there's discovery. Plaintiff's side works on things. Defense works on things. And we look at each other's work. Between Mr. Mellenberger and I, we went back and forth on the cow numbers and pretty much agree."

"Have you," Barry asked, "come on to any changes that would change the basic premise of your report?"

"No. It's been strongly supported."

Barry turned and went over to the exhibits. "Showing you 1377, Dr. Behr, is there anything about the animal turnover rate from 1975 until 1983 that would question your growth rate of 10 percent?"

"No. Because the cull rate over that period is almost right on 30 percent. I made a calculation: the average rate from '78 through '83 was 26 percent. If we put 26 percent in for a cull rate, we could actually grow at 14 percent."

Behr nodded. "If you look at the rate of change and compare the increase in the state average of milk production, about how many pounds can you look at?"

"Not more than 500."

"So," Barry said, "from a statistical point of view, can you tell us whether or not the change in the production level on the Allen farm is at all comparable to the state average?"

Dr. Behr shook his head. "No, not at all. The increase above the average there is around 3 percent. But the increase on the Allen farm was something like 40 percent over that same period."

Barry walked across the courtroom to the jurors. "Dr. Behr," he said, standing at the foot of the jury box, "in economics does the term 'maximizer' have a specific definition?"

The doctor grinned slyly. "It's a widely used term around at least 200 years."

"Does it have anything to do with level of damage?"

"Absolutely nothing."

"Explain maximizing in economic terms."

"Well," Dr. Behr said, "I indicated early in my testimony that economics was about use of resources which are insufficient to produce everything we would like to have. So in order for society to be as well as possible, we have to maximize use of resources to maximize social welfare. I've got quotes here from Nobel Prize winners—"

Judge McKay rudely cut him off. "Dr. Behr, we're not interested in other people's quotes."

Behr turned and looked up oddly at the judge. "And I'm adopting what they're saying which is that maximization is at the core of making choices to use resources in the best way."

Whether or not this was the intended meaning of "maximizer" in Behr's letter, I admired his skill in denying it had anything to do with payouts.

When Barry was finished, Trevor Will rose to re-cross and take his parting shots at the doctor. I doubted he would hit anything.

"Dr. Behr," Will said, getting up and passing Barry as the latter went back to his table. "On redirect you were asked about the maximizer comment, and it's true, isn't it, that your definition of the term in economics has nothing to do with winning lottery numbers, does it?"

"Not directly."

"It's also true that the person you were writing to was not an economist, was he, Nobel-Prize-winning or otherwise?"

This was another cheap shot. But what could I expect? The surprise was that there had not been more of them.

"He was a lawyer. I don't know what else he might have been."

"Would you agree that what you actually said was, 'Oh, if you want to see a maximizer in action, just keep your eyes open'?"

"Correct."

"You'd agree that someone who's not an economist might misinterpret that as meaning the way you go about creating it?"

"There's a large number that would," Dr. Behr said, and Trevor Will left it at that. He had no choice.

Then Barry stood for a last question and to make a final point to the jury. "At the lunchroom at the University of Wisconsin at Superior, in the economics department, was there a sign?"

Dr. Behr grinned, showing white teeth. "A sign," he said, "over the perpetually messy coffee area. Somebody had written, 'Be a maximizer, clean up after yourself. Your momma don't work here.'"

Judge McKay asked Will if he had any more questions. "No, Your Honor," he said. "I've never been in his coffee room."

I had no expectation of this sort of amiable banter and civility continuing for very long. Don Zipse was coming up, and the electrical engineer was not a man to suffer arrogance and contradiction gladly. He had spent part of his life abusing utilities. I wondered in what sort of predicament that was going to leave Kurt Meckstroth.

Chapter Ten

THE BIG GUNS FIRE: TWO

Doug Sutter was an educated, articulate, serious man. A former Brown County, Wisconsin, agricultural agent, Doug's main focus had been on applying university research on topics of importance to practical situations existing on farms throughout the state. He spent time holding public meetings, giving talks and disseminating information about such things as barn lighting, bag silos and mixing sites for calf feed.

Thirty-something and handsome, he was lean and fit, with sand-colored hair and light eyes. Aside from his warm and fuzzy personality, the thing I liked about Doug and what made him an important witness for me was that he was a disinterested, objective observer with a foot in two worlds: a university man and teacher as well as a farm boy and keen dairy-farm observer. His testimony, I hoped and expected, would contradict some of the WPS expert testimony soon to come from the utility's hired guns.

That morning, the fourth of June, another sunny, sparkling, Wisconsin spring day, again saw Barry Hammarback handling the direct questioning. Everybody in the courtroom had assumed their usual positions. Whether Will or Meckstroth would do the cross was anybody's guess, since both, at least for now, were seated at the defense table. But it hardly mattered.

Dressed nattily in a light-gray, glen-plaid suit, with white shirt and red silk tie, Barry opened, as he had with my other witnesses, by questioning Doug about his background and credentials, getting him to explain what the job of an agricultural agent entailed.

Doug smiled and adjusted the microphone in front of the witness chair as he settled in. "That would take a couple of days to fully explain. There's a lot of things. The overriding umbrella is that we're supposed to meet the needs of our clients. Primarily, that is farmers. But it could be a homeowner or somebody who owns land up north—anybody who wants information as it relates to agriculture."

"Do you have any interface with the university systems here in Wisconsin?" Barry asked.

"There are specialists we work with," Doug said. "Be it dairy, crops or other matters, there are specialists we work with for data."

"What's in your background," Barry asked, "that led you to this career as an agent?"

"My undergraduate degree was in agronomy. My master's was in integrated pest management."

This, I thought, clearly qualified him to deal with WPS.

"I'm born and raised on a dairy farm," Doug continued. "I worked for a consulting firm out of college, for a co-op after that, and was a tech rep for a company that produced organic fertilizers prior to coming here."

Barry asked to approach the bench and offered Exhibit 482 to Judge McKay—a copy of Doug's resume.

"The major area you list," Barry said to Doug, "is consulting regarding stray voltage, right?"

Doug affirmed this. Barry then wanted to know what particular areas of the stray voltage problem most concerned him.

"Initially, it was to try to resolve differences between what farmers are hearing from electricians, the utility and university research. More people would ask what I'd seen happen on other farms, to remove some of the grapevine rumor mill and have some verification of the reality of the situation."

At this point, Kurt Meckstroth, who I now knew would be crossing Doug, requested a bench conference. Whatever Meckstroth's complaint, McKay told Doug to go ahead and answer Barry's questions.

When Barry asked, "Would you say your role was primarily communication?" Meckstroth objected to his "leading" the witness. It seemed that Meckstroth was going to be taking potshots throughout the testimony.

"What," Barry went on, "was the predominant issue that arose with your work in stray voltage? Tell me what knowledge you gained."

Doug crossed his legs and adjusted his tie. "I was contacted by a farm with problems they couldn't identify anywhere else. Didn't appear nutrition-related, or cow-comfort-related, or—"

Meckstroth, frowning again, interrupted with an objection, and another bench conference took place. The interruptions were annoying; they made the testimony disjointed. But maybe that was the plan.

After the conference, Barry asked the witness, "When was the first time you were on the Russ Allen farm?"

"I was hired in June of '95. That summer I visited the farm."

"What was the purpose of that visit?"

"Russ was interested in remodeling the old dairy barn, and I had brought a specialist up from Madison to look at it and give some ideas. Russ had picked up a parlor from the farm show in Green Bay and wanted to change a stanchion barn into a free-stall to accommodate cow comfort issues."

Barry circled the courtroom, moving slowly from the judge's bench over toward the jury box and back again. "Did you have an understanding of the reason for that modification?"

Doug nodded. "Better utilization of the facility made it more usable. It was a 'special needs' facility to support the main dairy."

"The next contact you had with Russ Allen?"

"By the next summer he had concerns with stray voltage. I got involved at that point as he was looking for a loan to pay the cost to have the farm isolated. I had to run a computer program that would make projections as to what a payback would be on that change."

Still walking, Barry asked, "What information did you look at to figure out how it would impact cash flow or pay for itself?"

"Well, the numbers on investment return would come from Russ. The computer program shows costs to produce a 100-weight of milk, cropping history, and then, do I want 400 cows, 500, 600 cows, and does it break even at 20,000 pounds or 24,000 pounds. A big 'what if' computer game. If I invest $60,000 and get a 3,000-pound response per cow, does that pay for itself within a year's time, two years' time? It allows you to encapsulate all the variables on the farm and do some looking forward."

"As to the level of production expected or put in the plan, what was it?"

"Around 23,000," Doug told him. "When I run those programs, I do it as a service to the farmer. As everything with computers, garbage in, garbage out. So, I make sure farmers have ownership of the numbers they put in. If I think they're inaccurate I'll call them on it, but the number 23,000 would be a fairly average production level for that type of farm."

"Once you got done with the projection, did it seem feasible?"

Doug nodded several times. "Entirely."

Barry circled back toward the witness stand. "After Russ Allen found a lender to loan him the money for the isolation transformer, did you have another meeting on the farm?"

Doug nodded. "Prior to installation of the isolation transformer there was a meeting at the facility. An electrician would have been there, the nutritionist, the veterinarian, a couple of people that worked there, and Russ himself."

"What was the purpose of the meeting?"

"To establish a team," Doug said. "This was a vested interest to get off their flat milk production. So everybody needed to be on the same page on what expectations were, and to establish goals."

"What's the term 'limiting factor'?"

I thought Barry wanted to refresh the jury's memory of this term, which he had gone over in great detail with Dr. Johnson several days earlier. Understanding the phrase was important to understanding the desperation that had driven me into this courtroom.

"Something that would be in place," Doug replied, "that wouldn't allow you to go to a higher level in the case of a milking cow."

"Give me," Barry said, "a practical example of a limiting factor on a dairy herd."

Doug thought for a moment. "If you wanted 80 pounds of milk per cow per day and the ration was formulated to 60, the ration or the balancing of the ration was a limiting factor."

"What would happen if you changed other factors on the farm and didn't adjust the ration in your hypothetical?"

"In my opinion, you'd see very little production response. If the 60-pound limiting factor on the ration were still in place, you could improve any other factors and not see a dramatic increase."

Barry walked from the jury area and moved toward the front of the courtroom near the exhibits. "On Exhibit 1310, Dr. Johnson testified there was a list of rule-outs, differentials, to deal with on a dairy farm. Are these examples of limiting factors?"

"I don't know about records," Doug told him. "But certainly all the rest. Yes."

Barry nodded and went back over toward the jury. "Did you discuss the issue of limiting factors at that time with the folks at the meeting?"

Doug looked across the courtroom at Barry. "I don't know context, but everybody was asked, is there any reason we couldn't get up 75 pounds. Nobody felt, in their area, that they shouldn't be able to achieve that."

"What was your role at the meeting?"

"I was basically there for oversight. To document what had been going on—to track it through after isolation to see the response on the farm."

"Why was that?"

"Stray current was a hot topic. Quite a few people are concerned."

And more people need to become concerned. This menace does not merely harm animals; it harms humans in many ways and can kill them.

"I was getting calls from other farmers wanting some clarification—Extension, the agricultural agency, is supposed to serve as an unbiased third party. So I was trying to gather data and information, as other people were asking what was happening on a given farm, and to validate issues."

I watched to see the jurors' reactions to Doug, and noted that they were attentive. That had been the case with all my witnesses so far.

"And did you?" Barry asked.

"Yes. I did."

Barry nodded. "Did you follow up and go to the farm after isolation?"

He had.

"What information did you gather at that point?"

"I had conversations with Russ. He would say the cows did this or did that. My approach was to get a third-party verification. If he was talking about kicking, I might go talk to the person in the parlor and see what they felt, or look for myself. If talking to the electrician, I tried another party to get separate verification."

"Now," Barry said, "before isolation, I think you made a videotape?"

"Correct," Doug told him. "Upon Russ' request, I went through all the cows in one milking to videotape their feet, legs, hocks—just the condition of them."

"Before or after isolation?"

"Shortly before."

"And," Barry said, "I'll represent to you that the evidence in this case is that that tape is missing and cannot be found. Is that your understanding?"

"Right."

"If we go to Exhibit 1168 here," Barry crossed over to it, "which is a view of the parlor looking south, where was the videotape taken?"

"I had to walk in," Doug told him. "There were a dozen cows, so to video their legs and hocks, I would shoot cow one, go down to two, three and four and do the same thing on the other side."

"What's your recollection regarding the condition of the hocks or what you actually saw?"

Doug sighed. "Cows get swollen up at what I guess is the equivalent of our knee. Certainly, there was swelling in the hocks on Russ Allen's animals."

"Anything you would consider surprising?"

As Barry and Doug discussed the cows' miserable condition, I again began to feel the sense of melancholy, of irreparable loss. Because even if the farm were to be restored to its former health tomorrow, that couldn't bring back my dead sisters, or baby Jesse, or erase the suffering and deaths of so many poor animals. I felt a lump in my throat and blinked back tears. I tried to concentrate harder on Doug's testimony.

"I would say," Doug told Barry, "there were more cows with swollen hocks than you would like in any given barn. That's a debilitating thing to a cow, where she can't get around."

"What about the behavior of the animals in the parlor?"

"Well, that wasn't the intent of the video, but there's another farm where kicking was an issue. So, that was something that came up. I would say there was a fair amount of dancing, what we call rocking back and forth as they're being milked. That was apparent."

Barry thrust both hands into his pockets and came up near Doug. "After isolation occurred in December, did you make additional observations of the cattle?"

Doug nodded. "Water consumption became an issue. I did some visits to watch cows drink. Summer of 2002."

"Based on your evaluation, what happened to the production of the Allen herd after isolation? Did you go out and verify production, with the records?"

Doug had, and I remembered that he had.

"And," Barry said, "did the records indicate what was occurring after isolation, in December of 1997?"

Again Doug nodded. "There was a straight-line increase the first six months, around 20 pounds. Then, over the next year or so, he picked up another 10 pounds per cow per day."

"What's the highest benchmark that you verified?"

Doug told him, "87 pounds."

Rocking slightly back on his heels, Barry said, "At the level of increase prior to installation of the isolator, approximately what was their production per day?"

"I think," Doug said, "they bounced around between 55 and 60 pounds."

Barry, head down, now went slowly back toward the jury, stopped, turned and asked the witness, "After isolation, did you look at the Allen farm to see if there were any other management changes that could cause production to go up?"

This entire line of questioning, it seemed to me, was another thick layer of mortar and brick designed to reinforce the idea that stray electrical current on the farm, or its lack, was totally responsible for my gains and losses. Drs.

Johnson and Behr had testified to this, and now the county agent. I hoped the current's deadly effects were firmly etched on the jury's minds. After all, those 14 people had to live in Wisconsin and were potential victims of the hazard.

After a few half-hearted objections regarding foundation from Meckstroth, who looked as sour as if he were suffering from cramps, Doug said, "I noticed no change on that farm that would have indicated a 30-pound milk increase."

Meckstroth made a few more complaints to disrupt Barry, but Barry took it in stride and stayed focused on his Q and A. He knew Meckstroth's game.

"In 2002," Barry said to Doug, "you looked at cows and water. Why?"

"Water consumption has always been a big issue on farms that suspect stray voltage. Cows are unable to drink the amount of water they want or need. I identified four farms that would be representative of high production facilities, one of which was isolated, one of which was not, and two that were not or had been isolated but not responding, and Russ's facility, which was isolated, but not seeing high production continuously, had lost it. So I wanted to look at how cows approached the drinking water. Do they go right up and drink? Did they stand there and just look around? Did they lap? On those four facilities, with a video camera, we observed cows. There's a theory that they drink more on one farm than on another, I was trying to get visual observations."

"Did you make some of these observations on the Allen farm?"

"Yes."

"When?"

"Through the heat of summer—July, August."

"What did you see?"

"For the heat of summer, cows that would stand at the water and not drink, or put a nose down to the water and retract without drinking. As cows *would* drink, they would put their muzzles in the water, sucking in water—sounds like drinking through a straw. They'll stick their tongues out, and it's not lapping like a dog. They're testing the water, lapping."

They were getting ready for electric shocks, is what they were doing. Like Pavlov's dogs salivating to the sound of the metronome, my cattle had learned that drinking water brought pain. And so they avoided it as much as possible. And got sick. And had to be shot. And died.

"You saw that at the Allen farm?"

Doug sighed and drew a deep breath. "I saw more standing around looking than I did lapping. If a cow is thirsty, she would go up and take her drink of water and go somewhere else, or lie down. Our rule of thumb is you like to see them either eating or laying around. The fact they're just standing at the water in the heat of the day is contrary to what you'd like to see."

Barry next asked the witness to take a closer look at Exhibit 1010, a graph with rolling herd averages, among other things. With McKay's permission, Doug rose from his seat, pushed the microphone aside and walked over to the exhibit.

"After the first quarter of 2000," Barry said standing beside him, "there was a decline in production. Do you see that?"

Doug nodded. "Yes."

"Did you observe any changes in management or labor that would have caused that decrease in production?"

"No, I did not," Doug told him.

"In your opinion, Mr. Sutter, was the installation of the isolation device in December of 1997 a substantial factor in the increased production?"

Meckstroth suddenly leaped up from behind his table with a protest. "Objection, Your Honor! There's no foundation that he knows anything about electricity."

None of us around dairy farming had known anything about electricity until WPS reconfigured its distribution system and began electro-poisoning man and beast alike. We were *forced* to learn about electricity, more than we ever wanted to know, to survive. Of course Doug Sutter knew something about stray electricity. He was from Wisconsin.

But McKay, as usual, sided with Foley and Lardner. "Mr. Hammarback, with regard to your question, I don't have a clue what he knows about electricity. I mean, there is no foundation for that question."

Barry smiled and Meckstroth sat down. Barry turned and said to Doug, "Let me ask it this way: Assume there was testimony that the isolation device reduced ground current near the feed room. Based on your experience as an extension agent and your observations of the farm before and after verification of production records, do you have an opinion whether that decreased current caused the production to increase?"

Again Meckstroth sprang up, breathing fire. "Objection, Your Honor! *Foundation*! Mr. Hammarback has not laid a *foundation* for the witness to render such an opinion. He's got a background in agronomy and pest control and his history consists of working with seed and fertilizer companies!"

This experience with fertilizer might just come in quite handy when Doug got hit by Meckstroth's B.S. during cross-examination.

"He has not," Meckstroth went on, "testified to change in any measurements. He's not looked at any of the parameters that Dr. Johnson put on his chart."

McKay peered down at the attorney from the height of his bench and nodded sympathetically. "I understand your objection," he said in a soft voice.

McKay's constant coddling and favoritism toward Foley and Lardner enraged me; I thought about lodging a complaint with the state bar.

Barry said, "This goes to weight, Your Honor."

McKay thought for a moment. Maybe realizing he had gone too far, he said, "Well, I'm going to allow him to answer based on the fact that the question was posed as a hypothetical, and to the extent of your knowledge, you can answer."

Doug glanced at McKay and then looked at Barry. "I would go back to my earlier statement: There was nothing else that happened on that farm that would have caused that production fluctuation. I can't tie it specifically to the feed room."

"Or to the ground?" Barry said.

Meckstroth shot up like he'd sprung off a trampoline. "Your Honor, I object to Mr. Hammarback's question as leading and move to strike!"

Barry peered over at Meckstroth, grinned at him, and turned to the judge. "I described a hypothetical, Your Honor. He looked at the installation of the isolation system."

McKay shook his head. "I understand, but he's answered. Didn't you answer his question?" McKay said to Doug.

"I tried."

"Let me," Barry said slowly, "ask it to you this way, Mr. Sutter. In your experience as an extension agent and dairy farmer, and taking into account the actual evaluation you did both before and after isolation, and assuming the isolation reduced earth current on the Allen farm, do you have an opinion whether or not the reduced current I'm asking you to assume caused an increase in production?"

Meckstroth bobbed up again. For some reason he couldn't stand this question. "Objection, Your Honor," he sputtered. "Asked and answered."

This time Barry did not grin. He addressed Judge McKay but his eyes bored into Meckstroth like drills. "It goes to weight, Your Honor."

"I understand, Mr. Hammarback," McKay said, "but he's answered to the best of his knowledge. Objection sustained."

Barry did not respond at all to this. He said to the witness, "Was it your purpose in looking at the farm to have a complete checklist and check off every single item when you evaluated the change?"

"No," Doug said. "In hindsight it was too informal, but my information gathering was a conversation with the vet or the person milking cows. I would talk to people about what was observed and see if two and two added up to four."

Barry went over near his counsel table, glanced at Meckstroth, and leaned against the table, facing Doug. He had his back to the rows of spectators. "Is there a level of concern you used as an agricultural agent regarding electricity?" he asked.

Once more Meckstroth stood up and objected, claiming there was no foundation for the question. But this time McKay overruled him. Score one for the home team.

"I have a comfort level," Doug said sarcastically. "A zero. I mean I disagree with the current threshold that research would say is acceptable."

In other words, he felt exactly the same way about the subject as Dr. Johnson, the udder doctor, did.

With that answer hovering in the air, Barry finished his direct and gave way to Meckstroth's cross. I hoped Barry would interrupt him as many times as he had interrupted Barry.

"Mr. Sutter," Meckstroth said, standing and walking up toward the witness box. "When I took your deposition, you told me that you had been on the Allen farm at least 25 times. The only behavior you've seen that you attribute to stray voltage was that one day you saw an unusually high number of cows eating outside rather than eating inside in the shade, right?"

What was Meckstroth suggesting, that they were outside sweltering in 100-degree heat because they liked dining al fresco?

"I don't know," Doug said.

Then, from the blue, Meckstroth asked, "Is milk production a behavior? The fact they increased that much milk is a behavior, isn't it?"

I could see Doug stifle a laugh. "The only behavior, physical behavior."

Meckstroth said, "Mr. Sutter, please turn to page 226 of your deposition, line 9. Did I ask you these questions and did you give me these answers? Have you observed behavior on the Allen farm that you believed was prompted by stray voltage?"

Now Barry rose with an objection. I didn't care what it was about so long as it interfered with Kurt Meckstroth. "I object, improper impeachment. That's exactly the way he answered the question."

"You can," McKay said to Meckstroth, "explore his answer today or at the transcript of the deposition."

Meckstroth turned toward Doug. "Mr. Sutter, when I took your deposition I asked whether you had ever seen any behavior on the Allen farm you thought was prompted by stray voltage, right? Let's go to the transcript: *Have you ever observed any behavior on the Allen farm that you believed was prompted by stray voltage?*"

"We're talking physical behavior as opposed to cows producing milk?"

"Right."

Doug was silent. "I guess I would comment on that the fact that he had such a high percentage of cows outside feeding seemed contrary to what one would have thought they would eat on a hot day. You would assume they would come in the shade to eat."

"Did I ask you those questions and did you give me those answers?"

"Yes and yes."

"Now," Meckstroth said, "you didn't go inside the barn, did you?"

"No."

"And you would agree that if there was an excessive amount of ammonia inside the barn emanating from the pits beneath the slatted floor that could be a reason why cows would eat outside instead of inside?"

Barry objected to the lack of foundation for the question, and McKay instructed Meckstroth to pose it hypothetically, which he did. The judge also told Doug to answer it hypothetically. For the life of me I could never understand the difference between "hypothetical" questions and answers and "real" questions and answers; they always seemed the same thing.

"There's probably some level of ammonia that would at some point drive a cow outside," Doug said.

"And you don't know whether that level was present on the day you saw the cows eating outside, do you?"

"No, I do not," Doug said.

"When I took your deposition back in 2002, you made no mention of videotaping cow hocks and hooves, did you? And you didn't say you went down the line and looked at each individual cow, did you?"

"No."

"You taped milking for a couple of hours and weren't sure whether you taped even one complete milking, correct?"

"Not to say for certain, no."

"You concluded with respect to the cows' behavior there wasn't a major issue on Mr. Allen's farm, correct?"

Doug shrugged. "I don't know what it states in the deposition transcript but there was not excessive kicking on the day I was in there, no."

"And you told Mr. Allen that, didn't you?"

"I believe I did, yes."

"And he disagreed, right?"

Wrong. I only said that the cows didn't act squirrelly every second of every day.

"I don't know," Doug told him.

"He wanted the cows to stand firm while they were being milked?"

No, I wanted them kicking the bejesus out of everyone and everything. Of course I wanted them to stand still. And if they could've talked, they would've said that they wanted to stand still. And not be shocked.

"Yes."

Meckstroth paused. "By the time you asked Mr. Allen that question about the whereabouts of the tape you had made, he had filed his lawsuit against WPS, right?"

"That's possible."

"And you had known that he filed the lawsuit, didn't you?"

"At the time of the deposition I did, yes."

"You didn't press Mr. Allen at all about what he had done with the videotape, did you?"

Doug snorted. "Did I put him up against the wall and frisk him? I don't know…"

"You didn't even ask him who he loaned it to?" Meckstroth said incredulously.

"Probably not," Doug said.

Barry objected without rising. "Argumentative," he said. "And cumulative."

McKay sustained him. The videotape was misplaced and could never be found despite my best efforts. That was all there was to it.

Meckstroth sauntered over toward the jury box and stood several feet in front of it, facing the witness on an angle. "At the time I took your deposition in 2002, you didn't know whether Mr. Allen had a parlor camera or not, did you?"

"I get on a lot of farms," Doug said acidly. "I don't keep them all straight which ones have cameras and which ones don't."

But Meckstroth was nothing if not dogged. "When you learned that the videotape you made of Mr. Allen's cows had disappeared, you never thought to try to find out whether there were any parlor tapes available that would show the cows in the parlor?"

Doug chuckled. "Didn't know that it was an issue, no."

"Return," Kurt Meckstroth said, "for a moment to your observations of Mr. Allen's cattle drinking water. You don't have notes pertaining to those observations?"

"No, I do not."

"And you never prepared any, did you?"

"No, I did not."

Meckstroth's tone was inquisitorial, like Doug was on trial for something which hadn't yet been explained.

"Your observations lasted about an hour. You never counted the number of cows that drank water during that hour on Mr. Allen's farm, did you?"

"No, I did not," Doug told him.

"And you never counted the number of animals that seemed hesitant to drink water, did you?"

"No, I did not."

"And you never counted any number of animals that you believed lapped water?"

Barry stood up to object. "Cumulative," he said to McKay.

"Did you?" Meckstroth pressed Doug.

"I admit," Barry broke in and told the judge, "that they're very clever questions and they're all slightly different. But they're establishing the very same fact."

McKay held up a hand, palm outward, to Barry. "I'll allow him to answer the question," he said.

Doug said quietly, "There were no counts taken."

Meckstroth made an accusation couched in the form of a question. "You don't know whether it's natural for cows to lap water or not, do you?"

Doug grinned. "That's why I went to four farms. To try to establish some parameters."

"You don't do any electrical testing?" Meckstroth suddenly asked.

Doug said he did not.

"Matter of fact," Meckstroth said, "you've made a point not to get involved in electrical testing, right?"

Doug agreed.

"Because your background is crops?"

Again, Doug agreed.

"No one," Meckstroth said, "has ever told you that a cow contact measurement was made on Mr. Allen's farm that exceeded the Public Service Commission's level of concern, correct?"

"Correct," Doug told him.

"Mr. Sutter, did you not regularly receive from Mr. Allen production records for the farm? Production records which would be paper?"

Doug had not because I hadn't given him any.

"And you sometimes stopped by the farm and talked to Mr. Allen about milk production?"

"Correct."

"And Mr. Allen told you that his milk production increased after isolation, didn't he?"

"Correct."

"And he told you several times there were no management changes on the farm between the time of isolation and the time of his production increase, right? But you didn't play detective and try to verify anything he told you, did you?"

This was intolerable. Was he suggesting Doug was my pawn? That I lied and Doug swore to it? Meckstroth was some piece of work.

Doug also grew angry about the insinuation. "I stated this before," he told Meckstroth, and I thought I could see smoke coming out of his ears. "I would go there when Russ wasn't there and talk to the employees. I didn't bird-dog every statement made, no, but there were times I did independent verifications."

Meckstroth began to press Doug hard, triggering a stern objection from Barry that McKay had no choice but to sustain.

"In fact," Meckstroth said in his best bully-boy voice, "you didn't play detective and try to verify *anything*, did you?"

"Your Honor," Barry said, "I don't think there's any statement in the record that he's ever played detective."

"I'm going to sustain the objection," the judge said.

"You," Meckstroth accused Doug, ignoring both McKay and his ruling, "didn't try to verify *anything,* did you?"

At this Barry sprang from his chair. "Objection, Your Honor! He's already been asked and answered that!"

McKay nodded. "Sustained. He's indicated, Mr. Meckstroth, that he had his own method of verification. That's what he's indicated."

But Meckstroth, the pit bull, had his teeth clamped around Doug's ankle and did not intend to let go. "If that's his answer, can I read something from his transcript that's different, Your Honor? Page 238, line 20:

> Q: Did you ever try to find out whether anything dramatic changed during that time period?
> A: I asked if there were any other issues that would account for that increase. To play detective and try to verify anything, I did not.
> Q: And who did you ask that question of?
> A: Russ.
> Q: When?
> A: Probably several times.

Barry again spoke up. "Wait a minute," he said. "How about the rest of that?" Meckstroth obliged by reading further:

Q: That would have been sometime in 1998?
A: Sure. Yes.
Q: What did he tell you?
A: I don't know. Obviously there was an increase in milk production.

Meckstroth turned and went over to the defense table and dropped the transcript on it. Then he went back near the stand. "You would have considered starting to use BST to be a significant change during this period between isolation and the production increase, wouldn't you?"

Doug coughed and cleared his throat. "Not if it had been used prior to isolation, no."

Meckstroth went back to the table, picked up the transcript and quickly leafed through it. "Turn to page 240, line 3:"

Q: What would you consider a significant change?
A: The first one comes to whether BST was added or dropped in the management routine.

"Did I ask you that question and did you give me that answer?" Meckstroth asked.

"Yes and yes," Doug said.

"At the time I took your deposition, you did not know that Mr. Allen resumed using BST after isolation and before the production increase, correct?"

"I don't know when he would have gone back on BST."

Meckstroth lowered the transcript and tapped the thicket of papers against his thigh. "You would consider switching milking intervals from two to three times a day to be significant with respect to a milk increase after Mr. Allen's farm was isolated, right?"

"It would have an impact."

"At your deposition, you did not know that Mr. Allen began milking one alley of cows three times a day in the spring of 1998, did you?"

Doug shrugged. "That's possible."

Meckstroth paused. He stepped several paces closer to the jury box and the jurors, I noticed, were all watching him. I wondered if he was aware of 28 eyes on him.

"At your deposition, you did not know whether Mr. Allen had sold 32 cows one day in December of '97, did you?"

"I knew cows left," Doug said.

"In fact," the lawyer said, "Mr. Allen told you at one point that he kept cows he would have preferred to cull to satisfy obligations to his lender to maintain minimum cow numbers. And the reason he gave you for wanting to cull those cows was because of low production, right?"

"Certainly the production," Doug said.

Meckstroth again opened the deposition. "Please look at page 370, line 23.

> Q: Did Russell Allen give you a reason for why he would prefer to cull these cows?
>
> A: There may have been low production. If it was a cow with longer days in milk—if she wasn't bred and wasn't scheduled to be dried off—maybe she would have been shipped.
>
> Q: "Did I ask you those questions and did you give me those answers?"
>
> A: "Yes and yes," Doug said. "The key word there is 'may.' I was saying what he may have said."

Meckstroth lowered the transcript and again tapped it against his leg. "The typical cull rate in Brown County for a Holstein herd of 400 or more is 35 to 40 percent. Is that right?"

"Correct."

Meckstroth, mercifully, ended his cross and gave way to Barry for redirect. As Meckstroth took his seat, Barry stood, smoothing the front of his suit jacket and walking up close to Doug.

"Mr. Sutter, if the normal culling rate is 35 to 40 percent, how many cows a year would that be?"

"Whatever 35 percent of 400 is; 130, I don't know."

"Now, I want you to assume that the evidence has been that from '94 to '96, Mr. Allen used BST, and that from March 18 of 1998 and onward, he again used BST. Assuming those things to be true, would BST have had an effect on increased production after the change to the isolation system?"

"It could have."

"As a management practice," Barry said, effectively proving that it did not have *the* effect, "at least if it was used before and after, would that be a change in management practices?"

Clearly not, as Doug stressed. "If it was reintroduced, I don't know that I would call it a change in management, no."

Barry went over to the jury and stood with his back against the box rail. "On the effect of three times a day milking, when would a farmer start using it?"

"When cows are starting to see their udder—if we get under 70 pounds of production, it's pretty marginal. Above 80 or 90, to take pressure off that cow's udder you would implement that change."

"So," Barry said, "assuming that the three times a day milking change was made on alley three in April of 1998 and alley four about May of 1998, and assuming there were cows producing 50 pounds or less, do you have an opinion as to whether changing to three times a day would affect production significantly?"

Doug shook his head. "It would have minimal effect on anything under 50 pounds."

"Of the cow behaviors you believe were affected on the Allen farm by stray voltage, which one was?"

"If we're going to allow production to be accepted as a behavior, certainly milk production was affected."

And so there it was. The Brown County agricultural agent, with no dog in this fight, no axe to grind, who was an impartial observer charged with dispensing useful information to all involved in the Wisconsin farm community, had come down squarely on my side and against WPS in this fight over stray electrical current.

We were all about to obtain an education on electrical theory and practice, for next up were two authorities: Donald Zipse, a brilliant engineer and theoretician; and master electrician Larry Neubauer, who had been working hands-on with current, with juice, for most of his life.

I wondered what bed of nails Meckstroth and Trevor Will were now designing for those two.

Chapter Eleven

THE BIG GUNS FIRE: THREE

The biggest cannon of them all, a two-ton howitzer, was about to fire on my behalf.

Donald Zipse, a straight-talking, no-nonsense, forensic electrical engineer who was brilliant, creative and experienced, and who'd studied not only mine, but many others' problems with stray current, was taking the stand this morning.

I was excited for two reasons: First, Don was a heavyweight in the field whose testimony, I thought, would have great influence on the jury; and second, because Scott Lawrence was handling direct testimony of a witness for the first time in the trial.

Besides his first-rate legal skills and mastery of language, Scott had long since become an expert in the area of stray voltage; he'd been defending farmers against electrical poisoning for the past two decades. In addition to his degree in mathematics, Scott had also studied electrical engineering. No attorney involved in this case, for either side, had as much knowledge of the subject.

Bespectacled and affable, a portly but dynamic 69 year old, the top of his bald dome ringed by bushy hair on the sides, Don's direct testimony began with a summary of his background and qualifications, which were now up as Exhibit 1368.

"I started," Don said to Scott in his deep, clear baritone, "at Williamson Free School of Mechanical Trades in the power plant operation. After a bit with the Marine Corps, I ended at the University of Delaware where I obtained my degree in electrical engineering in 1961.

"At Williamson, a school that has no outside utilities, we generated our own electrical power, so I spent some time firing the boilers to generate electricity. I also spent some time as an electrician, working with an electrical contractor. Upon graduation, I entered Cutler-Hammer, a motor control company that also manufactures switches and panel boards.

"From there I went to Atlas Chemical, which was previously called Atlas Powder Company. They were bought out by ICI, and I worked for them as a senior electrical engineer."

Scott Lawrence, tall, thick-set, short-haired, with a brown mustache, dressed in a well-cut black pinstripe suit that could not hide his powerful physique, stood directly in front of the witness, his folded hands held low in front. I noticed what looked like a large, gold, college ring on the third finger of his right hand.

"Did you design electrical systems at that company?" Scott asked.

Don nodded at Scott. "Not only come up with concept, not only designed systems, but ordered the equipment and oversaw construction and installation, participated in start-up of the projects."

"What kinds of electrical systems or what kinds of applications?"

"Since we produced explosive powders, we had a distribution system within that facility. I designed a distribution system within a chemical plant at the bottom of the Delaware Memorial Bridge. We generated power in our own plants independent of outside utilities."

"What," Scott asked, "is the highest voltage distribution system you've designed?"

"The highest voltage distribution system I conceived was 138,000 volts, in Houston, Texas. Then I joined FMC Corp., moved up to engineering consultant, did the same as at Atlas Chemical. I was responsible for interfacing with utilities, setting up how and where and at what voltage we would get electrical service. And also the electrical rate structure."

"Since then what have you been doing?"

"Since then," Don said, "I've had my own company, Zipse Electrical Engineering, an independent consulting company doing forensic work. I also sell time to a company called Discovery Systems."

This was good stuff. I wanted Don to lay it on, fill the room with how much he knew, with his experience and innovations. Nobody on the other side, I felt sure, even guys like Roger Mellenberger, would be able to match him. He was going to dispute just about everything WPS asserted. I thought of Don like that old commercial for the brokerage firm, E.F. Hutton: *When we talk, people listen.*

"Do you investigate incidents involving electric shock as part of your consulting work?" Scott inquired.

"I do. Zipse Electrical has been in business approximately nine years."

Scott now went over to the jury box and took a turn at getting familiar with the jurors. "About how many incidents involving shock to living things have you investigated?"

"Over fifty."

"Do some involve cows?"

"They do."

"When did you first get into cows and shock?"

"November of '01," Don said.

"Would most of those incidents involve shock to humans?"

"Shock and electrocution," Don told him grimly. "Electrocution is death by electricity."

This menace is spreading steadily throughout the United States, metastasizing like a cancer. It's a new pollution, a new polio, a new plague. We're only just beginning to understand its deadliness.

Don next went on to mention his membership in the IEEE, the largest organization of electricity professionals in the world with 350,000 engineers divided into 36 societies, like Power Engineering, or Industrial Applications, a medical society, and a computer society, which is the largest. Don joined in 1957 and was elected as a Life Fellow.

Under Scott's questioning, Don discussed the National Electric Safety Code, for which the IEEE is the secretary. Its committees write the electrical code for the entire country. Don mentioned that he held a seat on the Grounding Subcommittee and explained the committee's purpose.

"The function is to approve changes in the code proposed by the public and to vote on those changes which have to do with grounding."

"Is that part of the code adopted as the electrical code in Wisconsin, with changes for Wisconsin?" Scott asked.

"Correct."

"What's the National Fire Protection Association?"

"That association dates back to before the turn of the 20th century. It's an organization that includes firemen, fire chiefs, fire inspectors, people interested in fire protection. It also includes electrical inspectors and insurance personnel. I am a member."

"What's the National Electrical Code, and how does it relate to the National Fire Protection Association?"

"The Fire Association produces standards in the same way the IEEE does. The National Electrical Code is a standard that the National Fire Protection Association administers. They're the secretary for it."

Scott nodded. "Is that code part of the electrical code in Wisconsin?"

"It is," Don told him.

"What part does it apply to?"

"To the installation of electrical equipment in residences, commercial buildings and industrial facilities."

"Are you a member of committees that write the electrical code? If yes, what are those?"

Don named several, such as the ones for hazardous locations, grounding and emergency/standby systems. These were listed in the exhibit with his resume. He then made mention of an important new paper he'd recently written that did not appear on the resume.

"The title of this new paper, which has been peer reviewed and will be presented at the Petroleum and Chemical Committee Technical Conference on September 16, is titled 'Hazardous Multigrounded Neutral Distribution System and Dangerous Stray Currents.'"

Sounded like it could be a good title for my biography, if I ever wanted to write one.

But Scott wanted to bring out still more of Don's accomplishments, so he asked, "Are you the author of a portion of an encyclopedia on electrical and electronics engineering?"

Don sat up straight in his chair. "I am, sir," he said proudly. "It is John Wiley, a large publishing firm. The volume I'm associated with is number eight. The editor is John G. Webster from the University of Wisconsin in Madison. Mr. Webster and his committee solicited my input for the grounding chapter. That chapter was peer reviewed."

Scott had him add the finishing touch for the jury. "Not to be modest, sir—are you an authority in that field?"

"I am, sir," Don replied.

Satisfied that the jury now had the correct picture of the witness, Scott moved on to asking what Don had observed at my farm during the course of his three visits.

"I've driven the complete line from the substation, the total circuit."

"And have you discussed with Larry Neubauer, the master electrician, the work he's done on the farm? And reviewed the records of Wisconsin Public Service in this case, and the transcripts of witnesses both for WPS and Mr. Allen?"

Don most definitely had.

"Did you request that data be gathered on the single-phase portion of the distribution system in the vicinity of the Allen farm?"

"Yes."

"What data did you want and why?"

Don inched forward in the chair and adjusted the microphone. "The amount of current flowing out the line, the phase current. And the other reading I wanted was the amount of current coming back on the neutral, known as the multigrounded neutral. Current flowing out should equal the amount coming back. If it doesn't, that's an immediate, simple indication of the amount of stray current flowing over the earth uncontrolled."

"And that's in evidence as Exhibit 1324," Scott said. "On 1324, can you explain the graphs and traces and their significance to your inquiry?"

"Yes. As I mentioned, primary current going out is in red, at a low of 5 amperes and goes up to a peak of 19 amperes. This was on a little over a day, and gives a good indication of the cycle. The blue chart or plot that you see is the amount of return current coming back on the neutral.

"I'm real interested in the red and blue. There is a difference: The amount of current flowing over earth is the difference between the blue and red line."

Scott walked over near the jury box. "And what's the significance of that?" he asked.

"When I see something like this it immediately tells me there's a problem with current flowing over the earth."

Scott remained standing near the front of the jury box. "Were you able to analyze how much is coming back compared to going out?"

Don's answer was an eye-opener for the jury; I saw a couple of them do a double take: "Approximately 70 percent of the current is flowing over the earth."

Scott was silent, allowing the courtroom to absorb this crucial piece of information. He then asked Don to explain the significance of the data on a new graph, Exhibit 1369.

"This," Don said, "is the amount of stray current at any moment in time during the period that this was recorded, starting on April 30 at 9:32 a.m. and concluding on May 1 at 1:06 p.m. The amount of current flowing over earth ranges from approximately 3.5 amps up to 15 amps."

These numbers, coming from this man, created a buzz, really seemed to make an impression on the jurors. They sure impressed the hell out of me. And my dead cows would've been impressed too, if they were still alive.

"Why," Scott asked, "do we see periodic ups and downs on the chart?"

"Well, that is the result of the amount of current being used by customers on the system beyond Mr. Allen's farm. As you can see, it occurs at approximately 6 p.m. on Wednesday and again at 6 a.m. or 7 a.m. on

Thursday. And that is the time of milking, when the electrical load is at its maximum."

"Mr. Zipse," Scott said, coming away from the jurors and walking slowly toward him, "have you seen similar graphs of current versus time in WPS' records taken at any time?"

Don shook his head. "I don't recall seeing a plot."

And why would he? Corporations that pollute the environment don't keep careful records of the poisons they release. I doubt chemical companies record the amount of toxins they dump into our water and air.

"Would it be important for WPS to take such measurements near the Allen farm for any purpose?"

Don nodded rapidly and opened his eyes wide. "Very definitely. Because it's a quick way of determining the amount of stray current that's flowing uncontrolled over the earth."

"The wire on the WPS system to the east and south of the Allen farm," Scott said "What's the nature of that conductor?"

"The type of conductor is what's known as a No. 8 copper weld. That wire dates back to approximately 1935."

Scott again paused to allow the jury time to absorb this.

"Did that surprise you?" he asked.

"Yes. Copper weld has a tendency to deteriorate and has a higher resistance."

"What, if anything," Scott asked, "should WPS have done about that 70 percent of the current flowing over earth at the Allen farm?"

"There's several things they could've done," Don told him. "One would be to insert a transformer which would start a new system, in between transformers. The substation had a transformer, and this happens to be the end of the line, so that is a system, but they could have put a transformer at this location, on this pole. Would be what's known as a 'one-to-one transformer.' We would start a new electrical system, and instead of stray current flowing all the way across the Allen farm to the substation, the current would then flow back to that transformer."

"What about good shorthand for the transformer you just described?"

"A transformer with a ratio of one to one."

"Where would this new system start?"

"At that transformer," Don said. "I would back up and say that the old system would be from the substation to that transformer. A new system would begin at that transformer. And since it's the last transformer on the circuit, it would then take in the rest of the distribution system there."

"What effect would such an installation have on the percentage relationship of the phase and neutral current back at the farm, where Mr. Neubauer measured them?"

"It's my opinion it would reduce the stray current to a very, very small amount, because the only stray current would exist from the poles." He pointed out the various down grounds on the exhibit.

"That would also go with this recommendation," Don continued. "That they change the conductor, use a semi-insulating conductor which is inexpensive, and the reason is to protect the line. It would not be grounded, so that would eliminate all those grounds there."

I eliminated them once, when I cut them down like diseased trees.

"Even if you left those grounds there," Scott said, "would such an installation improve the situation?"

"Yes," Don said.

"Is this costly?" Scott wanted to know.

"Not at all. It's one of the least costly solutions. The transformer itself is less than $2,000. I've estimated it would cost less than $10,000 to install, labor, et cetera. That's without installing new wire all over the system."

"Okay," Scott said. "What's the next proposed solution?"

"I would suggest rewiring the system."

"How extensive a job is that and what cost might that involve?"

As if money was tight at WPS. This utility was wealthy beyond dreams of avarice. Scott Lawrence knew this better than anyone, but wanted to show that our alternatives to the current situation were more than reasonable and fair, requiring minimal funding.

"One concept would be to add a single conductor, from the substation out. That we would call a 'neutral'—not a multigrounded neutral, but a four-wire system. That would require a more expensive transformer with two bushings. All the connections would be phase-to-phase. To make it simpler to understand, we would take the single-phase portion that's up on top. We now have a conductor that is grounded multiple times for safety purposes, and we would add a new, neutral conductor that would carry all the unbalanced current and would not be connected to earth. So there would be no stray current entering the earth. Convert it to what we call a four-wire system."

"What type of distribution system is this?"

"At the substation, it would be a wye system like we have now. However, it would be grounded only at the substation—the neutral would be connected to earth only at the substation. So it would be a four-wire system with a phase-to-phase transformer on it to differentiate it."

I watched the jury to see how they were reacting to these technical details and was pleasantly surprised by how attentive to Don they were. His deep voice, assurance, preparation and obvious expertise were a powerful combination.

"What would be the next possible solution?" Scott asked.

Don smiled. "That would be a five-wire system. A five-wire utilizes existing transformers. However, the transformer being used could be reduced in a five-wire system. With a five-wire, we want a single additional wire. That again becomes the neutral, where we have existing transformers connected from the phase to neutral. We connect this new wire, which carries only current from the transformer that is a neutral current. That's all it carries. It is not connected to earth. We're going to utilize the present connection to earth as a grounded wire.

"This system is what the utilities recommend for farmers. This is a unigrounded system. It's also been used in industry as long as I've been in the industry, since 1950."

"What would your next possible solution be, Mr. Zipse?"

"Rewire the system," Don said quickly. "Replace the copper weld wire which is a problem. When we rewire, use connectors that are properly installed so we can reduce current flowing back over a less than acceptable neutral. You would install an oversized neutral. However, that would still not eliminate all the problems with the system, not reduce it to an acceptable factor for Mr. Allen."

Scott walked over to the counsel table where Barry was seated and lifted a yellow legal pad, glanced at some notes on it, and handed it to Barry. Then he went back toward the witness stand.

"What would be the next possibility for solutions?" he asked Don.

"Well," Don said, "the next possibility is that down at the bottom of the circuit is another substation circuit, and there's been no reports from farmers of any problems on that line. I feel it would be worth a try to break the line up at Little Apple and Big Apple Road, and this circuit would then be fed to the other substation. That's the last possibility, three, in this order of acceptable solutions."

WPS could have done any of those things, and easily, if they'd cared to. Fact was that they didn't. That's why I was here. I found myself wondering about the kind of brutal cross-examination Trevor Will or Meckstroth had in store for Don. He'd already hurt them with his testimony; I figured he'd give them all they could handle during cross.

"Mr. Zipse," Scott said, "I'd like to discuss with you the concept of equipotential planes and shock. Will the installation of an equipotential plane in a dairy parlor or barn prevent cows from being shocked?"

Don made a sour face. He said emphatically, "No. It will *not* prevent cows from being shocked."

Here he was flinging a challenge at the WPS experts soon to testify, who would say that these planes were effective.

"What if you're not isolated?"

Don chuckled. "That's even worse."

"What's the idea," Scott asked, "behind an equipotential plane?"

"The concept," Don explained to him, "is that you have a section, an area that's got no differences in electrical potential. But that's not true, that is a misconception."

"Is it like the notion of a bird on a wire?"

"Yes, it is," Don answered. "We'll assume it's two feet, or very close together. There's no difference in potential as the current is flowing down that wire, and so the bird doesn't get hurt. If you were to grab hold of the wire like a bird and hang there with one hand and not come into contact with anything else—I don't recommend it—but you likely, like a bird, would not be hurt."

"Does installing equipotential planes on a farm do anything about the resistance in the farm as compared to the utility system?" Scott inquired.

"Yes, it does," Don said, and explained how. "An equipotential plane such as a milking parlor, where you put in a metal mat or rebar, connect this together, and it's connected to the grounding system. The grounding system is connected to the neutral system, and the primary neutral is connected to a secondary. So it gives an excellent path back to the substation.

"Now that equipotential plane, the metal, the mat that is not concrete, and the concrete has contact with the earth. Believe it or not, concrete is a semi-conductor of electricity. When we have a milking parlor with cows urinating and feces on the floor, and we wash that down after each group of cows come in, well, that concrete remains wet, and being wet it again decreases the resistance of the concrete. That is an equipotential plane."

"Does being wet have anything to do with cows getting shocks?"

"Yes, it does, because it lowers the resistance. Because of low resistance, the current, once in the vicinity of the equipotential plane or milking parlor, can easily get back onto the system through that circuit."

"Have you personally," Scott asked him, "investigated shock to humans in structures similar to equipotential planes?"

"I have," Don said.

Kevin Whaley then stood up to object on the grounds of relevance. Evidently, he didn't like the idea of talking about human beings getting zapped.

And as was becoming a bad habit, Judge McKay sustained him. Expressionless and without commenting, Scott switched to another line of questioning.

"Mr. Zipse, is the circuit theory involving shock to animals complicated? Can you figure out whether an animal is getting shocked by a straightforward application of Ohm's Law?"

"Unfortunately, you can't. The shock to a cow, because of continuously flowing current, will change the resistance of the cow."

"Can't an equipotential plane be useful if current isn't flowing on it continuously?" Scott asked.

"Well," Don answered, "it can be. It allows, should a fault occur, that such a wire, come into contact with the wet surface of a parlor, gives a low impedance path back to the transformer, allowing the circuit breaker to trip so that harm is minimized."

"In your professional judgment, was it good or bad when the requirement for equipotential planes was taken out of the Wisconsin electric code?"

Don nodded very slowly. "A good thing, sir."

"Have you," Scott asked, "seen any evidence of on-farm sources of stray current in this case prior to, or since, December of 1997?"

"No," Don said. "I have not."

Scott fell silent. After several moments he once again walked over to the jury. For a couple of long minutes the courtroom was completely quiet except for some hushed conversation along the back benches. I turned and looked over my shoulder and saw and heard the bailiff go over and tell them to stop the chatter while court was in session.

"Mr. Zipse," Scott said, "I want to ask you about issues I think make up the rest of this case. Explain to us in a little detail what a multigrounded system is, please?"

"The title in the Wisconsin State Electrical Code needs to be changed, and because normally we refer to this as a multigrounded neutral system, and the paragraph after the title begins with the neutral, we're talking about the neutral being multigrounded." Don opened his copy and read, "And it says 'It shall be a sufficient size and pass, and shall be connected at least four times per mile to the earth.'"

"Okay," Scott said. "A moment ago we were talking about electrical distribution systems. Is there a standard definition of system with respect to electrical power?"

"In the national code?"

"That's what I'm after," Scott told him.

"Yes, there is."

"How does it state the definition?"

"The IEEE standard 100 defines system electrical power as follows: 'Designates a combination of lines and associated transferring apparatus connected there with all connected together without an intervening transformer.'"

"Is the five-wire system you discussed with us earlier in that document?"

"It certainly is."

"Are the authors of that document authorities in the field?"

"Yes, they are."

"Do their recommendations have any relevance to minimizing stray current coming off distribution lines?"

"Yes," Don said. "This document says that the five-wire system eliminates ground current."

"Is the notion of ground currents being problematic anything new in the electrical industry?"

"Not at all."

Scott went to the table, picked up a document, walked over to the witness and gave it to him. "Mr. Zipse, I've handed you Exhibit 1142. Tell us what that document is, please?"

Don scanned it and said, "It's a United States Patent Office document entitled 'System Electrical Distribution.' It is specs forming part of a patent dated August 30, 1887."

"And who's the author of that patent?" Scott asked with a straight face.

"The author," Don said with a smile, "is Thomas A. Edison."

"Was he considered an expert in electrical distribution?" Scott asked, still straight-faced.

McKay broke in here to say, "I'll recognize Thomas Edison for what he is."

But apparently Mr. Kevin Whaley of Foley and Lardner, who bobbed up with an objection, didn't want to recognize the wizard of Menlo Park. "I'm not sure what the relevance of a patent from 150 years ago is, Judge."

Scott glanced over at Whaley and grinned. When it came to electrical knowledge, Whaley couldn't carry Scott Lawrence's surge protector.

"I'll lay a little foundation, Your Honor," Scott said to McKay. "Mr. Zipse, how are these pages from the Edison patent relevant to the issue of current in the earth?"

"This discusses what should be done about current flowing in the earth. Now, the original patent for an electrical distribution system was in

1880. In the intervening seven years, Mr. Edison had a problem, and the patent talks about that problem. The problem was—"

Whaley objected to relevance, McKay sustained him, and Scott moved on. But the relevance was this: the first Edison patent used the earth as a return conductor. In the second, Edison acknowledged this was a *bad idea*, and patented a three-wire system that didn't use the earth as a return. Scott's point was that this was a problem known 125 years ago to Edison, yet nothing had changed.

Scott then asked about the copper rings Larry Neubauer had placed around the farm.

"The buried ring is to detect current flowing through the earth," Don explained. "The surface ring also.

"The surface ring, 4,000 feet, is a loop, not a pure circle. It is in close proximity to the electrical distribution line, part of the three-phase line and part of the line running south from the Allen farm down Little Apple Road."

Scott slowly made his way back toward the jury. "Why does that result in induced current?" he asked.

"The large imbalance I mentioned earlier, the 15 amps. When we talk about alternating current we take a snapshot in time because current is fluctuating—we measure current going out. If current coming back is the same, the two will cancel the magnetic wave, the field generated around the conductor carrying current.

"Since there is an absence of return current, that is, we have the 15 amps difference—we generate a magnetic field equivalent to 15 amperes. That magnetic field dies down as a curve, as the square of the distance. As we get away from the distribution line, it will fade away. That is the magnetic field.

"As long as there is a magnetic field changing and a wire still on the ground, we're going to induce some current on it. That current is going to stay on that wire. That current has to complete its circuit."

"How about the buried ring?" Scott asked. "What are the sources of current on it?"

"Current flowing across the earth."

"They're induced on the buried ring also?"

"Yes," Don said, "but less than on the surface ring."

"Why?" Scott inquired. "Why did the buried ring, which is only a foot or two lower, have less induced current on it?"

"It's the distance," Don told him. "The distance from the source, the source being the two wires on the pole on the distribution line."

Scott spoke very slowly so the jury would get every word. "Is there any other characteristic of the buried ring that impacts the sources of current flowing on it?"

Don nodded. "Yes. There are ground rods that are driven alongside or attached to the ground ring. That ground rod acts as a collector of current going through the earth."

Scott paused, then said, "I'm going to hand you Exhibit 1124, in evidence, a document entitled 'Behavioral Studies of Dairy Cattle Scientists to Electrical Currents.' On page 1508 there is a table of how electrical resistances travel through various pathways. Is one of those pathways the mouth to all hooves pathway?"

"Yes, Table 3."

Scott nodded again. "Are you," he asked, "familiar with the test procedures for cow contact voltages set out by the Public Service Commission of Wisconsin?"

"Yes."

"What's the highest cow contact voltage reading according to that procedure that you're aware of on the Allen farm?"

"From my review of the data, it was over one volt when the instrument was pegged, it couldn't read the maximum."

"Would this be a straightforward application of Ohm's Law?"

Don lifted his drinking glass and took a sip of water. "Yes, sir," he said.

"For the lowest 10 percent resistance of cows, what would the current through the cow be at one volt and 244 ohms?"

Don made some fast calculations on a hand-held calculator Scott gave him. ".003, or, 3.3 milliamps."

"The 90 percent cows at 525 ohms?"

"That," Don told him, "would be 1.9 milliamps."

Any way you sliced it, this was a lot of juice flowing into the animals.

Next, Scott went into what I considered was an indictment of WPS' "see no evil" attitude toward farmers.

"Mr. Zipse, does the NESC have requirements with respect to inspection and tests of lines and equipment, and are they set forth on Exhibit 1370?"

"They are. There's five: 'Lines and equipment shall comply with safety rules; lines and equipment shall be inspected at intervals; shall be subjected to practical tests; any defects shall be recorded and maintained until the defects are corrected; lines and equipment with defects that endanger life or property shall be properly repaired, disconnected or isolated.'"

"Now," Scott said to him, and looked at the jury, "in your professional opinion, has Wisconsin Public Service Corporation complied with the requirements of Section 214 of the NESC with respect to the distribution system in the vicinity of the Allen farm?"

"No, sir," Don said loudly and clearly. "They have not. They have not complied with these requirements."

Scott waited. Then he said, "Do you hold that opinion to a reasonable degree of professional certainty, sir?"

Don looked him in the eye. "Oh, yes. I do."

Scott kept pouring it on. "Mr. Zipse, do you have an opinion to a reasonable degree of professional certainty as to whether Wisconsin Public Service Corporation has been negligent in the operation and maintenance of its distribution line in the vicinity of the Allen farm?"

"Yes," Don said forcefully. "It has been. Yes."

Still next to the jury box, Scott asked, "Do you have an opinion to a reasonable degree of professional certainty as to whether Wisconsin Public Service Corporation is providing adequate service and facilities to the Allen farm?"

"I do have an opinion: They have been negligent in meeting Rule 214 of the National Electrical Safety Code: Inspection and Test of Lines and Equipment."

Again, Scott paused to let the jury digest the last remark. "Mr. Zipse, if a solution to the condition of the single-phase line past the Allen farm is not implemented, do you have an opinion to a reasonable degree of professional certainty as to whether earth current will continue to access Allen livestock housing facilities?"

"I do," Don told him. "It will continue until the line is repaired, changed, modified in some way to prevent stray current flowing over the earth in the vicinity of the Allen farm."

And so there was my case in its technical details, and in no uncertain terms, presented by a leading expert on electricity and stray current. Scott had no further questions; there was nothing more to be asked. The jury had heard it all. Now it became WPS' time to take their whacks at Don and I was very curious to see how Foley and Lardner planned to attack and try to damage the man's credibility. I didn't have long to wait.

After a discussion with Judge McKay about the admission of certain exhibits into evidence, Kevin Whaley stepped forward to tackle the witness.

"Would you agree with me," Whaley asked Don, stepping up near the stand, one hand in his trouser pocket, "that for a dairy cow to be harmed by electricity, there must be current flowing through the cow?"

Don agreed.

"If there's no current flowing through a cow, electricity, it will not cause the cow's milk production to decrease, correct?"

Don agreed with that.

"And if no stray current—a term you like to use—flows through the cow, the milk production will not be adversely affected by electricity, correct?"

Don again agreed that this was correct.

I started thinking that I was maybe seeing the outline of a double branch of WPS' defense. It sounded like they might have their experts claim that stray current was not flowing through earth, or, failing that, that such a tiny amount was present that it couldn't possibly harm animals. But that remained to be seen.

"And a cow's ability to reproduce," Kevin Whaley said, "will not be adversely affected if no current flows through the cow, correct? And a cow's health will not be adversely affected by stray current if no current flows through the cow, correct?"

This seemed a weird way to ask questions. It was like saying if a bullet from a pistol was not fired at you, you would not be killed.

"That's correct," Don told him.

"A step potential is the voltage difference between two legs of a cow, correct? Typically, a step potential measurement is made between the front and back leg of a cow, correct?"

"Correct."

"When you're talking about a step potential, you are trying to determine how much current is flowing through the cow?"

"I am," Don said.

Whaley went and positioned himself halfway between the jury box and the witness stand. He folded his arms across his chest. "To determine if current is flowing through cows on the Allen farm due to earth currents, you take measurements, two points in the earth that the cow can touch simultaneously, correct?"

It was.

"One of the places is metal?"

"One of the places is a conductive surface, sir."

"And if," Whaley said, "you're trying to measure current flowing through a cow, you have to take into consideration the resistance of the cow, correct?"

"You do," Don told him.

"And when you try to determine how much current is flowing through a cow, you use 470 to 500 ohms of resistance, correct?"

"Correct."

"You have not taken any measurements on the Allen farm to determine how much current is flowing through the cows?"

"Correct."

The lawyer nodded and walked toward the witness stand. "Let's talk about the multineutral distribution system. You understand that every utility in Wisconsin has one, correct?"

"I understand that to be a fact," Don said.

"And you know that 49 of the 50 states in the US have exclusively or predominantly multigrounded distribution systems? And since 1950 the multigrounded system has been the predominant electrical distribution system in the United States, correct?"

"Unfortunately so," Don replied.

"You know it is an undisputable scientific fact that some current goes down a grounding wire that is connected both to a primary neutral and to a ground rod driven in the earth, correct?"

His tone was that of a teacher lecturing a backward student, which annoyed me, since Don had forgotten more about the subject than Whaley would ever know.

Don took it in stride and simply said, "That's correct, sir."

"And," Whaley went on, "it is an undisputable scientific fact that current flowing on the ground rod will then flow into the earth, correct?"

"Yes," Don said absently, as if to humor him.

"And that's what you call stray current, correct? And you believe the multigrounded neutral distribution system itself is defective, don't you?"

"I certainly do," Don told him.

"And you believe it is always wrong for utilities in Wisconsin to distribute electricity using a multigrounded neutral distribution system. You believe it is always negligent for a utility in Wisconsin to distribute electricity using a multigrounded neutral distribution system, correct?"

Don stared at him. "I certainly do from the information available in the public domain, sir."

"Because," Whaley said in a voice dripping with sarcasm, "that violates what you call 'Zipse's Law,' correct?"

"That's correct," Don said evenly.

"We have Ohm's Law," Whaley said, pacing slowly and continuing the sarcasm. "We have Newton's Law of Gravity. And you've named a law after yourself?"

"Yes, I have, sir."

And why not? Why shouldn't he? The man was 69 years old and a world-class expert, had made many important contributions and devoted his life to improving and innovating electrical systems throughout the country. As far as I was concerned, he was fully entitled. Zipse's Law was this: *In order to have and maintain an electrical installation safe from shocks and to prevent electrocution from stray current, all continuously flowing current shall be contained within a conductor insulated from earth, except at one place within the system, and in only one place can the neutral be connected to earth.*

Now this Q and A was starting to get seriously interesting. I wondered if or when Don would explode over Whaley's attitude. The jury, I could see, was paying close attention; some were grinning, while others looked wide-eyed and bemused. Dan Murray, sitting next to me, was having trouble stifling a grin.

Whaley continued, "And so even if WPS operated and maintained its system near the Allen farm in a satisfactory manner, your position would still be that the multigrounded neutral system is harming the cows. Correct?"

Don hesitated. "Well, sir, I've testified earlier that there is a possibility of having a multigrounded neutral system where the current is at such low value that it is not harming cows—but that does *not* include the Wisconsin system."

Take that, WPS!

"Let me," Whaley said impatiently, "ask my question again, Mr. Zipse. Even if WPS operated and maintained its distribution system that services the Allen farm in a satisfactory manner, your position would still be that the system is harming the cows on the Allen farm?"

Don sighed and drew a deep breath. "As I said, if you did it in a way that has been done elsewhere, where the current—and I described that previously—would not harm the Allen cows."

Kevin Whaley, I saw, was growing more and more annoyed and frustrated by Don's answers. He kept sighing loudly and shifting his weight from side to side. "But my question is in particular to the multigrounded neutral system in existence in the vicinity of the Allen farm. Do you understand that part of my question?"

Don smiled at the exasperated attorney. "I understand, yes."

"Now let me ask it again! Even if WPS operated and maintained its multigrounded neutral distribution system in a satisfactory manner, your position would still be that it's harming the cows on the Allen farm. Correct?"

Now it became Don's turn to get annoyed. "You have not listened to what I said. I said that if it operated as elsewhere in the United States, that the amount of current would not be harming the Allen farm with a multigrounded neutral distribution system."

Kevin Whaley shook his head and rolled his eyes. "Well, let me ask it differently. No matter what WPS did to the system in existence now, there is no way they can prevent stray current flowing over earth unless they remove all the grounds on their distribution system, correct?"

My point exactly.

"The way you have rephrased, correct," Don told him.

Whaley, though, continued setting traps. "In your opinion there is no way a multigrounded neutral distribution system can be operated satisfactorily, correct?"

Don was growing angry. I could see him biting down on his lower lip. "That is *not* what I said previously, sir."

"Is that," Whaley said accusingly, "what you told me at your deposition?"

Don slowly shook his head. "As I've said since the deposition, I have come across a system in the US that is a multigrounded system that would have been tested that appears not to have stray current of a magnitude that would harm animals."

"Did you tell me at your deposition there is no way that a multigrounded neutral system would be operated satisfactorily?"

"I certainly did."

"But in the 120 years since Edison, there are changes made?"

"I have changed my mind," Don said.

"WPS is required to follow the law, correct?" Whaley said.

Don agreed they were required to follow the law.

"And," Whaley said in his most sarcastic and arrogant tone, "when there's a conflict between the laws of Wisconsin and the opinions of Donald Zipse, author of Zipse's Law, you know that WPS is required to follow the laws of the state of Wisconsin and not your opinions. Correct?"

This, to me, was a really stupid question/statement/argument, whatever you want to call it. The law was always behind the times in catching up with new trends and new thinking, not to mention justice and morality. Was segregation right and proper during the decades that the law protected it? Was slavery?

"Unfortunately, that is true," Don said.

"You believe the law—requiring connection between the primary and secondary neutrals—is defective. That's the word you used at your deposition, correct?"

"Correct," Don said.

"With respect to the requirement under Wisconsin law that the primary neutral be grounded to earth at each transformer location, you think that part of the law is defective?"

"Correct," Don repeated.

"And with respect to the requirement that the primary neutral be grounded to earth at least nine times per mile, you think that part of Wisconsin law is defective, correct?"

"Correct," Don said once more.

"And with respect to this law that says the primary neutral shall be grounded at each neutral distribution pole since 1994: You think that part of Wisconsin law is also defective, correct?"

Don stated that he did indeed think so.

"And you have never designed a utility distribution system, correct?" Whaley said.

"I have not," Don told him.

"You yourself have never repaired a utility distribution system, correct?"

This was another dumb question. Lee Iacocca had never repaired transmissions but knew a faulty auto design when he saw one.

"No, I have not," Don said.

"And you have never built a utility distribution system for a utility, correct?"

Was Whaley kidding?

Before Don could answer, the lawyer continued on, breathless and overheated, "You have never been a licensed engineer in the state of Wisconsin, correct?"

Don smiled at him. "That is correct."

Looking suddenly pleased with himself, Whaley nodded and sauntered over to the jury box. "But for the purposes of this lawsuit, you think it's appropriate for you to term or declare Wisconsin law defective?"

Don looked up at Judge McKay and then across the courtroom at the attorney. "Not only Wisconsin, but the National Electrical Safety Code also," he snorted.

Whaley stared curiously at him for a moment. Then he said, "I'm going to ask about your five-wire system, one of your proposed solutions written on the chart over there by Mr. Scott Lawrence."

Don held up a hand. "I object to you calling it my system. It's been in effect a long, long time. I appreciate the compliment."

"What you aim to do," Whaley said, "is have 49 of 50 states switch from a multigrounded system to your proposed five-wire system?" He may as well have been asking, "And you aim to have eight of the nine planets in the solar system start orbiting Jupiter?"

Don nodded. "Over a period of time. Just the way that a receptacle were to change from two poles to three."

Whaley shook his head theatrically and threw out his hands. "And you want 49 states to change over to this system, even though you have no idea how much it would cost?"

Cost be damned, Don was talking about what was desirable and beneficial, what was best for life and living things. Just how much did Whaley and WPS calculate the lives of my dead sisters and brother and nephew were worth? I felt like jumping up and asking them.

"As I said, it would be a period of many years, therefore cost would be a nonissue."

"When I asked you if WPS were forced to change its system, if that would cost millions, you said you had no idea, correct?"

"I object to you referring to it as *my* system. It's been designed way before I got involved."

"I understand," Whaley said, full of impatience. "You are not aware of any utility that in the last 50 years has used the five-wire system of the type you are proposing?"

"Correct," Don said. "It is a unigrounded system."

"Your five-wire system is what you call a 'ground conductor ground?' And that's in accordance with Zipse's Law, correct?"

"No continuous return current on it."

Whaley licked his lips and moved a few feet nearer to the witness. "Your recommended four-wire system is a multigrounded distribution system, correct?"

"It is not," Don said. "When you say 'multigrounded distribution system,' it is understood that the word 'neutral' is in it. The law is sloppy. The National Electrical Safety Code is sloppy, because it fails to put the word 'neutral' in. It has the word 'neutral' as part of the text, so it can be confusing."

"If I'm keeping a list of adjectives about how you feel about Wisconsin law, we have: Defective, vague, ambiguous, sloppy, confusing. Correct?"

It looked as if Whaley had a personal stake in defending what he seemed to feel was the holy writ of Wisconsin law. And maybe he did at that. Maybe his job with Foley and Lardner was on the line.

Now he began to spar with Don over words and their definitions, which told me he had nothing of use to undermine Don's testimony or refute his expert opinions. The thought encouraged me.

"Let me ask this," Whaley said to him. "Public Service Commission 114 requires in Wisconsin that you have a multigrounded system and the system has to have a primary neutral, correct?"

Don smiled and shook his head. "It has no primary neutral, and it's not prohibited by Wisconsin code or the NESC."

"Your four-wire system violates the NESC, doesn't it?" Whaley said.

Don's eyes flashed with anger. "No. It does *not* violate the NESC."

"The NESC requires that the secondary neutral be grounded to earth at each service entrance, correct?"

"Only," Don told him, "if there's a neutral involved, sir. When you say 'a four-wire system' as we've been talking about, this is an ungrounded system."

Seeing there was nothing in this for him, Whaley now tried impeaching Don by referring to his deposition. He directed him to a certain page and read aloud: "'Q: Is it fair to say your four-wire system violates the NESC because the secondary neutral is not grounded to earth at the service entrance?'

"'A: It does.'

"Did I ask you that question and did you give me that answer?"

"I did, sir," Don replied, deadpan. "And I stand by that answer. Because when you ground at the service entrance, you are grounding a second time. A second time allows stray current to flow just like it does on the primary system where the neutral is grounded many times. So, it's a problem, one I'm striving to eliminate."

"So," Whaley said, again sarcastically, "now we know that in your opinion, Wisconsin law is defective and the NEC is defective, correct?"

Don laughed. "That's correct! The NEC is revised every three years, and the reason is so that when we discover defects in the code, we can change the code and the public can act. We write a proposal to change the code. So, yes, the code is defective."

Whaley chewed on that in silence. "Well, have you recommended to the NEC that it eliminate its requirement that the secondary neutral be grounded to earth at the service entrance? And has the NEC rejected your request?"

Don spread his hands out in front of him. "It takes time to educate the NEC. It took me 21 years to finally get the ranges and dryers in homes to be wired correctly. So I don't expect it to be accepted on the first go-around. It was rejected, and I expected that."

"Another solution you have," Whaley said, "the idea to tear down the line, the single-phase east of the Allen farm. If WPS moved the single-phase service over to the west mark substation, it would continue to be a multigrounded neutral system. And stray current would continue to flow in the earth in all directions, correct?"

I couldn't help but think that Whaley was conceding the fact that WPS' electricity was flooding my land. I sure hoped the jury had that impression.

"Correct," Don said.

"And all that stray current would flow over property owned by people other than Mr. Allen, correct?"

"That is correct," Don replied.

Whaley then pointed out that this proposal would redirect current from me to my neighbors, which was true enough. Don told him that he'd made the suggestion because there were no complaints from other farms as yet, and because the solution would be temporary as well as inexpensive. And it was only one of a number of solutions he'd offered.

Whaley went off on yet another tangent, and I could see the jury's attention starting to fade. Most of them were looking around absently or fidgeting in their seats. I hoped he was losing them for good.

"Before the farm was isolated," Kevin Whaley said, "there were two problems. One, the primary neutral is bonded to the second. And the primary neutral is multigrounded to earth, correct?"

Don agreed that this was true.

"Current, once it enters the earth from WPS' distribution wires, goes in all directions?" Whaley asked.

"We can assume it goes in all directions, according to the resistance of the circuit that it takes. Currents flow in all directions, but if one direction has high impedance—same word, 'impedance,' is electrical for AC—less current would flow in that direction."

Once again the attorney moved over close to the jury. I wondered if he noticed what little attention they were paying him.

"You told us earlier," he said to Don, "that the theory in a farm facility is to have an area with no differences in voltage, correct?"

"That's the theory. But it doesn't hold."

"Under Ohm's Law, current equals voltage divided by resistance?"

"Correct, sir," Don said.

"If there is no voltage, there is no current. And to determine whether there is voltage in a farm with an equipotential plane, you need to take electrical measurements. And Larry Neubauer took measurements in the milking parlor from the Allen farm in 1997. And even though there was an equipotential plane in the free-stall barn, it's still a good idea to take electrical measurements, correct?"

Don affirmed that it was.

"You have not seen electrical measurements from before April 1997. So you can't say one way or another whether Mr. Allen's on-farm electrical system contributed stray current before April 1997, correct?"

"Yes, sir."

Whaley fell silent, maybe assuming he'd scored a point. "You know," he resumed, "that the buried ground ring was installed on Mr. Allen's farm in May 1999? And the surface ring in May 2002?"

"Yes, sir."

"And it's your professional opinion that the buried ground ring diverts some stray current flowing across Mr. Allen's property? And it was your professional opinion that 100 percent of the current on the buried ground ring came from grounding wires on the WPS circuit servicing the Allen farm, correct?"

"Correct," Don said.

"And it was your opinion that most of the stray current came from the single-phase portion of the system as opposed to the three-phase? And you thought it was an excellent idea for Mr. Neubauer to install the buried ground ring?"

Don nodded. "I did."

"And the surface ground ring was another invention and development that you were enthralled with, correct? And, in fact, you thought Mr. Neubauer should patent his idea of buried or surface ground rings to divert current, correct?"

"To detect," Don said. "The amount it diverts is insignificant."

Whaley turned and faced the jurors. "What's your hourly rate in this matter?" he asked Don.

"My hourly rate is $150."

"And you've billed over $50,000 in this case, correct?"

"No, sir," Don said. "I've billed under $50,000."

"You had put in well over 150 hours in this case, correct?"

I couldn't wait for Scott and Barry to start asking WPS' witnesses about how much they were making for their testimony. I knew for a fact that Roger Mellenberger had made well over a quarter of a million dollars off the utility in one year.

"I probably did," Don said.

"And in that time, it was your professional opinion that 100 percent of the current on both the buried and surface ground rings were from WPS' grounding wires?"

"That's what I said at that time. And I'm now more knowledgeable about the subject."

Kevin Whaley folded his arms and leaned with his back against the post of the jury box, facing toward the witness. "Now," he said, "you are aware that Mr. Allen has cut WPS grounding wires, correct?"

"Yes, I am."

"In fact, when WPS' lawyers faxed a letter to Mr. Lawrence stating that it was inappropriate and dangerous to cut grounding wires, you happened to be in Mr. Lawrence's office when the fax came through, correct?"

"That is correct."

"And WPS' lawyer said in that letter, 'tell us who is cutting ground wires.' Correct?"

"Yes."

"And you laughed at that part of the letter. Correct?"

I heard some giggling from the spectators behind me.

If Don didn't laugh, he should have. He should've laughed with satisfaction at my lessening of the electrical pollution that poisoned the dairy cows and wrecked my family and business. No one else was helping me, least of all WPS, so I had to help myself.

"I don't know whether I laughed at that or not. Don't recall."

"Did you tell me at your deposition that you laughed?"

Don shrugged. "Doesn't sound like something I would say. But you can look it up if it will save you time. I accept that."

Kevin Whaley was frowning heavily, as if Don had committed some kind of sacrilege by laughing over that stupid fax. What was Don supposed to do, weep and wail for poor, abused WPS?

Whaley just couldn't let it go; this supposed laughing of Don's infuriated him.

"And you *laughed* because you thought it was funny, because we didn't know who was cutting the ground wires. But *you* did, correct?"

Now I wanted to laugh.

"No," Don said, leaning back away from the microphone, and I could tell he was struggling to suppress a smile. He leaned forward again. "If I laughed, it was because we were preparing an affidavit at the same time you sent in that fax. So I was probably laughing at the coincidence. I guess you had clairvoyance, knew we were preparing this affidavit and kind of cut us off at the corner."

"You assume it is illegal for somebody to cut utility ground wires?" Whaley asked.

"I can assume that."

"And yet you know that Mr. Neubauer cut grounding wires on WPS transformer poles, correct?" Whaley's attitude seemed to suggest he thought of us as terrorists who were terrorizing defenseless WPS.

"I'm not aware of that," Don told him.

"You were aware of it," Whaley countered, "when I took your deposition, so let me take you—"

"That was then," Don snapped. "I don't recall it now."

Whaley read aloud the relevant portion of the deposition transcript, which Don didn't dispute.

"And," Whaley continued, "you *applaud* Mr. Neubauer for cutting the WPS grounding wire on the transformer pole, correct?"

Damned terrorists that we were, we always laughed insanely and applauded when destroying things.

"Yes," Don admitted.

"And you were aware that Mr. Allen cut grounding wires from the intersection of Little Apple Road and Big Apple Road southeast of the farm and northeast of the farm?"

We were waging a *jihad* against WPS, all of us, Whaley was trying to say.

"I am aware, sir," Don said. "And that resulted in increased production."

"That's your professional opinion?" Whaley asked.

"Yes, sir," Don said breezily.

"And you *applauded* him for cutting ground wires?"

It was as though the so-called applause and laughter were far worse offenses to Kevin Whaley than the actual cutting of the wire. Very strange.

Don grinned. "I did."

"You were aware that Mr. Allen had ditch wires laid next to the WPS distribution poles? And if ditch wires were connected directly to WPS' ground wiring that would create a parallel path?"

"Correct."

"It's your professional opinion," Kevin Whaley said, "that transient electricity is not a problem on the Allen farm today?"

"They're not a major problem," Don told him. "The major problem is the stray current. And the stray current—if it's eliminated, the transients will go with it."

Whaley walked over to his defense table, looked at his pad a moment and then went up near the witness. "I want to talk a little about this law of physics you've named after yourself, okay?" This naming of Zipse's Law, I could see, rankled Whaley. As much as the applause and the laughing bothered him.

"Thank you," Don grinned.

"You are a firm believer in Zipse's Law, correct?"

Dumb question. Why wouldn't he be?

"I am, sir."

"And you think Zipse's Law is a fundamental law of physics?"

"It's a fundamental law of electrical distribution."

"You consider Zipse's Law as scientifically sound as Ohm's Law, correct?"

"For electrical distribution, I certainly do, sir."

"And Zipse's Law is one of the fundamental beliefs upon which you have staked your reputation, correct?"

"It is, yes," Don said.

"And you've gone around the country telling people about Zipse's Law, correct?"

"It's the only way they're going to learn about it."

"You agree your ideas on grounding requirements for utility distribution systems are outside the mainstream of electrical engineering?"

"At this point in time, yes."

As were Columbus, Galileo, Fulton, Einstein and a host of others in their times. They all thought "outside the box," as the saying goes.

"When I took your deposition in 2002, you told me that both the NESC and NEC had rejected all your ideas on changing grounding requirements, correct?"

Don shrugged and smiled. "As I said, it took 21 years to get ranges and dryers, which are in everybody's homes, to be wired correctly. So I expect it will take 21 years. And since I've been working about three years on this, I expect it will take another 18."

"You submitted to the NEC four to six new proposals relating to what you call 'stray current.' And these new proposals all related to what you call 'Zipse's Law'. And you consider Zipse's Law related to the issues in this lawsuit, correct?"

Don nodded several times. "Oh, I do. Yes."

"The NEC voted on your proposals in 2003. You made a proposal to Panel Five regarding your theory on Zipse's Law, correct? And between January 2003 and April 28, 2003, when I took your deposition, you could have called any member of the code-making panel to find out results of the vote, correct?"

"I was willing to wait until the publication was final, because the committee vote is not final until the document is published in July."

Whaley went up and handed Don a document. "I'm going to show you what's been marked as Exhibit 1738. Do you recognize it as an NEC test report prepared by the Edison Electric Institute, a member of the NEC?"

Don recognized it.

"This document," Kevin Whaley said, "sets forth your proposed changes to the NEC, correct? And in connection with those proposals, you submitted written substantiations, correct?"

"You have to."

"And the substantiation you submitted to the committee specifically refers to Zipse's Law, correct?"

"Yes, it does," Don said.

"And you know with this document in your hands," Whaley said, in full gloating mode, "what the vote of the committee was. Don't you?"

"It doesn't surprise me at all," Don said evenly.

"They rejected all five of your Zipse's Law proposals by a vote of 16 to nothing. Correct?"

"Yes," Don said. "But recognize that all five were one paragraph in the documentation. I addressed each sentence of the paragraph, so they rejected my rejection of the paragraph—that's correct."

"So," Whaley continued to gloat, "by a cumulative vote the NEC has voted on Zipse's Law and rejected it 80 to nothing. If we add up five votes of 16 to nothing, that equals 80, correct?"

Don grinned. Then he made a memorable remark, maybe the most memorable of the trial, one that perfectly summed up WPS' approach to the epidemic of stray current and to my case. "*Liars can figure,*" Don said. "*And figures can lie.* I would just say it was rejected by the committee."

"And then there is the panel statement of rejection, correct?" Kevin Whaley said. "The statement begins, 'The submitter has failed to provide meaningful and specific substantiation for his recommendations. The substantiation he does provide is not based on existing science or engineering principles but rather anecdotal and personal opinion.'

Did I read that correctly?"

"You did, sir," Don told him. "That's their opinion."

"And," Kevin Whaley went on, "the first sentence of the second paragraph says, 'The submitter's use of farm and livestock examples are not only anecdotal in nature, they are off the mark.' Did I read that correctly?"

"Yes."

"Then the next paragraph says, 'The submitter's claim that earth currents associated with grounding is hazardous is overstated and leaves out reputable evidence to the contrary. Reference is made to a study commissioned by the Minnesota Public Utilities Commission which asks the question of whether or not earth currents can be harmful to livestock. The answer was provided in a report prepared by a team of scientists known as the Minnesota Science Advisors. The team worked four years and spent nearly one million dollars reviewing the existing body of relevant research. Among other things their final report states, 'We have not found credible scientific evidence to verify the specific claim that currents in the earth or associated electric parameters such as voltages, magnetic fields and electric fields are causes of poor health and production in dairy herds.' Did I read that correctly?"

"You read it correctly," Don said.

This notorious report was bogus in just about every way. Several scientific advisors who participated in the study later said so. This study, to which Kevin Whaley was referring, was conducted at 19 farms in the state of Minnesota. But there was no record of *which* farms or the names of the farmers, as Mark Cook of the Public Service Commission testified to in an earlier deposition. Scientific?

A field study to examine the effect of stray current on dairy farms was noteworthy because of its unscientific nature and the harsh dissent from its conclusions by a few eminent members of the panel, including Drs. Abraham Liboff and Charles Polk. Alex Furo was involved but resigned from the study because he found it completely slanted in favor of the utilities. The methodology was faulty from start to finish.

Dr. Liboff—a physics professor at Oakland University in Rochester, Michigan, and the author of many papers on the biological effects of electricity and electromagnetic fields—has stated that he thinks the utilities got to the legislatures and got the funding yanked. Suddenly, they were told that the money ran out? They were not allowed to test for what they wanted. Dr. Liboff said he believed that something other than management of the farm and diseases were affecting their cows, and felt bad that they were not allowed to help the farmers who he thought were very nice people. After the funding suddenly and mysteriously ran out, the study was sent back with Douglas Reinemann, a University of Wisconsin shill, to be finished in his laboratory. Dr. Liboff stated that they knew that the study would be in trouble when it was sent back with Reinemann. He had said that you couldn't call what Reinemann did 'research;' that wasn't how research was done.

Douglas Reinemann knew ahead of time the results he would get: That electricity did not harm cows. The project was rigged from the start according to Dr. Liboff, who has said, "Reinemann keeps doing the same old things he's done for years, which show nothing. He should've looked for changes in biochemistry. All he looked at was if the cow's tail switched. The study was political. And working with cow variables—housing, feed, et cetera—is an insane way of conducting earth current experiments."

Utilities drag this obviously flawed field study into court and cram it down the jurors' throats. Is all this stuff legal? Should the attorney generals of our states be looking into this?

Both Dr. Liboff and Dr. Polk—and—others refused to sign the final report.

I have to confess that my heart sank on hearing the panel's conclusion. I was afraid it would severely hurt my case, since the finding was issued by a commission that the jury would probably regard as a high authority. Still, it was only one opinion. But what I wanted was a slam dunk win, though I knew it wouldn't possibly be that easy. The main thing was to win. A win in football by the score of 3-0 is just as good as a 41-0 blowout. A win was a win was a win.

At this point Kevin Whaley had nothing left to ask Don, and Scott Lawrence moved quickly into redirect. I hoped he could repair the damage I felt had been done during the final part of cross.

Scott rose and buttoned his suit jacket as Kevin Whaley passed him on his way to the defense table.

"Mr. Zipse," Scott said, "you made reference to three-pronged outlets in the homes in response to one of Mr. Whaley's questions. Do you recall that? Did the principles involved in three-pronged outlets on a 120-volt circuit in homes also relate to utility systems?"

"They do," Don told him. "In your home, you know, you have a three-pronged receptacle. That used to be a two-pole receptacle, and in the 1950s the NEC—which we've just been talking about—said it was safer if we had an equipment-grounding conductor. That's the green wire. The utilities have taken the two wires and twisted them together into one to save money and save installation cost. For the lineman's safety it's required that it be grounded four times per mile so the voltage doesn't build up on that combination wire. I submitted a proposal to the NEC, in jest, to do just that for your home wiring, so they would realize the significance."

"Are you requesting that the code require utility systems to do what's done routinely in homes?"

"Yes," Don said. "That's the extra wire we've talked about being added, the fourth or fifth wire."

"By the way," Scott said. "Was one of your proposals to the code committees contained in Exhibit 1773 accepted?"

"Yes, sir. That had to do with the messenger. The messenger is the bare steel conductor that runs from the pole to Mr. Allen's farm initially on the single-phase service and also runs to all conductors. In the case of Mr. Allen's service, it supported two black-insulated conductors and that was all. However, my proposal is that the messenger cannot be used as a current-carrying conductor. In Mr. Allen's farm, the messenger is used as not only the green ground wire but also as the neutral carrying conductor. So they've combined the messenger into three functions: a messenger supporting other wires, a current carrying conductor and the equipment-grounding conductor, that green wire."

Scott went over near the jurors. "Does this proposal relate to your concern to separate the grounding and neutral?"

"Yes," Don told him. "And it's been accepted."

This was excellent; I felt better now that the jury realized that Don had this idea adopted by the committee.

Scott paused to let the courtroom absorb the information. "Now," he resumed, "Mr. Whaley talked with you about the use of split bolts. In order to have the use of split bolts continue to be acceptable, is it necessary to inspect and maintain them?"

"Yes. Because, with herds, there is a vibration, and that vibration can loosen them up, if the connection is not properly made. To properly make the connection, you have to follow the manufacturer's instructions. In today's world we have a paste material known as No-ox, which helps prevent oxidation. It's usually used with aluminum but can be used with copper and Copperweld. By applying that paste, it reduces the possibility of corrosion. And the other thing is, it has to be torqued up to the manufacturer's specifications."

Still standing close to the jury, Scott asked, "To periodically inspect split bolt connectors, what does one have to do?"

"Normally," Don said, "go with a torque wrench and check the torque on the bolt not only looking at that, but making sure that it doesn't show signs of corrosion."

"What effect will failure to maintain connectors like split bolts eventually have on the amount of current going into the earth?"

Don's forehead furrowed as he considered the problem. "Well, a good example was the pole 16W11. There was a loose connection into that pole. A split bolt, if improperly made, will have a higher impedance and that will force more current into the earth, which is a cause of the current flowing over the Allen farm."

"Have you seen several examples of that occurring at that pole in the logger data Mr. Allen has given you?"

"Yes, sir. That data was taken on August 4, 2002."

Unbuttoning his coat, Scott left the jury area and went over near the witness stand. "You weren't at that farm on that particular date. You were on other occasions that summer, however. Correct?"

"Correct."

After some objections from Whaley about hearsay and a bench conference with Judge McKay, Scott said, "Mr. Zipse, assuming the ditch wire was not connected when that chart was made, what does the data tell you about the status of connections in the vicinity of that pole?"

"Well, starting on August 2 and up until sometime between 12 a.m. and 12 p.m. on August 3, current was extremely low on the pole 2221-16W11. However, after that period of time it rises for approximately twelve hours to, we'll say, .6 amperes."

"Mr. Zipse," Scott said to him, "are you on the NEC committee that writes Section 96 of the National Electrical Safety Code?"

"I am."

"And portions of that document are adopted in Wisconsin law?"

"Portions are," Don said.

"Mr. Zipse, with respect to Section 97 E-2 as adopted in Wisconsin, does that have to do with interconnection of neutrals between utility and customer on both grounded systems?"

It did.

"We now have on the screen Section PSC 114097 D-2 of the code. Does the section describe exceptions to the requirement of neutrals to be interconnected?"

Don looked at the exhibit, squinting a bit. "It does," he said. "It describes two types of devices. One is a spark gap, two points close together that an arc can jump over. It's normally referred to as a 'lightning arrestor type.' Another type is a solid state which has a variable switch that closes depending on what the parameters are—voltage, et cetera."

"In general," Scott said, "not necessarily in relation to farms, what would be the purpose of separating the neutrals with these devices?"

"Well, since the primary and secondary neutral is connected, that means that primary neutral current can return to its parallel path. It can return through the farm, the home, et cetera, and this device—this allowance that's here—allows them to put a spark gap in there to prevent that continuous flow of current into the house.

"Now, the reason why the spark gap is there is because originally they connected the lightning arrestors to the incoming neutral, and from the original concept over 70 years ago, it's progressed to this point where the primary neutral and secondary neutral are connected together, allowing lightning current to flow in. So if lightning current starts to flow in with this device, it closes and grounds it."

Scott nodded once or twice and then paused. "Do any of your suggested solutions to the stray current problem on the Allen farm that we wrote down involve the use of a delta system?"

"No," Don replied. "However, they can be confusing because the four-wire system has only three wires going out. If somebody was out on the line, away from the substation, and just looking at the wires and seeing only three wires, and everything connected phase to phase either from A to B or B to C or C to A, they may misconstrue that as a delta system because back at the substation, it is a Y—that is, it is a Y with a neutral and a neutral is ungrounded."

"Why," Scott wanted to know, "do you call it a four-wire system if only three go up?"

Don leaned closer to the microphone and adjusted it. "Because the fourth wire is what we would call in low voltage the 'equipment-grounding conductor'—that green wire, that wire that allows a safety path of low impedance back to the substation. Should a phase fall to the ground, the current would flow through the ground back to the substation, allowing sufficient current to flow to trip the circuit breaker. You have to have a low impedance path so a lot of current will flow, and that will cause a circuit breaker to trip."

"A grounded Y," Scott said. "Can you either carry that equipment grounding wire out in the system or just leave the grounding wire at the substation and not carry it out?"

"You can," Don explained. "But then you're relying on the earth, which is a very poor conductor and not recommended."

"With respect to ground rings and current going down the grounding wires, if the current stops going down the grounding wires on the WPS poles in the vicinity of the Allen farm, what happens to the current on the rings?"

"The current on the ring would cease also."

"Has that been shown in the power outages?"

"Yes, it has."

Scott looked over at the jurors and then slowly crossed the courtroom toward them. "Mr. Whaley asked you about your ideas on grounding being outside of the mainstream. In the city, in industrial applications, are your ideas out of the mainstream?"

"No, not at all," Don told him, and I was glad of the question. This was something I felt the jury definitely needed to hear. "They've been used since 1950. That is the five-wire system."

"Mr. Zipse, given the electrical configuration of the systems at the Allen farm as of April 1997, have you seen any evidence indicating that the Allen farm wiring system was creating stray currents?"

"None at all," Don said to him.

"Mr. Zipse," Scott said, saving what I thought was the best question for last, "who are the members of the grounding committee subsection of NESC who rejected your proposals? What industries do they come from?"

"The majority comes from the utility industry, sir."

And it was in that majority's own best financial interests not to change the system, not rock the boat, to maintain the status quo, to reject new, progressive ideas, no matter the harm the old system was doing to farmers. I was glad Scott had brought this out.

"And has that been true generally during your 50 or so years on that committee?"

"That is correct," Don said.

"That's all I have," Scott said, letting Don's last answer reverberate through the courtroom.

As I expected, Kevin Whaley stood up. He was not about to let Scott and Don have the last word, especially so strong a last word.

But his few questions were duds; there was nothing to them.

"It's your testimony that a messenger wire is part of the secondary neutral as we've used that term in this lawsuit?" Whaley said.

"Correct."

"And if there is a messenger wire on the Allen farm, that belongs to Mr. Allen, correct?"

"Yes," Don said. "It belonged to Mr. Allen, but it was prior to the rewiring."

"And there hasn't been a messenger wire on the Allen farm since 1997, correct?"

"That's correct," Don told him.

Kevin Whaley had ended with a whimper instead of a bang. "Nothing further," he muttered, then walked slowly away from the witness box and over to the defense table where he slouched down in a chair next to a nervous-looking Trevor Will.

Chapter Twelve
THE BIG GUNS FIRE: FOUR

I figured that Foley and Lardner would do their damnedest to make a whipping boy out of Larry Neubauer. Larry was a crucial witness for me, had spent more time on the farm helping me than anyone. Trevor Will indicated in his opening statement that Larry and I had conspired together to hatch this court case to get a big financial windfall out of WPS, so it was pretty clear that they intended to go after him hard.

Certainly, I'd never tried to hide the fact that I wanted a settlement to pay for all the damage they'd inflicted; that would be justice. But I also hoped the jury would decide that not only was WPS liable for monetary damages, they must be forced to correct the deadly electrical menace so I could continue to run the family dairy business with a prospect of success. More than anything, I wanted the business to thrive.

This fine morning, Larry, a friend as well as an excellent electrician and idea man, was approaching the witness stand for his direct testimony. Looking like the proverbial Packer lineman—big and burly—Larry seemed ill at ease in his too-small blue sports jacket and crooked tie. And who could blame him? I felt exactly the same way about going up there, sitting in that fishbowl and testifying. If Larry felt nervous now, I could only imagine how he'd feel after Will or Meckstroth or Whaley started kicking him around. Hell, I could imagine how I was going to feel. I just hoped I'd be able to keep my cool.

Sensing his discomfort, both Scott Lawrence and Judge McKay attempted to put him more at ease. After the court clerk had sworn him in, McKay said gently to Larry, "Mr. Neubauer, if you'll just come over here and have a seat, please. Make yourself as comfortable as you can. Adjust that microphone. There, that should be good."

Scott began by asking him about his occupation and background. An electrical contractor as well as a consultant, Larry held a master's license in Wisconsin as well as North and South Dakota, and a journeyman's license in Idaho. He'd started in the business in 1982, served a four-year apprenticeship through the International Brotherhood of Electrical Workers, and ran mostly industrial jobs in paper mill technology, installation of drives and process control. In 1993, Larry created his company, Concept Electric, Incorporated. The company had grown and become successful, and Larry was currently doing consulting work on a nationwide basis. He went on to say that problems at dairy farms with stray current had begun to take up more and more of his time until it became the business' focal point. To date, Larry had performed stray electricity identifications on over 600 farms in 25 states.

Scott asked him how he'd gotten involved with me, and Larry explained that Mike Hoerth, my equipment dealer, knowing that he specialized in identifying electrical phenomena affecting dairy cows, had put us in touch. I spoke with Larry in 1997 and hired him to come out to the farm and conduct some tests.

"I explained to Russ," Larry was saying, "that I needed to monitor his electrical system, power steel voltages, currents, access and the grounding network.

"I had Russ explain how his operation worked. I looked at a layout of the farm and commenced a design to come up with a testing procedure." Larry unbuttoned the neck button of his blue dress shirt and loosened the collar.

Scott asked what Larry was looking to find.

"Attempting to get an overview of electrical phenomena that would travel not on wire but through dirt or the parlor feeder, or through the parlor floor. If a connection issue is present, it's bad because then the electricity would have to take the earth. A connection issue would be if the secondary neutral of the dairy, feeding the power to the farm, is broken. You could shock cows or electrocute them under the right conditions. I found no connection issues on that dairy."

"Did you," Scott asked, "test various motors and electrical equipment in the free-stall barn?"

"Yeah," Larry said. "I set my machine on low parameters, but the filled memory meant something was happening electrically. Whether on farm or off, at that point I didn't know. Then I started checking equipment and panel wiring.

"I found large current-flow numbers present on the meter display screen on the grounding conductors. And it raised my alarm that my first premonition was right—that we had primary neutral, secondary neutral entries throughout the dairy due to the service laterals. Question was, was something on farm wrong that was causing the current I was measuring on the grounded conductors, or was it just the loop from the primary neutral?

"I went through every circuit in the back and front equipment rooms one by one to identify if there was on-farm leakage."

"Did you find any?" Scott asked.

Larry shook his head. "No on-farm leakage, no defective equipment."

"How long did it take to do all that?" Scott asked, inching up a little closer to Larry.

"Standard test procedure in '97 was about a day. This dairy took two."

"Why?"

"Because I was confused by what was going on with the large amounts of current on the grounding conductors traveling through the dairy. And Mr. Allen was explaining his concerns about the adverse effects on his cows—a lot of them were in very poor shape."

This was a big understatement. Thinking about those emaciated, crippled animals even now breaks my heart.

"After checking out electrical equipment," Scott said, "what did you do?"

"Well," Larry said, turning his head side to side as if the shirt collar bothered him, "I used a four-and-a-half-inch-diameter clamp that I use for checking currents on parlor steel, where the cows would be. You use a scope meter with that. You set that on the one-volt amp range so you can get into milliamps; 1,000 milliamps equals an amp. Clamping the parlor steel, I found 800 milliamps. So I shut the incoming electricity off, and it remained. So it alarmed me."

Scott nodded. "Is a finding like that unusual in investigating farming?"

"Yes," Larry quickly answered. "In this dairy, it was highly unusual from what I'd seen in the past."

"What conclusions did you draw?"

"What I found," Larry told him, "was that current wasn't taking the neutral wire back to the substation. That current was taking the direct connection into the dairy's grounding field and traveling through the dairy into the earth."

"And who owns the primary system?" Scott asked.

"Wisconsin Public Service."

Scott paused for several long seconds. "What did you do next?"

"Well, I told Mr. Allen that I wanted to focus on removing the two services from this dairy and getting it down to one. Because, if we don't, this dairy isn't going to be around much longer."

"Did you tell him that would solve all his problems?"

Larry smiled. "Mr. Allen didn't believe me at first. He had a lot of good questions, and I told him to go get a second opinion. I told him to call WPS and have them test the dairy. So he did."

"What did you do after that?"

"Talked to Mr. Vern Peterson from Wisconsin Public Service, and asked him about bringing this dairy down to one service so that the primary neutral system wouldn't have adverse effects upon Mr. Allen's cows."

"What did he tell you?"

Again, Larry smiled. "That Mr. Allen didn't have a problem. He was below the level of concern set by the Public Service Commission of Wisconsin."

"Where did you leave it with Mr. Peterson after this conversation?" Scott asked.

"I had a meeting with him at the dairy. Basically, Vern Peterson does not believe in anything people like me do," Larry told him. "Because it's contrary to his protocols. I explained to him that my concern on this dairy was high current levels accessing the grounding conductors.

"He kept telling me that it takes two milliamps or so to affect a cow. I said we needed to get this dairy to one service—it had to be done. He kept reiterating there wasn't a problem. So, after a lengthy discussion, he offered Mr. Allen an isolator. My question to him was, where would you install the isolator? Because without relocation of the three-phase service out in the yard, an isolator would most likely fail."

Scott put both hands in his pockets and slowly paced back and forth in front of the judge's bench. "Why?" he asked.

"The proximity of the primary neutral down ground was less than eight feet from the side of the dairy wall which has an 8-foot-deep pit."

"Manure pit?"

"Manure pit. That's important because manure pits, beneath slatted floors, are highly conductive."

Both Scott and Barry felt that WPS would try to blame my equipment for leaking current, so were interested in having as many witnesses as possible refute this point ahead of time.

"If you were to install an isolator on a farm with two services," Scott said, "would you need one isolator or more?"

"We put two transformers into the dairy," Larry said. "We put all those loads in separate panels, fed them with the transformer and checked them: No on-farm electrical issues."

"Was it ever shown to you by anybody at WPS that there was any sort of on-farm electrical issue that you determined was real?"

"No," Larry said flatly.

That took care of that. Or so I hoped.

"Did Mr. Peterson agree to cooperate in getting the Allen farm down to one service?"

"Yes," Larry said. "He did."

I couldn't wait to hear Peterson's version of these conversations when he took the stand, which would be very soon.

During one conversation with Vern Peterson out at the farm, I said to him, "Why don't you go in there and see if you can milk those cows?" Larry and I were discussing with Vern the moving of the pole and other of Larry's findings.

"Did you," Scott inquired, "have the opportunity in 1997 to test whether the isolator was effective?"

"We did: The isolator was bypassing. An isolator is supposed to block voltage and current flow. Putting a current clamp on the neutral conductor of the secondary, basically when Mr. Allen didn't run electrical loads, it went near zero. But when loads were run, it increased dramatically. We measured over 60 milliamps coming in with an isolator installed. It meant the isolator couldn't handle the load or block aspects of the primary system."

"Is it possible," Scott asked, pounding away at what was sure to be a WPS theme, "that the 60 milliamps was from anything on the farm's wiring system?"

"No," Larry shook his head. "Because the load box is hooked directly to the transformer."

"Did you share your findings with Mr. Peterson or anybody from WPS?"

"Actually," Larry replied, "Mr. Peterson shared his opinion with me."

"Okay," Scott said. "Tell us what he told you."

Oh, boy, I thought. Scott's cross-examination of Vern Peterson promised major explosions.

"I received a phone call from Mr. Allen after they had tested the isolator, and Mr. Allen was upset. He said we had miswired the dairy and that WPS wanted to check out his house, because they felt it was an open neutral in the house that was causing the voltages they were measuring when they tested their isolator."

Scott stopped pacing, went over to the jury and stood near the first juror, the foreman. "What's an open neutral?" he asked.

Looking more at ease than he had earlier, maybe because of Scott's relaxed style of questioning, Larry's knowledge and expertise, I thought, came through very well. "An open neutral is a situation where all the current has to return through the earth because it cannot take the insulated conductor back to its source."

"Essentially, a broken wire or a missing connection?"

"Correct. At that point, since we had Vern Peterson there we asked how come he wouldn't relocate the three-phase service. I had done a graph with a flute meter. And I actually measured not 60 milliamps, but well over 400.

"And I showed him the graph of the isolator bypassing. So we pressured him if we could get this down to purely one service and relocate the three-phase. Because otherwise Mr. Allen's dairy, in my opinion, was going to fail."

Standing beside the foreman with one hand on the rail of the jury box, Scott asked, "Did he agree to do anything to help Mr. Allen at that point?"

Larry glanced toward me and we made eye contact for a moment. Then he looked over at Scott. "After repeated discussions, he said, 'I'll agree to relocate but only where I want to put it.'"

"Did he tell you where?"

"Well, I requested that he relocate it across the street. He refused. He located it near the well casing on the farm."

"What happened next with respect to resolving the electrical concerns at the Allen farm?"

Larry sighed. "Well, at that point I told Russ to take it. I said, 'We've got to get this away from the parlor because we had an isolator on and it was being bypassed.'

"Mr. Allen was complaining things were worse for the cows. They were bad enough when I walked into the dairy. I knew he was in deep financial trouble."

The understatement of the year. I was going down for the third time, and figured there wasn't going to be a fourth.

"We knew," Larry continued, "the isolator wasn't holding because of conditions on the WPS line. The fact was, even if it did hold at some point, it would close and open depending on load orientations. What we needed was to get the three-phase transformer as far away from the dairy as we could. From that point, we installed the isolation transformer."

And production shot way up.

"Have you personally installed such systems on other WPS farms since then?" Scott asked.

"Within a year after that time," Larry told him, "WPS granted me two services of that nature on two dairies. They were also large dairies."

Scott came away from the jury and moved toward the witness stand as Larry turned and quickly sipped some water from his glass before answering the next question. "About when did you actually put a three-phase isolation transformer in the center of the Allen yard?"

"I believe November of 1998." Larry looked at the glass and then set it down.

"Had the three-wire ungrounded feed been granted by WPS, would that have saved Mr. Allen money?"

"Yes. Isolation transformers run from $1,200 to $2,700 a unit for a single phase. On the Allen dairy we needed a 480-volt system primary. We wanted to create a 480-volt isolation system. So we needed a transformer we would call a '1 to 1.' Those are special ordered, cost about $6,000. That's what we ended up having to purchase. Plus, you've got to fuse the primary of that transformer. You've got to buy disconnects, buy rain-tight equipment that sits outside in case the transformer gets destroyed by lightning or melts or something happens to it, that you can run the farm on a generator. The cost just continued to escalate."

Along with my sleepless nights.

"We did end up," Larry said, "installing two transformers, one for the house and one then for barn one and barn two, and then a separate transformer for barn three so that none of the buildings would interact with any other structures."

Scott next asked about some of the photographic exhibits of farm structures and utility wiring. "Mr. Neubauer, tell us if, when you're making stray measurements, ones you consider important, if you double check your measurement with two different types of meters?"

"You should double and triple check."

"Is the farm not isolated in Exhibit 1331 when that measurement was taken?" Scott asked.

Larry leaned forward, squinted at the photo, and then took his glasses from his coat pocket and put them on. He shook his shaggy head. "It is not isolated. This is actual current coming into the dairy."

"What's Exhibit 1331?"

"A picture of the same meter."

"What's the condition being photographed?"

"It says power off."

"And what's being read?"

"183.8 milliamps," Larry told him.

"Do you recall what time of day this was?"

"No."

"Will such measurements vary with time of day?"

"Yes," Larry said. "The utility system is dependent upon customers using electricity, when, where and how. So, if down the street, customers all turn their lighting on, current on the Allen dairy would increase. If they all turned down their power usage, then the primary neutral would have less returning to the substation. It's dynamics."

"Exhibit 1332," Scott said, "another photograph of a meter. What are the conditions under which measuring is being done?"

Larry craned his neck forward. "The utility isolator is now installed in the three-phase service pole which is the only service feeding the dairy. They've turned the load box on creating an artificial secondary power source, create return current going on the neutral."

"And what are you measuring?"

"We're reading 60.4 milliamps showing isolator bypassing installation."

"How is it," Scott asked, "that you have current coming in with the load box on and the isolator in this place?"

"That was a tough one," Larry said. "But the down ground on the three-phase service is located less than eight feet from the dairy barn. There was over 180 milliamps accessing the dirt right there. The isolator is not blocking everything. It's allowing this amount to get through."

"Explain what's failing and why, please?"

Larry cleared his throat, took a drink and held on to the water glass with his right hand. "Even though we spent $57,000 plus on the isolator, when this piece of equipment hooked up and the transient at this level showed up on a grounding conductor, not even electrically connected to the dairy, yet I was able to measure it in the dairy, that told me it was traveling through the earth into the dairy.

"I mean, it's precise. Looking at the well casing next to the utility transformer, at that point I went out and cut the utility down ground and relocated it to the road with the other service equipment to get them away. So, we couldn't be affected anymore."

"Did you do anymore testing after that?" Scott asked him.

Larry took another gulp of water and set the glass down. "I hooked up my dranetz machine one more time, didn't find the same aspects on to the barn neutral. Then I went through the monitoring system and double checked it for current flow."

Scott questioned him about the monitoring system, and Larry told him he was more concerned with it than down grounds. "The monitoring system," he said, "was supervising the electrical system and it was ground fault protection. I wanted to make sure that nothing on farm was causing these, which I ruled out. Again."

Larry Neubauer was nothing if not cautious, careful and thorough; it was why I trusted him.

"Mr. Neubauer," Scott said. "I'm going to shift gears a little. Did there come a time when you had discussions with Mr. Allen about installation of what's called a ground ring?"

"Yes. Primarily in '97, '98. We did a lot of dairy farms that had electrical problems. And Wisconsin Electric used to, when they had an electrical issue on a dairy, dig around the dairy and drive a bunch of ground rods and connect a ring of wire. But, then they would also bond it to the primary neutral."

"On how many farms have you installed such a device, roughly?" Scott asked.

"Well over 30 or more; 50."

"Do they fully resolve issues of current in the ground?"

"No, no, they do not fully resolve them."

Scott walked over toward Larry and stood halfway between the witness and the jury boxes. "How was it that one became installed on Mr. Allen's farm?"

"After we'd put in several," Larry said, "a lot of customers were happy. Mr. Allen knew this. He called and said he wanted one. He was complaining that spring had come, his cows were re-kicking in the parlor after all the work that had been done. So we put the ground ring in."

And after all the expense, and the grief. Ever since the spring thaw, the soft, damp earth became so much more conductive of the stray ground currents.

"Do you have information as to whether Mr. Allen cut ground wires on the WPS system in the vicinity of his barn at around that time?"

Larry looked over at me. He hesitated a second before saying, "When Mr. Allen saw the amounts of current on his ground ring, after that he removed WPS' down grounds."

"Does cutting ground wires on the WPS system in the immediate vicinity of the Allen farm affect current on the ground ring?" Scott asked.

"Yes."

"How do you know?"

"It's not only the current, it's the voltage, too. But the current is reduced because the down grounds are no longer able to inject current near the farm. It makes the neutral do more of its job."

Scott meandered up near Judge McKay's bench. "At some point a ditch wire was installed in the vicinity of Mr. Allen's farm. Did your company do some of that work? Tell us what the ditch wire was all about."

"It was actually a bypass wire," Larry told him. "I helped drive the ground rods. To reduce earth current, Mr. Allen ran a ditch wire all the way down to the corner about a mile south of his dairy to build a neutral. It was an experiment. It was connected to three or four ground rods on either end. And it was an isolated number two copper conductor."

"And is this providing a parallel path for WPS return current on its neutral?"

"Correct."

"Was this your suggestion?"

"To a degree."

"Is the installation of that wire causing any damage to WPS' system?"

"No."

I assumed that Trevor Will or his underlings would say that it did cause damage.

Scott nodded and walked over near his counsel table. "Eventually, was the three-phase transformer moved to the north side of Little Apple Road? Was that the location you had originally requested?"

"Correct."

"How did it come to be that that finally occurred?"

"Russ was highly concerned that issues were getting worse again. I knew that loads were increasing on the system and that the system wasn't maintained. I told Russ we should get that transformer as far away as we can. If we can get it across the road, we'll gain another 25 feet on the grounding conductors.

"Mr. Allen agreed," Larry said, glancing at me out of the corner of his eye. "And we re-contacted Wisconsin Public Service who said, 'If you want it there, we'll move it there.' Whereas the first time they refused."

"Was Mr. Allen charged for the move across the road?"

Charged? 'Raped' would be a better word.

"Yes," Larry said. Then added, "He was upset with that."

Enraged and outraged was more like it.

"But he paid both times, correct?" Scott asked.

"He paid both times."

"In retrospect," Scott said, "do you think moving the transformers helped?"

Larry smiled. "I wasn't prepared for what they did. And, no. It didn't."

"What happened?"

Shaking his head but smiling, Larry said, "They put in two new poles directly to the west to raise the overhead, and then they added down grounds. So now they had to fight the grounding near the road in front of the dairy."

"What would be the effect of adding down grounds?"

"The effect is: You already have a poorly maintained neutral. Now you have it grounded even better, in front of the dairy, injecting more current and affecting the dairy even worse."

Scott paused. "Did Mr. Allen put yet another ring of copper around his farmstead?"

"Yes."

I felt like guys must've felt in the Middle Ages, digging moats and filling them up around their castles to keep out marauders. But it's harder to keep out electricity that you can't see, and that flows 24/7.

"Did you have involvement in doing that?" Scott asked Larry.

"To a degree. Mr. Allen knew that when the surface conductor—surface condition of the soil was wet—that he could see adverse effects to his cows because electricity travels better on moist soil than dry.

"He attempted to borrow my trencher. He was going to cut a trench around the dairy, fill it with rocks and put in a pump to keep the water off his property; he was that concerned.

"After he went about 30 feet, he realized it was a very tough task. I told him, 'Why don't you put a surface ring on and see if it helps?' So he did."

"Will a surface ring result in reduced current flowing through?"

"Definitely."

"Why is that?" Scott wanted to know.

"Kurkoff's Law. It's basically a continuous loop similar to a rubber band. Break it, it's gone. Induction is purely magnetic fields, current flowing through or across the conductor creating it."

"If all the WPS return current stayed on the neutral conductor, what would happen to the current buried in the surface rings of the Allen farm?"

"It would go to zero!" Larry replied.

"Do you have any data," Scott asked, again hitting the idea that nothing on farm was responsible for the ground currents, "to establish whether Mr. Allen's on-farm system is contributing to this current?"

Larry nodded. "Two days after Mr. Allen put a monitoring device on this ring, the power went out in the area. Mr. Allen then ran the entire dairy on a generator, with the same electrical loads of milking equipment, the feeding operation, everything that was running before was running again, but from the generator. *And the ring measurement remained at zero with the power company's line down.*"

"You've got the dairy running on the generator?" Scott asked.

"You bet," Larry told him.

This, to me, was ultimate proof of how the WPS distribution system had wrecked me. I didn't know how the utility's hired guns would refute this evidence, only that they would try their hardest.

"Do you," Scott now asked, "incorporate cow contact voltage measurements, according to the guidelines of the Public Service Commission of Wisconsin, in your typical test routine on dairy farms?"

"Sometimes. I generally try to measure currents more than resistances."

"Why?"

"Because there's so many electrical aspects. Radio frequencies, DC components, distribution lines. The electrical industry is so complex you've got to be willing to use what works, at times go non-conventional to determine where problems exist and where the data comes in."

Larry was another person who, like Don Zipse and Mike Behr, could think "outside the box." He was conscientious as well as creative. I was proud of these men and their work on my behalf. I found it impossible to imagine that the jury would reject their opinions. Maybe that was wishful thinking on my part, but I sure felt that way.

"In your experience at over 600 dairies," Scott said, "do cow contact voltages measure according to the Public Service Commission of Wisconsin protocol?"

"The answer to that," Larry told him, "is that it doesn't meet those guidelines. All farms are unique, different. They all vary."

"Do farms with equipotential planes and parlors never have stray electricity problems?" Scott asked.

Larry pursed his lips. "Some of the worst," he said.

"By the way," Scott said suddenly. "Back in July '97 when you were at the farm with Mr. Held of WPS, Mr. Peterson and Mr. Allen, did you or Mr. Allen ever deny those folks the opportunity to make measurements of the barn?"

That was what WPS claimed, and would claim again when its boys on the payroll got up there to start earning those huge fees. But the idea was ridiculous, as Larry quickly pointed out. "We couldn't," he said. "With an isolator installed, the power company has explicit rights to monitor the grounding system of the dairy. It's one of the protocols set up in their agreement."

"Did you ever, for any reason, tell them they couldn't go in there and measure?" Scott asked.

Larry made a face. "I have never told a power company representative that he was unwelcome on anybody's dairy. That's the last thing I would do."

Next, Scott guided Larry through a number of photographic exhibits depicting Larry taking measurements and collecting data all throughout the farm at many different points in time. Again and again, the tests proved there was no on-farm source of electrical pollution, only that which came from the WPS system.

Then Scott asked, "Were you present when the WPS people, including both their electrical and farm expert witnesses in this case, came out to the farm?"

Larry was there all right. He was with me.

"Were attempts made by WPS personnel to recreate the situation with regard to connection of the primary neutral, the farm grounding system, before 1997?"

"Yeah, they recreated it," Larry said. "I explained to Mr. Schott, we'll go to the back feed room where the existing service was, because you'll want to tie in there. Fifteen steps up the driveway, he asked me if a panel existed in the front equipment room where the new service was, and I said yes.

"He stated that the potentials were the same and that he was going to hook up there. I requested that the original situation was located in the back feed room, not in the front, northern part of the parlor. They still hooked it up in the equipment room." Larry shook his head.

"Why would that make a difference?" Scott asked.

"He basically took all the return current that would have accessed the dairy and put it for the north end instead of allowing it to travel through the dairy to recreate the original situation."

"Did you," Scott asked, "advise Mr. Schott that it would be more appropriate to make the connection in the feed room where the meter had been back in 1997?"

Shaking his head and shrugging, he said, "That's exactly where I first took him. Until he decided otherwise."

"After WPS had hooked up the connection, did you make any measurements on the wire they used?"

Larry raised his eyebrows. "I was a little concerned right off the bat that they didn't hook up where it should've been hooked up if they wanted to recreate the system that existed.

"They broke for lunch, we had a sandwich or two, and curiosity got the best of me, so I clamped back on the bonding jumper. I'm not a good math guy, so I like to use meters that measure voltages and currents. I put my clamp on and basically found no current."

Scott seemed surprised. "Did you find any current at all?"

"Less than seven milliamps."

"Did that strike you as unusual?"

"Well, I alerted Mr. Allen's counsel—you—and told you gentlemen that I felt the connection was improperly made. Or intentionally not made. I wasn't sure which."

But I was.

"What happened then?" Scott asked.

"I got another two meters out to reconfirm that reading. They locked up the same reading."

"And did you have photographs taken of those meters?"

"I did."

"Are two of those photographs depicted in Exhibit 1335?"

"Yes," Larry said.

"What happened after these measurements?"

"I talked to counsel for Mr. Allen and explained that in no way could this be right. Nothing I've ever seen on any testing procedure—this could not be correct. I told counsel we should notify WPS immediately, which counsel said to do."

"What happened after that?"

"I contacted Steve Schott. Him and Mr. Lavern Stetson, who is an agricultural engineer, then came into the dairy. I showed them my readings. They went and got their own meter. They came up with the same reading. Then they started looking at their connection to the panel, most likely to see if I had tampered with it. They replaced the wire, rolled it up and took a new spool out. And I reclamped it."

Scott turned and once again walked slowly over toward the jury box. "Had that initial test lead stayed in place," he asked, "would valid test results have been obtained?"

"No," Larry replied.

Scott then asked Larry about the amount of work he'd done for me, which was quite a lot, and which was shown in spreadsheets marked as exhibits. "In accordance of this lawsuit, have you or your office staff dug out all your telephone records and logs related to that business? And have those all been provided to WPS?"

"Correct," Larry told him.

"Have you also estimated in the spreadsheet the amount of money that could have been saved in this work had you received more cooperation from WPS?"

"Correct."

Scott went through the invoice amounts with Larry for his work. "Cooperation" and "WPS" were mutually exclusive words. They did nothing, ever, to help solve my problems. But then, they were the ones who'd created them in the first place.

To provide the jury with a good example, Scott asked, "What does WPS' cooperation or non-cooperation have to do with a farm generator?"

"You can't put an isolation transformer inside a building. That means all the generator and transferring equipment, should a power outage occur, has to be placed outside. That means it's got to be made raintight."

"Was such equipment in the feed room before?"

"Correct. We located the generator outside but then had to buy all raintight equipment. And spent $11,000 to relocate the service, the second time, across the road."

There were a dozen or more other things, major expenses, no thanks to the utility.

"Let me ask you," Scott said, "what are the total charges your company has made to the Allen farm between April '97 and today?"

"$220,461."

"A lot of money, isn't it?" Scott said.

"A lot," Larry agreed.

"Does this include materials?"

Larry nodded twice. "Materials, parts, labor and all the maintenance."

Scott ended his direct questioning of Larry by identifying the problem that was nearest and dearest to my heart at this moment.

"Mr. Neubauer, is WPS current still reaching areas where Mr. Allen's cows are housed?"

"Yes," Larry said emphatically.

Scott paused so the jury could think about this.

"When you installed your first isolation system on this farm in late 1997," he said, "did that prevent some of the WPS current from getting to Mr. Allen's cows?"

"Yes," Larry told him. "It prevented some of it."

"Thank you, Mr. Neubauer. That's all I have," Scott said, and went and sat at his counsel table next to Barry.

It turned out to be Kevin Whaley, who was tabbed to do the deed with Larry. This guaranteed a ton of sarcasm and flare-ups, although Larry was a calm and nonconfrontational individual. Pairing these two could mean an irresistible force (Whaley) would collide with an immovable object (Larry), the results of which would be highly interesting.

Predictably, Whaley began by trying to make Larry look bad. "You've been a licensed master electrician since approximately 1995," Kevin Whaley said. "And you took the test four times and flunked it before you became a master electrician in Wisconsin?"

The faintest trace of a smile appeared on Larry's lips. He could see right off the bat how this was going to unfold. "Actually, five times, and passed it." Larry was one of those people who did not test well. But he knew his business.

"Right." Kevin Whaley nodded, happily. "You flunked the first four and passed on the fifth time, correct?"

Larry reaffirmed this.

"And after that, you took the same test for a master's electrician twice in the state of Minnesota, correct?"

"Not the same, a different test, correct."

"And you flunked both times?"

"Correct," Larry said.

"And then you went to the state of Michigan and took a test there to be a licensed master electrician. And you flunked that one?"

"Correct."

"So," Whaley said, clearly pleased with himself, "you've taken the licensing test to be a master electrician and you've flunked it seven out of eight times. Correct?"

"Correct," Larry told him.

Larry was blind in one eye—that was why he flunked timed tests. Just try it for yourself. He really knew his stuff.

Whaley nodded. Then he said, "You know that the unit of measurement for current is amperes, correct?"

"Correct."

"But when I asked you at your deposition, you told me it was a jewel. Correct?"

"Well, correct."

"And you know that the unit of measurement for resistance is ohms, correct?"

Larry said he knew.

"And when I asked you at your deposition, you said 'R', as in the capital letter R. Correct?"

Again Larry agreed.

"And you know," Kevin Whaley said, "that the unit of measurement for voltage is volts, correct?"

Larry said he knew that, too.

Kevin Whaley grinned. "When I asked you at your deposition, you said that the unit of measure for voltage could be watts, correct?"

"Could be," Larry answered.

"And," Kevin Whaley, apparently believing himself to be on a roll, said, "when I asked you where in the ground earth currents flow, your answer was, 'Truthfully or scientifically.' Correct?"

Larry affirmed this.

"You distinguish between scientifically and truthfully, don't you?"

"I don't understand what you mean by 'scientifically,'" Larry said.

On his way over to the jury, Whaley stopped dead in his tracks, turned and went back over toward Larry. "When I asked if you distinguish between scientifically and truthfully, you said you definitely do, correct?"

"Well, I would agree."

"And you believe in Ohm's Law?"

"Correct."

"But," Whaley said, "you do not believe Ohm's Law is always applicable in electrical situations?"

"That can be the issue," Larry told him.

"You think there is a real world versus a theory world?"

"Correct."

"You are not a licensed professional engineer and you are not an electrical engineer, correct?"

This had nothing to do with anything. It seemed to me that Whaley's strategy was not so much to refute Larry's findings as it was to try to make a monkey out of him. I hoped the tactic might backfire by gaining sympathy for Larry from the jurors.

"Correct," Larry said.

Now Whaley made his way over to the jury box. "From April of 1997 through February of 2002, you charged Mr. Allen some $230,000 for work on his farm. Correct?"

This was true.

"And that was certainly the most you ever charged any single customer during that time frame, correct?"

Larry again agreed with the statement.

Kevin Whaley looked at the jurors, smiled at them, and then looked across the courtroom at Larry. "Would you agree," he said, "that your work on Mr. Allen's farm, given all the time and money and effort you've put into it, represents your best work?"

"That's correct," Larry replied.

"And it's your best idea on how dairy farms should be wired in terms of safety?"

"No," Larry told him. "I would not like to have the isolation system. I'd like it utility supplied."

"But for all the work you did, some $230,000, you tried to make that farm as safe as you possibly could?"

"That's correct," Larry said. His forehead wrinkled, as if he were trying to figure out how Kevin Whaley intended to mousetrap him.

He would find out in a hurry.

"Mr. Neubauer, I'm going to show you a document we've marked as Exhibit 1606. Exhibit 1606 is an order from the Wisconsin Department of Commerce, entitled, 'Notice of Violations and Orders.' And you know who Tom Garvey is, a state electrical engineer?"

"Correct," Larry said.

"As an electrician licensed by the Department of Commerce in Wisconsin, you have to follow the rules and orders of the department?"

Larry said that he did.

"And you know that last month Mr. Garvey went on Mr. Allen's farm and conducted an investigation of the electrical system you are responsible for?"

"Correct."

"During and after that investigation, Mr. Garvey issued this report we've marked as Exhibit 1606. Mr. Garvey, on behalf of the Wisconsin Department of Commerce, found 22 code violations to the electrical system on the Allen farm that you're responsible for. Correct?"

"Correct," Larry said.

Whaley nodded with satisfaction. "As a master electrician, you are familiar with the fact that a code violation raises a safety issue?"

Larry shrugged. "Could go both ways. It does, generally."

"And so it's fair to say that when Mr. Garvey inspected this system last month, 22 issues regarding safety for the electrical system, that you're responsible for it?"

Once more Larry shrugged. "I'll agree with that."

I shot a look over at the jurors to try to gauge how they were reacting to all this. The lady in the floral print dress was looking at the girl next to her; both appeared to be interested in something going on among the spectators. Two male jurors were glancing about the courtroom and looking intermittently at the ceiling, and one lady was writing on a notepad. Several others were looking at Larry, Kevin Whaley or the bailiff.

"On behalf of the state of Wisconsin," Kevin Whaley said, "Mr. Garvey ordered Mr. Allen to make certain changes to the system to comply with these 22 code violations. Have some of those been done?"

"All," Larry told him.

"Were you involved in that work?"

"No."

Kevin Whaley came back toward the witness stand and stood silently for several moments. Larry took a drink from his water glass.

"Now," the attorney said, "with respect to the buried ground ring and the surface ground ring, you mentioned something called 'Kurkhoff's Law,' is that right?"

"Yeah."

"Your testimony was that Kurkhoff's Law causes induced currents on the surface grounds, correct?"

"It's one of the basic principles of why induction is present, to make current flow across the grounded conductor."

"Would you agree that Kurkhoff's Law applies to the buried ground ring just as it applies to the surface ground ring?"

Larry nodded at the lawyer. "In retrospect, yes."

Whaley mentioned several of the chart-exhibits that Scott Lawrence had discussed earlier. "We saw the logger chart showing current on the ring of life on top, the surface ground ring. And then it disappears when the power goes off. And we saw the same with respect to the buried ring. When the power went off on WPS' distribution system, the current on the buried ring went to zero, correct?"

Larry agreed.

"If the power goes off," Kevin Whaley said, "that means there's no current on those distribution lines. And if there's no current on those lines, there's no magnetic fields associated with current. And if there's no magnetic fields, there would be no induced current from those distribution lines. So the fact that current on the rings went to zero when the power was off is consistent with the idea that current on those ground rings is induced current, correct?"

"No!" Larry said, and Whaley ignored his answer.

"With respect to the costs you went over with Mr. Lawrence—for example, the ground ring—you claim that WPS is attributable for the $8,000 that Mr. Allen paid you for the ground ring?"

"I agree with that," Larry said.

Whaley laughed. "That was your idea," he said. "Correct?"

"That's correct," Larry said.

Apparently trying to portray Larry as a money grubber as well as an incompetent, Kevin Whaley folded his arms across his chest and turned to face the jury. "With respect to the testing equipment you're trying to get WPS to pay for, that was equipment associated with this litigation?"

"Right, but Mr. Allen didn't pay for those."

"Are you," Whaley asked accusingly, "trying to recoup the price of those loggers from Mr. Allen?"

"I charge so much per week for them," Larry replied.

"Those loggers were purchased in connection with this litigation, correct?"

"And for my own testing purposes in the future."

Taking the harsh, accusatory edge off his voice, Kevin Whaley said softly, "Now, you agree that in order for current to flow through a cow there must be a voltage differential between two points of the cow contacting simultaneously?"

Shifting in his seat, Larry nodded. "Correct."

"You agree that for a dairy cow to be harmed by electricity there must be current flowing through the cow? And you agree that electricity will cause a cow's milk production to decrease only if there is current flowing through the cow?"

Again Larry nodded. "Correct."

Still eyeing the jury, Whaley said, "A step potential measurement measures the voltage difference between two places in the earth and cow contact. And a step potential is a measurement of voltage between front and back legs of the cow, correct?"

Larry said it was.

"Now, before we talk about the step potential measurements you took on Mr. Allen's farm, I want to talk about how many times you've been on that farm. You were on the farm in 1997, 1998, 1999, 2000, 2001 and 2002 in connection with Concept Electric?"

Larry stated that he was.

"I'm going to show you now, Mr. Neubauer, Exhibit 667. We've listed year by year the invoices that Concept Electric and then Concept Services sent to Mr. Allen in connection with your work and Concept Services' work on the Allen farm. From the last page, Concept Electric invoiced Mr. Allen

$219,211.41. When you add what Concept Services did, the total came to $229,846.97. Correct?"

"Correct."

"And invoices for your work since March 2002 total approximately $14,000. So that brings the total to just under $245,000, correct?"

"Correct."

"And Mr. Allen has paid all of that to you but several thousand dollars?"

"Correct," Larry told him.

I wondered how Whaley was going to expand on the idea that Trevor Will had first introduced: That Larry and I had cooked up this conspiracy theory about stray current so we could swindle WPS out of millions. It was coming all right; I just didn't know when or in what form.

Whaley went to the jury box and leaned back on the railing. If he'd gotten any closer to the jurors, he'd have been in their laps. "Clearly," he said, "given the amount of money you've invoiced Mr. Allen, would it be fair to say you have worked *very* closely with Mr. Allen over the last six years? And you've been on the farm maybe a hundred times?"

Larry shook his head. "Not correct. That's high."

"Seventy-five times?" Whaley ventured.

"I'd have to say that would be pretty close."

"You can go on Mr. Allen's farm pretty much whenever you want?"

Larry nodded. "I'd agree with that."

"And once you're on the farm, you can pretty much do anything you want?"

Wondering what Whaley was up to, I leaned forward in my seat, elbows resting on my knees and my chin on my fist. Whatever it was wouldn't be to my advantage, that was for sure.

"No," Larry told him.

Whaley smiled faintly. "You can pretty much take whatever electrical measurements you want?"

"Correct," Larry said.

"And for all the time you were on the Allen farm, you took step potential measurements only twice?"

"That I can recall."

"And you used a 470 ohm resistor for those measurements. But with respect to step potential measurement, you have to use a resistor for step potential measurement if you're going to represent current going to a cow. And the two measurements you took were, in your words, very low. Correct?"

"That's correct," Larry told him.

"So low you didn't bother to write them down because those measurements didn't concern you, correct?"

Larry spread out his hands, palms upward. "They concerned me, but it wasn't part of my investigation to figure out how to correct those issues."

Larry didn't write very much down. But on at least one occasion, he showed Vern Peterson what he was picking up on the scopemeter when WPS' isolator was closing. Certainly, if Larry was preparing for a lawsuit, he would have written more things down. I was with him that day.

Whaley suddenly straightened up and walked briskly across the courtroom toward the witness. "The second time you took measurements was the day before WPS came to Mr. Allen's farm to take measurements in connection with this lawsuit, correct?"

"Yes," Larry said.

"You took four measurements. And for all four, you bypassed the isolation transformers you had installed, correct?"

"I believe I did it both ways."

"For testing purposes, you essentially reconnected the primary and secondary neutral. With respect to all of the photographs Mr. Allen's attorneys will show the jury in this case, none of them will relate to step potential measurements taken by you, correct?"

"Correct."

Whaley nodded and continued on. "Now, you know that the impact of using a smaller resistor is that you get a higher current reading, correct?"

"That is," Larry said.

"Now," Whaley said. "Cow contact measurements. The only time you have taken them on the Allen farm were on April 1997 and June 2002?"

Larry affirmed the dates.

"Do you always take cow contact measurements in the same fashion?"

"Yes."

"Do you take cow contact measurements the same way on farms in Wisconsin, Idaho, South Dakota and California? And you have been on the Vierstra farm with Mr. Allen in Idaho, correct?"

It was true. Larry and I had been there.

"The Vierstras have sued the utility in Idaho. And you were an expert witness in Vierstra's case?"

"Correct," Larry said.

"And you have taken cow contact measurements in the Vierstra case in Idaho?"

Larry said that he had.

"Mr. Neubauer," Kevin Whaley said, "I'm going to show you what we've marked as Exhibit 1719." He went over to it. "This videotape that you had an opportunity to look at last night, of you taking cow contact measurements on the Vierstra farm in Idaho, correct?"

Larry further loosened his already loose shirt collar. "All sorts of measurements. Cow contact also," he said.

Whaley now played a snippet of the tape, which lasted about ten seconds.

The video ran, with a voiceover saying: "Okay. Cow contact test. Generator running. Shunt resistor in series right here. Tied to the parlor steel, tied to the foot."

That was it. Whaley then said, "Mr. Neubauer, the voice we heard was your voice?"

"That is correct."

"You know, don't you, that in order to take a legitimate cow contact measurement, to determine how much current is flowing through a cow, the resistors have to be placed in parallel, not in series?"

Larry agreed that this was so.

Whaley moved a few paces closer to the witness stand. "We've already gone through your step potential measurements on the Allen farm, Mr. Neubauer, so I want to talk about your cow contact measurements before this lawsuit was filed.

"You claim the highest measurement was .914 volts, correct?"

Larry nodded. "Off the bunk feeder."

"And your second highest was .25 volts?"

"Correct, I think."

"Both were taken in the free-stall barn?"

"Correct."

"You didn't take measurements in the parlor because there was an equipotential plane? With an equipotential in the milking parlor, that means there was no voltage differential between the metal and the floor?"

"No," Larry told him. "It just means it's low."

Kevin Whaley looked down at his left hand and adjusted the silver bracelet of his wristwatch. "Is it your testimony," he asked, looking up at Larry, "that the .25 volts was from the feed bunk hood to the floor or from the free-stall divider to the floor?

"The .9 was from the bunker feeder to the floor."

"You claim that this measurement from feed bunk to floor is a cow contact measurement. That's because you believe cows' heads touch the feed bunk hood while they are eating?"

"That is correct."

"Even though the feed bunk is six feet off the ground?"

"Roughly six," Larry said. "Correct."

Whaley stared at him. "You were testing by yourself when you claim the test resulted in .914 volts. With respect to this measurement of .914 volts, you fully loaded the system—turned on everything that could be turned on?"

Larry said that he had.

"When you had standard loads on, your reading was only .25 volts, correct?"

It was.

"With respect to the .914 volts, you couldn't determine if the on-farm system contributed to that?"

I couldn't believe Whaley was still whipping this tired old horse, the "on-farm" system.

"I couldn't find the on-farm system contributing," Larry said.

"Because," Whaley said in a smug, *gotcha* tone of voice, "to make that determination, you would have had to shut off the farm and take that same measurement again. Correct?"

"That is correct," Larry admitted.

"And you didn't do that on April 15, 1997, when you took this measurement of .914 volts?"

"No. I focused on the current flow in the parlor."

Kevin Whaley went over to an exhibit. "267, Mr. Neubauer, is your report of the testing you did on Mr. Allen's farm in April 1997. There is a fax line on the bottom, because you had obtained this form from a friend named Spark?"

"Spark Burmaster, yes."

"And Spark told you to fill out a testing tracking form for data collection?"

"Well," Larry chuckled, "thoroughness is the key to success."

"So you know how important it is to record your data," Kevin Whaley said. "The first five pages of Exhibit 267 are in your handwriting?"

"Yes."

"But nowhere do we see any reference to .914 volts. In fact, .914 volts isn't seen anywhere on Exhibit 267. And nowhere do we see a reference to .25 volts, correct?"

That was correct.

"You didn't write down anywhere a measurement of .914 volts. Because the measurements were so low they didn't concern you. So, you didn't bother to write them down. Correct?"

"Wrong!" Larry said, and his voice rose with anger. "Incorrect!"

"But you have no notes or printouts of that measurement of .914 volts?"

"Correct."

Whaley now scooted back to the jury area and stood with his arms folded, looking over at Larry.

"WPS was out at the Allen farm within 10 days after you took that measurement. And you never said a word about measuring .914 volts in the free-stall barn. Correct?"

"Correct."

"And you have no letter to WPS about the volt measurement? And no notes of conversations?"

Larry said he didn't. Whaley was insinuating that Larry had made up numbers to help me, and make WPS look bad. All of that was true except the part about making up numbers. The numbers were depressingly accurate.

"And you *claim* to have shut off the dairy farm on April 15, 1997, in connection with your testing?"

Larry said that he had indeed shut it down.

"This measurement of .914 volts you *say* you took on April 15, you didn't write that down on April 16, when you were still doing testing on the Allen farm. And you didn't write it down any time in 1997, 1998 or 1999, correct? And you didn't write it down any time from January through April 2000. Correct?"

"I can't recall," Larry told him.

"The first time you ever wrote it down was on May 2, 2000. Correct?"

Larry shrugged. "Could be, yes."

Kevin Whaley now showed Larry Exhibit 266, which included a letter he'd written to my lawyers in which he wrote down his 1997 measurement of .914 volts.

"This is a letter," Whaley said, "to attorneys representing Mr. Allen who were preparing to sue WPS. Correct?"

"Correct."

Whaley paused for effect and returned the exhibit. "Now," he said. "Under your direct testimony, you told us that you measured in April 1997, 800 milliamps on parlor steel. You used a clamp meter. You did not have one test lead hooked to metal and another lead five or six feet away?"

"Not five or six feet away. It was to a remote rod."

"That measurement of 800 milliamps on the parlor steel was not a measurement of current flowing through a cow, correct?"

"No, it was current flowing through a pipe," Larry told him.

Whaley shot back, "You did not write down that measurement, did you, Mr. Neubauer?"

"No."

The attorney nodded. "Now, with respect to some other of your measurements: Measurements of current on WPS grounding wires are not measurements of current flowing through a cow. Measurements on the surface and buried ground rings are not cow contact measurements. They are not step potential measurements. Measurements of current on a primary neutral are not measurements of current flowing through a cow. Measurements of current on WPS' hot wires are not cow contact measurements, not step potential measurements. Correct?"

Larry agreed that all these statements were true.

Whaley nodded, clearly satisfied. "By April 16, 1997, you told Mr. Allen he had a stray voltage problem attributable to WPS. And you told him that had to do with the fact there were two services to the farm—that what needed to be done was eliminate or relocate one service across the street. You were telling Mr. Allen the transformer pole had to go across the street. You never told Mr. Allen the transformer pole should go to the middle of the yard?"

Larry stretched his arms out wide. "That was WPS' decision. Not mine."

"Was the on-farm grounding system bonded to the metal work in the parlor and free-stall barn?"

"Yes."

"You agree with me that Mr. Allen owned the service panels? And that Mr. Allen was responsible for grounding the secondary neutral to earth at each panel? And for bonding the grounding system to the metal work in the buildings?"

Larry agreed.

"You know that every dairyman is responsible for their own wiring from the meter socket on, correct? And if the source of stray voltage were on-farm electrical leakages, it is the farmer's responsibility and not the utility's. Correct?"

"That is correct," Larry said.

If I was forced one more time to hear this idiotic concept of "on-farm leakage," I thought I would scream. There never was any—never, ever, at any time, anywhere on the farm. And though I knew tossing out the idea was just a tactic in WPS' defense, it infuriated me. I could feel the muscles tensing along the back of my neck and my fists start to clench.

Whaley now began to pace. "You told Mr. Allen in 1997 that to get rid of what you called 'loop current,' you needed to eliminate the single-phase service to the parlor and then rewire the farm?"

"Correct."

"You charged Mr. Allen $9,000 for that work. After you eliminated the single-phase service, you told Mr. Allen you had reduced the current flowing through the parlor and free-stall barn. Correct?"

"Definitely," Larry said.

Continuing to pace, Whaley said, "Have you seen any production records that show that milk production actually went down after you did this work in May 1997?"

"No."

The attorney stopped moving and faced Larry. "Assume for me that there are such records. If that were the case, would you attribute the decrease to the work you did?"

Larry frowned. "No!"

Kevin Whaley's lips twisted into a small smile. "You would agree, though, that eliminating what you call 'loop current' had no positive benefits?"

"Correct."

"Is it fair to say you misdiagnosed the problem?"

"Yes," Larry confessed. "It was the three-phase unit needed to be relocated. The isolator was bypassed."

"Did you refund Mr. Allen that $9,000?"

"Nope."

Again Kevin Whaley paused. "When you isolated the farm in December of 1997, how much did you charge Mr. Allen for that one?"

"$55,000," Larry told him.

"Did you take any cow contact measurements before that work to determine how much current was flowing through the cows?"

"No."

"And after that work, did you take any cow contact measurements to determine how much current was flowing through cows?"

"The only thing I addressed," Larry said, fiddling with the knot in his tie, "was current flow through grounding conductors."

Kevin Whaley nodded and walked toward the jury, stopping well short of the railing. "After installing the isolator, in 1998 did you take cow contact measurements to determine if the work had effect on how much current was flowing through the cows?"

"Other than the wave form analysis, I knew the current was pretty much absolved from the dairy. So no cow contact or step potentials."

"Mr. Allen cut 13 grounding wires on April 21, 1999, correct? And you knew Mr. Allen was cutting the wires, didn't you?"

"After he had cut them. Yes."

"You knew Mr. Allen reconnected those wires at some point. Did you take any cow contact measurements after he cut the wires to determine if that had any effect on current flowing through the cows?"

"When Mr. Allen reconnected them, no. But when WPS reconnected them, yes."

Kevin Whaley shot a puzzled look at the jury. "What you're talking about is, Mr. Allen actually cut grounding wires a second time, didn't he? And after WPS sent him a letter telling him not to. Correct?"

"I believe so," Larry said softly.

Whaley said nothing, and came back over to the witness. "You know," he said, "that WPS is regulated by the Public Service Commission. And that the PSC has established a level of concern of 2 milliamps of current flowing through a cow?"

Aha, so there *was* current flowing constantly? The only question was how much? Did Whaley want the jury thinking like that? I sure did.

"That's their level," Larry agreed.

"With 500 ohms of resistance, two milliamps is one volt, correct?"

That was right.

"So, let me make sure," Whaley said. "As of June 26, 2002, you had never taken a cow contact measurement on the Allen farm of one volt or more. Correct?"

"That I can recall."

Looking pleased with himself, Kevin Whaley now said, "You told us earlier about how you believe WPS hooked its lead wire to the wrong panel on June 27, 2002. And you told that to Mr. Lawrence and Mr. Hammarback, didn't you? But you never told WPS. Correct?"

"That's not true," Larry said to him.

Whaley nodded. "Now, you know that an equipotential plane is designed with the intent of making all contact voltages in a dairy equal, so if there was electrical leakage, the cows would not perceive it. Correct?"

"It's *designed* for intermittent electrical contact voltages, should they occur, so cows could not perceive it. I'll agree with that."

What Larry clearly meant was that what it was designed for and how it actually worked were two completely different things. Basically, this was the same argument Don Zipse had made in different words.

"Have you never recommended to Mr. Allen that he disconnect the plane in the free-stall barn from the on-farm electrical system?"

"It would be impossible," Larry told him.

"With respect to the plane in the free-stall barn, is it possible that parts of the slatted floor broke? And that Mr. Allen, who fixed it, didn't bond it in?"

"That's possible," Larry said. He sounded skeptical, as well he should have. I'd done the job properly.

Kevin Whaley stuck both hands in his pockets and rocked slowly on his heels in front of the witness stand. "You've had farmers tell you that the ground rings you've installed have done absolutely no good, correct?"

"I would agree with that," Larry said.

"When I took your deposition, you told me you thought the current on the surface ground ring was from the grounding wires on WPS' distribution system except for 1/1000 of a percent. Correct?"

Larry hesitated a moment before answering. "Well, yeah, uh, there would be a natural current of earth to some degree, but current from the down grounds of 15 amps will give current on the ring, whether earth current or induction. It's caused by electromagnetic imbalance."

"What percentage of the current measured on the surface ring is induced from the magnetic fields from WPS' distribution lines?"

Larry thought for a moment. "I believe close to 84 percent."

"With respect to induced current on the buried ring from WPS' lines, does it get on and off the ring before it can be measured?"

This time Larry answered without hesitation. "Induction is induction," he said. "Generally, it doesn't flow off. To flow off, it has to return to a source. That would be return current."

Kevin Whaley quit his rocking movements and stood still in front of the witness box with his feet apart and his hands clasped in front of him. "So it's your testimony that there is no induced current from the magnetic fields getting onto the buried ground ring?"

"No," Larry said. "Some of it is on the buried ground ring because it's a continuous loop, but it is earth."

Kevin Whaley crossed his arms over his chest and walked toward the jury. He stopped midway and turned toward Larry. "Assume," he said, "that the evidence will show 95-99 percent of the current on the surface and buried rings is induced from WPS' distribution lines. Would you agree that all the measurements have no relation to milk production on the Allen farm?"

Larry shook his head vigorously. "I disagree."

"You would agree, though, that that would mean those ground rings are not capturing and diverting earth currents at the Allen farm?"

Larry shook his head again in contradiction. "I disagree with you," he told the attorney.

Whaley now proceeded to the jury box. "Do you remember that during your deposition I asked you about the surface ground ring, and you told me that Mr. Allen was being electrocuted through the earth?"

Those last four words, *electrocuted through the earth*, were the true heart of all matters before this court. Either I was being electrocuted or I wasn't. There could be no middle ground. That was what the jury had to decide, and all that it had to decide. I was glad Larry had phrased his words this way, and glad that Kevin Whaley had brought them to everyone's attention.

"If you looked up in *Webster's Dictionary,* the term electrocution, that's defined as that."

This answer didn't sit well with Whaley. He scrunched up his face like he'd just swallowed a mouthful of sour milk. "And when I asked you what proof you had to back that strong of a statement, you mentioned that a light bulb could be lit off the surface ground ring. Correct?"

"Right," Larry said. "Which is uninsulated."

"The fact," Whaley said, his tone skeptical and testy, "that a light bulb as big as the eraser tip on a pencil can be lit off the surface ground ring is your proof that Mr. Allen is being electrocuted?"

Larry sighed, took a deep breath and looked hard at the attorney. "If a wire laying on the surface of the earth can light a light bulb, there's energy behind it! That light bulb is a resistance between that surface."

Whaley now took his own deep breath, as if trying to match Larry in heavy breathing. "And now you've testified that 84 percent of current on the ring is induced. Correct?"

"It's induced," Larry said with impatience, "by the imbalance of the WPS neutral. If it was balanced, it wouldn't light, sir!"

"Not what you call 'earth currents'?"

"Well, the buried ground ring is earth-current based. The surface ground ring is more induction."

Whaley now finally went and stood beside the jury box. "When Mr. Vern Peterson of WPS told you that measurements at the Allen farm showed that the WPS Dairyland neutral isolator was functioning properly, you did not disagree with that, correct?"

"I disagreed. I didn't disagree that it wasn't functioning to his specifications."

This was a nice way of saying that the device was a piece of crap that di d not work. Which is exactly what it was.

"Now, you took measurements on the Dairyland Isolator for three days, but never after. And you didn't demand that it be removed?"

"I wanted a new service installed at the Allen farm, and I wanted it removed. Mr. Allen couldn't afford to have it taken down at that point."

Not unless I felt like torturing the animals some more.

"After it was installed," Kevin Whaley said, "cows were not sicker, correct?"

"Well, the first day or two they weren't. They were after."

"Were the cows more hesitant to come into the parlor after the Dairyland Isolator was installed?"

"At first, no," Larry told him. "Then it got bad again."

Still standing beside the jurors, Whaley said, "You refer potential clients to Mr. Allen to talk to him about the kind of work you do, correct?"

"Correct. Anybody can call any of my clients."

"And Mr. Allen has traveled with you to Idaho to visit dairy farms there?"

"Yeah," Larry said. "In about a two-week period, I think."

"Has Mr. Allen physically helped you test on some of these dairies?"

"Roll out a few wires for me and roll them up, yes."

"So he has actually assisted you in tests?"

"Correct."

"And you and Mr. Allen have traveled to California together to test dairies, correct?"

Wondering what in hell he was driving at, I leaned farther forward to make certain I caught every word.

"I took him along," Larry said.

"Is it your claim," Kevin Whaley asked, "that all of Mr. Allen's problems are due to electricity from WPS' distribution system?"

"All of his problems? That would be impossible."

It might not have been Larry's claim, but it sure was mine. Personal problems, no, were not WPS' fault, of course not. The problems for which I was in this courtroom? Most definitely!

"Are you claiming that, in connection with this lawsuit, the problems regarding electricity on the Allen farm are all attributable to WPS' distribution system?"

Larry looked across the courtroom at the attorney. "Yes," he said forcefully, finally agreeing with my feelings. "In this lawsuit."

Kevin Whaley now read a portion of a letter that Larry had written me back in 1997, Exhibit 668, when I was looking for that $58,000 loan to purchase an isolation transformer.

After introducing it, Kevin Whaley said to Larry, "At the very end of this letter you wrote: 'Remember, any product WPS issues to solve a problem is always designed around their system's functioning and capability, and in that message lies their interests, friends, stockholders, costs, legal department, et cetera.' Did I read that correctly?"

Larry acknowledged these words that he'd written.

"What you were telling Mr. Allen was, 'You can't trust WPS.' Correct?"

That was absolutely what he was telling me. With good reason.

"No," Larry said, probably figuring that at this point discretion was the better part of valor. "I was telling him that the Dairyland Isolator that was going to be installed would likely fit their characteristics."

Kevin Whaley now whipped out another letter, Exhibit 1764, from my lawyers. "This letter," he said, "is dated January 18, 2000. And what this letter references is a meeting that's going to include Mr. Allen, yourself and Mr. Scott Lawrence regarding conditions on the Allen farm, and to prepare for the lawsuit to be filed against WPS. Correct?"

Did Whaley think he had some kind of scoop here? And that it was somehow conspiratorial for clients to meet with their lawyers and expert witnesses?

"Correct," Larry said.

"By July 28, 2000, you had received money, $5,000, for being an expert witness for Mr. Allen, correct?"

"I believe half that amount," Larry said.

"And this is five or six months before this lawsuit was filed, correct?" Whaley asked accusingly.

"I believe so," Larry told him.

"That's all I have, Judge," Kevin Whaley said. He turned on his heel and walked briskly over to the defense table. He had a smug look on his face.

Why, I didn't know, because I thought Larry had testified very well; I didn't see that Whaley had hurt him at all. Therefore, I was a bit surprised when Scott stood up to do a redirect. I had to assume there were some weak points that Scott wanted to shore up.

"Mr. Neubauer," Scott said, coming up toward Larry while he buttoned his suit jacket. "Mr. Whaley asked whether you took cow contact measurements according to the PSC code, both before and after various things happened on the farm. Why didn't you?"

Larry responded immediately. "Because it would have been pointless. I was trying to identify what needed to be corrected, not sit there and deal with voltage and current going through cows. It was an electrical issue. And I dealt with it that way."

"Mr. Whaley asked about your being critical with respect to the three-phase service brought to the Allen farm back in the mid '70s. What are you critical about?"

"Well," Larry said, "it's apparent that if you wanted a transformer relocated back then, you couldn't do it. It had to be site-located by a power company engineer. The engineer should have put the service up front but put it in the back, ran overhead conductors and supplied single-phase service for the other structures."

"You removed that dual service in 1997. That resulted in an improvement of the dairy herd?"

"At that point," Larry said. "But it didn't help."

Scott nodded slowly and sadly. He knew the whole pitiful cow story by heart. "With respect to Mr. Garvey's report on code inspection on the farm, at whose request was that done?"

"Wisconsin Public Service."

"What's that about with respect to the code?"

Larry edged forward in his seat and folded his hands together in front of him. "The code states that at a building service or near the first panel you would run a grounding conductor from the ground bar—in the city it would be a water meter—and put a jumper on it. Therefore, you're tied into the plumbing system, so electricity could travel.

"On a dairy, if you have a building and a metal water pipe comes up, the code says you have to bond it. That was one of the issues—through the electrical system, they want a properly sized bonding conductor, copper wire connected to that pipe."

"If you have copper water lines not bonded with the wire, what can you do to solve that problem from the standpoint of the code?"

"At Mr. Allen's, they cut the pipes off as they entered the building, put on a piece of plastic, bonded it to the copper or water line to make it inaccessible to human touch, and it met the code."

"Can that lead to stray electricity concerns?" Scott asked.

"Yes!" Larry told him. "Because the water line is similar to the ground ring, it's return current that travels on the water line."

"Is the same true of equipotential planes if you're bonded back to the utility service?"

"They lower the resistance of the dairy to the primary neutral and increased current flows."

This was precisely what Don Zipse had said.

"When was it," Scott asked, "that you relocated the WPS down ground from the transformer pole that they put in the middle of the yard?"

"January 7, 1998."

"With respect to the video of the Vierstra farm, Exhibit 1719, could you explain what you were doing with the resistors in those pictures?"

"Okay," Larry nodded. "The test performed on the videotape is a bucket which is stationary to the cow platform. When you heard my comment on seriesing my cow contact resistor, this is for my records when I get back home from Idaho, so I can get a layout of the dairy."

"In April of 1997, when you first met with the WPS reps, what measurements did you tell them about that you had made on the Allen dairy?"

Larry took a quick, short sip of water. "Told them we had loop current horrendously high and that the cows were being affected by the dual services. And it was reiterated then—by them—that there wasn't a problem, there was no current."

Here was a further instance of the WPS see-no-evil syndrome. If the facts proved inconvenient, deny them. I had to hand it to WPS; their MO was at least consistent.

"How did it turn out with respect to the dual services?"

Larry shrugged his shoulders. "Didn't solve the problem."

"What didn't?"

"Relocating the service to the center of the yard and relocating that down ground to finally push it out to the road, and extensive on-farm wiring to meet codes."

Scott walked over toward his counsel table but stopped well short of it and turned back to Larry. "Last summer, on the day WPS was testing on the farm, at what point in time did you tell them they should connect to the feed room, the back panel?"

"I was asked by Mr. Schott, sitting there—" he pointed to him in the gallery, "where can I connect to the neutral. Mr. Schott was well aware that the service existed back in the feed room. Everybody was.

"I explained to him that back in the feed room is where the neutral was bonded originally. He asked me again, 'isn't the panel in the front also connected to that? We'll just go there.' I said, 'That's your choice,' and away he went."

"So," Scott said. "This was before he even made the connection. It was his decision. Mr. Neubauer, I'm going to show you the order of the Public Service Commission of Wisconsin on Exhibit 1024. Is the first sentence a suggested range of resistors?"

Larry looked at it a moment. "Yes. 350 to 560 ohms."

"And the ones you used to do that sort of test in that range?"

"360 generally, and a 470."

With that, Scott ended, but Kevin Whaley wanted one last crack at Larry. At this point, I couldn't imagine what he expected to accomplish, but he got up quickly and eagerly, buttoning his suit jacket as Scott sat down in his chair.

"Mr. Neubauer," Kevin Whaley said, "do you have any documents that would establish the date in 1998 that you claim you finished work on the Allen farm?"

Larry told him he would have to look through his old invoice records.

"Is that work about installing an alarm system for motors?"

"Actually," Larry said, "an alarm system for detecting leakage of on-farm electricity."

"I'm going to show you part of Exhibit 1025, PSC docket 115." He went up to the witness stand and leaned over Larry's shoulder to let him see the document.

"Does the PSC in 115 say that the use of a 500-ohm resistor to stimulate the resistance of a cow, the concrete barn floor, and the cow's contact resistance is reasonable? Using a 500 ohm resistor in the testing protocol is conservative because it calculates stray voltage levels that equal or exceed the actual amount of current a cow in the barn would experience?"

Larry nodded. "Yes, sir. That's what it says."

Whaley nodded and ended. That was all he had.

As Larry stepped down off the stand, I drew a deep breath and tried to take stock of the proceedings. My important expert witnesses, my big guns, were through testifying. I thought they'd presented themselves incredibly effectively and proven beyond doubt my contention that WPS' stray current was responsible for destroying the Allen dairy business. I could only hope and pray that this jury of my peers would see things in the same light.

It would now become the turn of WPS to present its case and its witnesses. I'd seen several of them milling around the courthouse, and knew that WPS had brought in its own heavy artillery in the persons of Steve Schott, Vern Peterson and Roger Mellenberger. My job, when I got up on the stand, might be to try to undo whatever damage they'd caused my case; I hoped I'd be up to the challenge. *Just tell the truth,* I kept thinking, trying hard to believe it, *and you'll be fine.* I reminded myself that I had Barry and Scott on my side, two of the best, now preparing to switch over to defense.

Cross-examination of those slicksters, I thought, was going to be like one of those WWE over-the-top-rope battle royales. If my life hadn't been on the line, I would've sat back and enjoyed my ringside seat at what promised to be more fun than a barrel of monkeys.

Chapter Thirteen

METERS CAN LIE, LIARS CAN METER

When Vern Peterson of WPS took the witness stand for the defense, I felt we had a golden opportunity to win the jury over permanently.

Peterson was a snake oil salesman who'd say anything, no matter how contradictory of the facts. What was true and right meant little; all that mattered to him was shunting blame from the utility and laying it off on me. I'd seen that up close and personal when he was out at my farm. In a sense, I understood; he worked for the company and had to toe the line. However, when there's a conflict, ethics should always trump loyalty.

This day, the Big Kahuna himself, Trevor Will, was handling Peterson's direct, which indicated that WPS considered him a high-value witness. That cut two ways; he would also become a high-value target for Scott Lawrence.

But target shooting would have to wait. For now, Peterson, currently WPS' customer service manager, was telling Trevor Will about his prior position in the company, when he was electric distribution engineering project supervisor. That was in 1997, when he first visited my place. He mentioned that he held bachelor's degrees in science and electrical engineering, and that he was a registered professional engineer in Wisconsin.

Under Will's questioning, he talked about his initial visit to my farm, when he'd come out with that notorious document falsifier Jerry Held, who'd since been fired.

"What was the reason for going out there at the time?" Trevor Will asked, looking thin, nervous and elegant as ever in a gray, three-piece suit, wearing a pale blue,French-cuffed shirt with silver cuff links.

"Russ Allen and Larry Neubauer wanted to discuss options and Russ Allen's stray voltage concerns. Russ had concerns that his cows were jumpy and that he had overall low production."

Trevor Will walked slowly toward the jury box and stopped just short of the railing. "Did anyone tell you that Mr. Neubauer had measured .914 volts over a feed bunk? Did anybody tell you that Mr. Neubauer had measured 800 milliamps of current on the parlor steel?"

"No, they did not."

"Did they give you any test results?"

"No," Peterson said.

"Did you have a response to their request about moving the transformer out to the road?"

"Yes. That we were not able to, as a result of a declaratory ruling—an order from the Public Service Commission."

Trevor Will nodded. "Is it typical to have discussions with the Public Service Commission staff about orders?"

Vern Peterson nodded back at Trevor Will. "They're our regulators, control what we do as a utility with the stray voltage program. I talked to Dan Dashal, head of the stray voltage program at the PSC."

Aha, so there was such a thing as stray voltage after all? It wasn't just a boogeyman that Neubauer and I conjured up to swindle the utility? Couldn't have been, not if the Public Service Commission had a special *program* in place to deal with it.

As for Dan Dashal, he was sued in another case where he was accused of cooking the books. He was let off the hook because of a "civil rights doctrine"; the judge said he qualified for immunity. We believe he quit sometime in 1997 under pressure to do so.

"Under what circumstances," Trevor Will asked, "was WPS committed to move transformers at a customer's request?"

"We had to use sound engineering judgment," Peterson told him.

"Did you discuss with Mr. Dashal the concept and meaning of sound engineering?"

"Yes."

"When you were out on the Allen farm in summer of 1997, did you make reference to this PSC order, to Mr. Allen?"

"Yes."

"What did you say to him?"

"That the order told us we could not move the transformer at his request."

"Did you have options to offer him?"

"Isolation on demand," Vern Peterson said. "Where we install a neutral isolation switch otherwise known as a Dairyland Isolator."

Which is practically useless, as it proved all too fully.

"What did you say to Mr. Allen or Mr. Neubauer about the Dairyland Isolator?"

"That it was a very good solution," Peterson told him.

He actually said that with a straight face.

"Mr. Peterson," Trevor Will said, walking over to the exhibits. He picked one up and took it to the witness. "If you look at Exhibit 151A, seventh page, this was a load box test on the three-phase setting at the Allen farm, as part of putting up the isolator. Why did you use the three-phase at that point?"

"That's the setting that serves the parlor and free-stall area."

"Is there anything about those readings," Trevor Will said, "that indicates to you a stray voltage problem on the Allen farm?"

"No, there is not."

The surprise would've been if he'd said there was.

"What readings support that conclusion?"

Peterson glanced at the document. "The .0022 which is our cow contact reading with our load box on its highest setting."

"What happened after you installed the neutral isolation switch?"

"We did another load box test."

"Using the data here," Trevor Will said, "can you calculate the percentage of the current that's returning on the primary neutral?"

"I would take the primary neutral—I brought it down and that's 1.95 amps. And divide that by the 2.5 amps that are on the primary line." Judge McKay leaned over and handed him a small calculator on which he quickly punched some keys. "That's 78 percent," he said.

"Are there any values or readings in this study that indicate there might be a stray voltage problem on the Allen farm?"

"No, not at all," Vern Peterson said, and I chuckled out loud.

Trevor Will walked over to the jury. "What numbers do you rely on for that opinion?"

"Our cow contact voltage, which is .001."

The attorney was silent for several moments. Then he said, "Was Mr. Allen present when you were installing the Dairyland Isolator?"

"Yes."

"Was Mr. Neubauer?"

"No."

"At any point during the installation did you say anything to Mr. Allen about what you were finding?"

Peterson nodded. "I told him we were able to get isolation but I was concerned about the numbers."

So were my cows. They didn't like them one bit.

"What did you tell him you wanted to do?"

"Well, we tried to isolate the single-phase service. I was able to, and get appropriate readings. We spent time going through some wiring, trying to figure out where that location was on the single phase on the services, to try and figure out what that interconnection was."

"Interconnection," Trevor Will said. "Are you talking about something on the utility system or the farm system, or somewhere else?"

"It would typically be through the farm system."

So, through all this, Will was again approaching his theme of "on-farm leakage," that the stray current was somehow caused by my faulty equipment and wiring.

"Did Mr. Allen indicate to you that he was upset or unhappy?"

Although I have to admit that I wasn't nearly as upset and unhappy as my animals.

"Well, he was concerned. And that was appropriate."

"How did you leave it with Mr. Allen at the end of the day?" Will asked him.

"I told him we would come back and try to pursue it further, or he could get his electrician involved."

"And when you left, where was the Dairyland Isolator?"

"On the three-phase setting," Peterson answered.

"And was it functioning?"

"Yes, it was."

Will fell silent. He walked slowly toward the judge's bench and then up near the witness stand. "What happened after that?" he asked.

"We received a call either from Mr. Allen or Mr. Neubauer the next day. The idea was they'd found the source of the problem and asked us to come out to the farm. We went out right away."

"Did anybody go with you?"

"Jerry Held came with me," Vern Peterson said quietly.

I'd have been quiet too, if I had a statistics forger for a colleague.

"When you got to the farm, what did Mr. Neubauer say?"

"That our reference rod was over a pipe that ran from the back of the old barn to a manure pit, and he felt that was affecting our reading."

"And what did you find?"

"That he was correct," Peterson said. "We moved our reference rod, and readings looked better."

"What did the readings show about whether the Dairyland Isolator was working?"

"It showed it was working."

Of course it did. The isolator worked until it didn't.

"Did Mr. Neubauer raise any other issue?"

"Yes. He claimed the isolation switch was closing." In other words, failing to prevent current getting through and onto my farm.

"Did he offer evidence of that?"

"He showed me a scope, a current level he'd measured over several hours."

"Was there any way of determining when and where that reading had been taken?"

"No, there was not. I was concerned, but I also questioned the reading."

As though Larry was faking it. Maybe Peterson had a guilty conscience because his good buddy Jerry Held had been nabbed doing precisely that. The psychologists call it 'projection' when you accuse others of your own shortcomings.

"Why were you concerned?" Trevor Will wanted to know.

Peterson thought for a moment. "I guess I was concerned at where he took it and in what time frame."

"Did he indicate where he had taken it?"

"On a secondary neutral on the farm."

"What would be the significance—assuming the reading legitimate—of taking it on a secondary neutral?"

"Neubauer claimed that because there was secondary neutral current flowing on the farm, the isolation switch was not working."

I loved these words: *legitimate, claimed.* It wasn't working, period. The only way it could have been working was if there was current flowing from my farm equipment. This was the point these two collaborators were trying to drive home.

"Now," Trevor Will said. "You had questions about the reading itself. Why?"

Vern Peterson smiled. "Well, I was very concerned because Mr. Neubauer had a reputation of taking readings—"

At that moment, Scott Lawrence cut him off, and I wished he hadn't. I was anxious to hear how Peterson intended to trash Larry.

"I object to this, Your Honor," Scott said to McKay.

Trevor Will retorted, "Your Honor, reputation of the defendant goes to how he reacted to what was presented to him."

McKay called both lawyers to him. After a bench conference, McKay allowed—what else?—the witness to answer, but limited him to stating his own perceptions and nothing more. A partial overrule was better than a total one, I supposed.

Peterson said, "Mr. Neubauer had a reputation of taking claimed readings that were elevated and no one else would be able to duplicate those readings. He also had a reputation of taking readings in locations that didn't make any sense."

"Did he," Trevor Will asked, "specifically show you where he took the reading that he claimed was on the scope?"

"No. I told him I wanted to do a complete load box test and a 24-hour chart."

"What was his reaction?"

Vern Peterson opened his eyes wide. "He was very upset by that."

"Was Mr. Allen present at that point?"

I was indeed present.

"After Mr. Neubauer got upset, what did you say?"

"That I wanted to test," Peterson told him. "To verify those readings myself. We got into a conversation about the Dairyland Isolator—the claim that the isolator was closing didn't make any sense to me.

"The device will not close unless the voltage across it exceeds 31 volts, and the testing we had done previously was nowhere near that level on our primary system. I found it very unlikely that the isolation switch was closing. Then the conversation turned to Mr. Allen."

"What did Mr. Allen say at that point?"

"Mr. Allen was very upset," he told Trevor Will. "He came over and said, 'You know, Mr. Neubauer is telling me that this isolation switch has a load on the feeder, that the switch cannot handle the voltage and current coming from our system.'

"I told him I didn't believe that was true. He got very upset and said he was done dealing with us. He was going to go with Mr. Neubauer. He was going with an isolation transformer, and he wanted us off his farm immediately."

"Did Mr. Held have to get any equipment?"

"Yes."

"What happened with Mr. Held's retrieval of his equipment?"

"He was told he couldn't go near the feeder equipment by the main panel. He could only go to where his lead was and remove that."

"Almost six years ago," Trevor Will said, "how is it that you can remember so clearly?"

"It was the first and only farm I have ever been kicked off of."

"And how did that make you feel?"

Looking as hangdog as a scolded puppy, Peterson muttered, "I was very upset."

All we needed now to complete this sad picture of a victimized and abused Peterson was a quartet of violins to start playing.

"Why was that?"

"Because Russ didn't believe me. I also felt I had been set up, and it was upsetting to get kicked off that farm."

Because I didn't believe him! My opinion of him and his so-called findings pained him to the depths of his soul. If only I'd known!

"When was the next time you remember being back on the Allen farm? And who was involved in that meeting?"

"Late summer, early fall. It would have been Mr. Neubauer, Mr. Allen. Jerry Held was with me."

So I didn't kick him off the property? Or I did, but then thought better of it and invited him to return? Peterson couldn't decide how he wanted to play this.

"At the point when you had agreed on the location of moving the transformer, what seemed to be the atmosphere?" Trevor Will asked.

"Everyone was pretty happy we had come to an agreement."

Yes, we were like one big brawling but loving family.

"Mr. Peterson, we were talking earlier about when the subject of an ungrounded 480-volt delta system to service the farm came up."

"Yes. Mr. Allen and Mr. Neubauer requested we provide a three-wire, 480-volt service."

Trevor Will unbuttoned his jacket and crossed his arms. "What did you tell them about their request for 480-volt ungrounded service?"

"I denied their request."

"Why?"

"I did not feel it was appropriate voltage for a dairy facility. I was concerned about safety."

"What are those safety issues?" Trevor Will asked.

Peterson coughed, cleared his throat and gulped from his water glass before answering. "The ungrounded service—if you do get the first ground fault, nothing, no breakers trip. Problem is, unless you clear that fault, before you get a second one, that second is a 480-volt fault, which is very high energy. In that wet dairy environment, it's very unsafe, both for humans and the cows."

Unsafe! WPS was worried about the safety of animals and people on the Allen farm! There was a good one. If anybody believed that, Peterson had an igloo in Alaska he wanted to sell him or her.

"Did there come a time when WPS did provide 480-volt ungrounded service to a farm in this area?"

"Yes. We viewed that farm as a different type, large dairies, over 1,000 cows. A 24-hour operation. We felt there would be a responsible person there around the clock.

"The other thing was, they worked with an electrician on a regular basis. These 480-volt services, farmers typically do some of their own wiring, and there are risks and dangers unless you have qualified people working on those systems."

I had a master electrician working closely with me, but apparently that didn't count. And I was on the farm 24/7, but I guess that didn't count either.

"How many farms actually got the 480-volt ungrounded delta service?" Will asked.

"Two."

"Have other farms requested it?"

"Yes."

"Did you see Mr. Allen again before this lawsuit was filed in November of 2000?"

"Yes," Peterson told him. "In November of '99."

Will thrust both hands in his pockets and walked a few paces nearer to the witness stand. "Did Mr. Allen ever approach you to talk about possibly resolving the claim he had?"

"No."

Of course not. Why would I? Given WPS' constant denials, evasions and outright lies, the problem could only be resolved in one way: Through the court.

"How many farms have you been on in your career at WPS for stray voltage testing?" Trevor Will asked.

"Over 1,000."

OH MY GOD, over 1000. That should throw up a red flag for any company when you have to investigate the same problem over 1000 times. That should certainly tell them that they have a problem with their system. Mark Cook from the Public Service Commission of Wisconsin did over 5000 investigations. I was with Larry Neubauer testing a farm in California. The farmer wanted to rewire his farm, so

Larry and the farmer decided that they should move the transformer at the same time. Larry called the utility and asked to speak with their farm consultant who tested for stray voltage. The utility said, 'We don't have one.' Larry said 'why not.' The utility said, "We don't need one. I think WPS has about 8 or 9 Farm consultants testing for stray voltage.

"Based on your experience, did you see anything on the Allen farm that indicated a stray voltage problem?"

"No, I did not."

"Have you found stray voltage problems on other farms you've tested?"

"Sure."

"And when you found them?"

"We've corrected them."

I have never heard of them ever finding a problem on any farm, and that's no secret among the farming community

"That's all I have," Trevor Will said to Judge McKay. He must've thought that answer sealed the deal for this witness.

It didn't, and Scott Lawrence rose to begin what I expected would be a sharp cross-examination. I slid forward to the edge of the bench, leaned my elbows on my knees and inclined my head. I didn't want to miss even one word.

"Mr. Peterson," Scott said, standing close to the witness and clasping his hands together below his waist. I could see the big college ring with the blue stone on Scott's right hand. "On those 1,000 cow dairies where this ungrounded secondary was provided, the purpose was to reduce current through the dairy?"

"That was what Mr. Neubauer requested, yes."

Scott nodded. "In fact Mr. Neubauer was the electrician for both of those farms?"

"I believe both."

"And you considered him a qualified electrician?"

"No," Peterson snorted. "I did not."

"But you provided the ungrounded 480 regardless?"

Peterson shifted uncomfortably in his chair. "They were working with their regular electricians," he explained.

"But Mr. Neubauer performed the isolation work," Scott reminded him. "Correct?"

"Yes, he did," Vern Peterson acknowledged.

"And those farms have parlors with equipotential planes?"

"Correct."

Scott went up closer to him. “The cow contact voltages in those parlors were very, very small, were they not?”

“That’s correct.”

“And yet,” Scott said, “the owners of those 1,000 cow dairies thought they had a problem. Correct?”

“In one situation, yes.”

“Was the subsequent course of milk production at either large dairy reported to you after the isolation work was done?”

“I was not given specifics.”

That was a cute way of not saying that production had increased on those farms after isolation.

“Now,” Scott said, walking over to the jury box, “in the system Mr. Neubauer installed at the Allen farm, isn’t it true that by design, customer current on that secondary neutral was totally eliminated?”

“Yes.”

“And he did that by installing sub transformers toward the load in the barn that made the farm system look like a phase-to-phase load to the utility system, correct?”

“Correct.”

Scott nodded again. “The second item in the Public Service Commission’s order in which you felt you had to deny Mr. Allen this service was that placement should be outside animal confinement areas, correct?”

This, also, was correct.

“And,” Scott continued, “Mr. Allen and Mr. Neubauer were not requesting a transformer that was in the cow yard, correct?”

“No, they were not,” Peterson agreed.

“The last of the Public Service Commission concerns,” Scott said, “is, and I quote, ‘The customer’s preference must also be considered in placement.’ Correct?”

“Correct.”

“And there was no doubt about Mr. Allen’s desire for placement, correct?”

“That is correct,” Vern Peterson answered.

“So,” Scott said, “we’re three for three. There’s no problem moving the pole?”

“Yes, there is a problem moving the pole.”

Scott did a double take and raised an eyebrow. “And in your estimation that’s the Public Service Commission’s position?”

“Correct. And that’s who we have to answer to.”

"And who has no jurisdiction over farm wiring. Correct?"

Vern Peterson shrugged. "In the declaratory ruling, which said we could not rely upon the customer's devices to take care of that secondary neutral drop. In the future, that system would be changed back to a more normal operation and secondary neutral voltage drop wouldn't be a concern."

Scott leaned both hands on the rail of the box and faced the jurors. "Mr. Allen was not party to that proceeding, was he?"

"No," Peterson said. "He was not."

"But you permit this kind of move now?"

"Yes, we do."

Next Scott showed Vern Peterson Exhibit 151A; this document recorded various contacts WPS had with me. Scott dug through a number of entries until he was able to unearth some real gems.

"Third entry down, step information," Scott said, "refers to meters being set in the milk parlor. Do you know when that happened?"

"The entry date is 4-22-97. I don't know which one that refers to."

"The next entry," Scott told him, "says 'pulled meter.' And you weren't involved in that either?"

"Correct."

"The following entry says, 'Customer notified, data discussed with Russ. He wanted farm isolated on 7-10-97. Vern P. and Jerry H. hooked up isolator and tested switch. It was functioning properly.' Did I read that properly?"

"That's correct."

Scott nodded and continued, "Page 11, first entry says, 'Required action by WPS, none.' Correct?"

"That's correct."

Again, Scott nodded. "And the next entry is 'order complete.' Correct?"

"That is," Vern Peterson said.

Scott nodded once more, then left the jury area and went over to the witness. Then, seemingly from nowhere, he said, "And neither on page 10 nor 11 do we see an entry indicating that *anybody* was refused permission to test. Do we?"

Vern Peterson took out a handkerchief and wiped his forehead. "No, you do not."

"We don't see an entry indicating that anybody was kicked off the farm. Correct?"

"No."

Scott's brow furrowed as if he was confused. "And yet," he said, "there are items entered on here that occurred the very same day you indicate that happened. Correct?"

"That's correct," Peterson admitted.

This was strong stuff because if, as Peterson made out, he was so traumatized and upset by getting thrown off the property, why hadn't he recorded it in the WPS log? Especially since he'd made detailed records of a bunch of trivial, mundane items. I watched the jury to see if they were reacting. The lady in the floral print dress was nodding, but I had no real idea of what that meant.

"And that to you was a very memorable event, being kicked off the farm," Scott said to him. "Correct?"

"Yes, it was!" Peterson exclaimed, and I almost burst out laughing.

Credibility was everything here, and if it became clear that WPS lied about being thrown off the farm, to make me look like a bad guy, the jury would wonder what else they were lying about. It would start doubting all their assertions, even true ones.

Then came another gem from Scott, which, unfortunately, he didn't get to finish: "Now, what's Mr. Held's reputation for truthfulness in reporting voltage readings?"

"It's very good," Peterson said with a straight face, and again I nearly laughed out loud. This was the same Jerry Held that WPS had fired for falsifying records. If he didn't get caught by Barry Hammarback, WPS would have probably given him a raise. The guy with the "good record" according to Peterson was a liar and a forger. And the company knew it.

"Is there a reason," Scott asked, "that he's no longer employed by WPS?"

Trevor Will suddenly leaped up and objected. Somehow he had to find a way of defending Held's dishonor. And the only way was to prevent the jury's hearing about it at all. Judge McKay, in full coddling mode, sustained him.

So Scott went a different route. "Isolation on demand is provided in part to give farm customers answers on whether eliminating electricity below the level of concern may help them, correct?"

"That is correct," Vern Peterson told him.

"And in the case of the large farms you discussed a moment ago, they wanted to isolate with a Dairyland transformer, correct?"

That was exactly what those farms had requested.

"And that's a fast device, will close on short impulse?" Scott asked.

"The duration of impulse will determine how fast it will close."

The thing was a piece of garbage, bottom line. Kept shutting down when you needed it most.

"Have you ever seen any measurements by Mr. Neubauer in January of '98 after the isolation system was put in?"

"No."

Scott moved a step nearer to the witness. "Your company never received an order from the Public Service Commission directing that it could not change transformer locations on the Allen farm, did you?"

Peterson shook his head. "Not specifically related to that fact, no."

"Isn't it true that the system Mr. Neubauer proposed to install included an alarm system?"

Which would've eliminated Peterson's safety concern about not having a 24/7 operation with someone in attendance. Maybe safety wasn't the real reason they refused to help me?

"It could have. I don't know."

Scott hammered him on this: "The systems installed on those other two farms have an alarm system, do they not?"

"I don't know that for a fact."

He knew all right. He was just parsing words.

"In addition to providing the ungrounded service, you did some other things to help those other farms. You loaned one farm a generator. Correct?"

"We paid for a generator," Peterson said.

"Mr. Peterson," Scott said, again going over to the jurors, "before you went to the Allen farm, you didn't get any records from WPS on the condition of the line in the vicinity? You didn't get records on the age of the line? Or records of inspections of connections on the line? Or any measurements of phase and neutral current recorded on the line?"

"No, I did not," Vern Peterson replied.

Here was a perfect example of their attitude toward the problem: Neglect it, run from it and deny it.

"You've never seen copper weld conductor on a customer service order?"

"Correct."

"And you haven't seen knob and tube wiring from the 1930s installed on a 500-cow dairy?"

"I would say no."

Scott went over to his counsel table and looked at some notes on his pad. Then he went back to the witness. "Mr. Peterson, there's no regular inspection program where a state inspector will come out and do a line inspection on a WPS distribution system, correct?"

Peterson said there was no such program.

"And there never has been. Correct?"

"Not in my career, no."

"Mr. Peterson, you're familiar with the principle of four and five wire systems on farms?"

"Yes, I am."

"Those systems separate the functions of the ground from the neutral?"

"Correct."

"The hardware you have to use for high voltage is different than for low voltage?" Scott asked.

"Correct."

"The concept of a four- or five-wire system can be applied to a utility system as well as a farm system?"

"There's certainly a lot more consideration on the utility system."

Scott turned sideways to the witness stand and placed his hands in his pants pockets. "You have no experience with four- or five-wire systems that separate ground and neutral?"

"No," the witness said. "I do not."

"Keeping connectors in good repair is a good thing to do on any electrical system?"

"I would agree."

"Keeping the conductor size adequate on any system is a good thing to do? And upgrading the conductor size as time and experience and electrical loads dictate is important? And all of that applies to utility systems as well as farm systems?"

"That's correct," Peterson told him.

And all these things were precisely what WPS had ignored, for years.

"That's all I have," Scott said, and Trevor Will rose quickly to start his redirect.

He seemed in an awful hurry and more nervous than usual. A sign, to me, that he felt he needed to rehabilitate this witness.

"Mr. Peterson," Trevor Will said, buttoning his jacket and going up close to the witness stand. "You were asked a series of questions about the order and docket 106 as it relates to the metering pole. Who owns the metering pole on the WPS system? The customer or the company?"

"The customer," Peterson told him.

"The transformer pole," Will said. "Who owns the pole there?"

"The utility."

"When Mr. Allen and Mr. Neubauer were talking about moving a pole on the farm, which pole were they talking about?"

"The transformer pole," Peterson told him.

"You were asked before by Mr. Lawrence about providing a generator to a farm. Why was it the company provided that?"

"That was part of experiments we were doing on the farm." (10,000 dollars' worth of experiments—*phew.*

Will nodded. "You were asked about the isolator closing. After it was removed from Mr. Allen's farm, was it tested?"

"Yes, it was."

"What were the results of the test?"

Peterson craned his neck forward to get closer to the microphone. "The isolator closed in at the 32 volts just as the specifications call for."

No surprise there. The fault was not in its operation; it was the design. There sure must have been a lot of stray current to close the Dairyland Isolator.

Trevor Will nodded and walked over near Judge McKay's bench. "You were asked if you got any line inspection data before you went out to the Allen farm. What was the practice of WPS before you went out to do a stray voltage investigation?"

"We did not do that."

"And you were asked whether you completed additional testing in July. Why didn't you complete the phase-two testing and primary profile in July 1997?"

"Because," Peterson told him, "I was kicked off the farm."

Again with this? But I guess that was his story and he was stuck with it. By this time he might even have begun believing his own lies.

"That's all I have," Trevor Will said and went back to his seat.

Scott Lawrence, though, wanted another crack at Peterson. He rose and went up to the witness stand.

"You indicated the company provided a generator to one of the large farms as part of an experiment. The farmer was also an author of a column on the farm page of the paper?"

"I know there were a few articles," Vern Peterson said.

"He wrote about this situation fairly extensively, correct?"

"That's correct."

The articles had documented the stray current problem and skewered WPS for wrecking that farm's animals.

Scott had nothing more to ask, but Trevor Will wasn't yet ready to release Peterson. He wanted to fire a parting shot, which, to me, was a dud.

"What was the practice at the company," Will asked, "with respect to whether you tried to keep the meter pole and the transformer pole close or separate?"

"We would try to keep them close."

And that was it for Peterson, a witness who went out with a whimper instead of a bang; Trevor Will had nothing else. Maybe he was saving his best for the next witness, the quarter-million-dollar man, Roger Mellenberger. Mellenberger was a key witness for the defense, an intellectual heavyweight who would present their arguments and theories with the same kind of force that Mike Behr and Don Zipse had done for me. Nervously, I wondered how Scott Lawrence would attack him; Mellenberger had the reputation of being as slick as they came, as well as articulate and personable. I felt my heart begin beating faster as the man slowly rose, walked to the witness stand, took his oath and sat down in the chair.

Chapter Fourteen

MILLION-DOLLAR MAN

Whatever else he was, and he was many things—huckster, intellectualoid and showman—Roger Mellenberger was a first-rate mind. Of that there was no doubt. After all, WPS wouldn't pay somebody hundreds of thousands of dollars every year for testifying unless he really knew his stuff and could make them look as kind as Mother Teresa.

Mellenberger knew his stuff. He'd published lots of articles and papers on the dairy industry, agriculture and stray voltage. From my vantage point, the problem wasn't that he lacked expertise; it was that he placed his expertise in the service of a bad purpose: He drew all the wrong conclusions from his studies. Just as his paymasters demanded.

For me he was a sell-out, a traitor, because he'd been raised on a dairy farm and should've known better. He was a prostitute. A high-class one, certainly, but a prostitute nevertheless.

Kurt Meckstroth, Mr. Arrogance himself, was handling the direct, and it would be interesting to see which of the two would prove more insufferable. Obnoxious as Meckstroth was, Mellenberger gave new meaning to the word "conceit." He was as proud as Lucifer and almost as slick.

Meckstroth began with the usual: Having the witness present his *bona fides* and report his background. Roger Mellenberger had received his bachelor's degree in agriculture from the University of Wisconsin, had been raised on a 50-cow farm and performed every variety of work there, including milking, tractor driving and feeding the animals. He'd recently retired from teaching dairy and animal science at Michigan State University and become a "dairy consultant."

This sounded a lot nicer than "corporate shill."

"Are you familiar with something called the Kellogg Biological Station?" Kurt Meckstroth asked him.

Mellenberger was, intimately. "That's one of three research dairy farms we had at Michigan State University. There was a dairy herd there that W.K. Kellogg started back in the 1930s. We replaced that in 1986 with a 140-cow Holstein herd and built new facilities. From '86 to '98 I was faculty supervisor for that facility."

"What did that job entail?"

"I would sit with the farm and dairy manager and make decisions whether we'd milk three times a day, use BST, how many acres of corn we needed, what research project might get done. And I used Kellogg to train people, from milk inspectors, field personnel and laborers."

"You mentioned that Michigan State had three dairy farms," Meckstroth said. "Did you have responsibility with respect to the other two?"

"The one on campus, of course, I used extensively for teaching purposes, for milking equipment or mastitis areas. And I also did a herd in the Upper Peninsula—if they had a health problem I would advise them."

"While a professor at Michigan State, did you do any work related to the state of Wisconsin?"

"Yes," Mellenberger told him. "A year before I retired, I took a sabbatical and did a couple of months at the University of California, and then did work at the University of Wisconsin for the other 10 months."

He went on to explain that he'd written numerous dairy bulletins on various aspects of dairy farming and herd management, and belonged to the American Dairy Science Association and the National Mastitis Council, where he served for a number of years on the editorial board. He also mentioned that his extension duties consisted of bridging the gap between intellectual research and actual problems encountered on farms.

"What does the term 'peer review' mean to you, Dr. Mellenberger?" Kurt Meckstroth asked.

His answer would be interesting, because peer review was a stacked deck. Only articles and papers sympathetic to the powers that be, conforming to the prevailing wisdom and company line, would be accepted for publication. It was a closed system, a communistic system. Since the utilities funded most of the university research, independent thinkers and intellectual mavericks beholden to nothing but the truth, men like Don Zipse, were frozen out. Had it been possible for the utilities (and might someday be), these individuals would've been declared enemies of the state.

"Peer review," Mellenberger said, "means that the article had been submitted to somebody that has the same knowledge base that I do. My article has to go to something like *The Journal of Dairy Science* to two different reviewers, sometimes three. They have to review the article to see if it's sound."

In other words, to see if it conformed to the party line. If it deviated, it went straight to the bonfire. *The Journal of Dairy Science* was warned to tighten up its peer review process. By the way, Gustafson, another corporate shill, controls the peer review process.

Mellenberger continued, "They will either reject it or send it back with suggested revisions. If I were the author, I would have to make the revisions and they would either accept it or reject it."

The "they" in this case functioned like the old, hack Soviet Writer's Union that Stalin ruled in the 1930s, which existed to glorify the regime. Officials all thought alike and made sure their writers did too. Or else.

"Have you authored any articles that have appeared in peer-reviewed literature?"

"Yes, several."

Of course he had. It was all part of the WPS job description for a salaried university flack.

"Dr. Mellenberger, you were retained by Foley and Lardner in this case. What were you asked to do?" Meckstroth went away from the witness and over to the jury.

"Look at the Allen farm—the time of the complaint period. Look at records, just see what happened from a dairy herd management standpoint over those years."

As soon as I heard the word "management" I knew where this was headed: Toward my "incompetence." Toward blaming the victim one more time.

"What years were you looking at?" Meckstroth asked.

"1976 through 2002."

"What methodology did you file to conduct this analysis?"

"My normal methodology is to take all records from the farm or businesses that have sold to that farm. Enter the information into the computer and develop a spreadsheet so that at any point in time I can evaluate the mastitis situation, reproduction and milk production. Look at what happened to that farm and herd over time. When I was first called in February of 2002, there were 46 boxes of documents waiting for me, records about anything: milk checks, feed invoices, vet records, herd production records, bank records, sales slips to stockyards—everything imaginable."

"How were the records organized?"

Mellenberger sighed. "Totally mixed up—the years all mixed up. No order whatsoever."

"What did you do?" Meckstroth asked in a tone of sympathetic surprise.

Mellenberger wagged his head theatrically. "Painstakingly went through page by page. And I arranged it category by category. Had to take all the years and sort by category, by year, box by box by box."

What a he-man! But no work was too tough for this hero. What he failed to mention was that our side had offered to put the records in chronological and alphabetical order before sending them to the defense. But they were all in a lather to receive them and declined our offer for the sake of speed.

"Did you encounter problems collecting data from the documents?"

"From a couple of standpoints," Mellenberger whined. "For instance, from 1984 through 1989 we had no cow numbers. No records. Same for '96, '97. Same thing from September of 2000 to February 2001. And then when he went on his computer system in 1997, there was no way to extract the information back out, so I had to hand enter all of that. That was very laborious." At first when it was convenient I had boxes and boxes and boxes of records, and now, no records—which was it?

But the big lug persevered, he did.

"Were there other records that were obtained after the depositions of Mr. Allen's employees began?" Kurt Meckstroth asked.

"Yes. The Robert Allen records on treatment came pretty late, on health and reproduction, other things that happened to cows."

That was a nice touch, *things that happened.* I liked that. It sounded much better than saying *the animals got fried by electricity and croaked.*

"Dr. Mellenberger, in conducting your analysis why didn't you just rely on the information presented by Mr. Allen and his expert witnesses?"

Mellenberger made a face. "If I'm asked to assess what happened to this herd over time, I need specific numbers, based on numbers from the farmer's records."

"What information were you provided by Mr. Allen and his experts in that regard?"

Mellenberger smiled across the width of the courtroom at Kurt Meckstroth. "None that I can think of."

"You referred to information you took from records and put in a database, into an Excel spreadsheet," Meckstroth said.

He went over to Exhibit 591, asked McKay's permission, and then brought it to the witness. It was one of several notebooks filled with data that Mellenberger had compiled relating to feed purchases, invoices, histories of particular cows, animal reproductive status, health status and veterinary records. Meckstroth asked him about his work when he visited my place.

"Took feed samples, observed what the cows looked like, how they're housed, how they're cared for, viewed the milking routine. I counted cows, the number of stalls, all just to have some idea how they're being handled and taken care of."

"Dr. Mellenberger," Meckstroth said, going back over toward the jury, "have you been paid for the time you've spent on this case?"

Did Bill Gates make a profit from Microsoft?

"I charge $110 an hour for regular work. $150 an hour for deposition and trial testimony."

"How much have you charged WPS in this case for your time?"

"$140,000."

Nice work if you can get it, and like I said earlier, you must be able to lie with a straight face and still be able to live with yourself.

"Has WPS paid your invoices?"

"Yes."

Meckstroth needed to get this on the record before Scott and Barry did. He probably realized my lawyers were going to beat Mellenberger like a drum on the issue. And rightly so.

"Have you worked on any other stray voltage cases? If so, how many?"

"At least 60."

"How does the amount of time you've spent on this case compare to other cases?"

"The maximum I've spent on any other farm is maybe 400 hours. So it's three times higher, 1200 hours."

"Why have you spent so much time working on this case?" Meckstroth asked.

I didn't really think Mellenberger was going to say he wanted to take his wife on a vacation to the French Riviera.

"Because," he said, "the records were in such disarray. I had to put them in a form that I could use them."

Obviously, he had not been in any hurry to fix up the records. After all, the south of France is an expensive place.

He rushed on, almost out of breath, nearly hyperventilating over all the backbreaking labor he'd performed on my Stone-Age-era farm. "If I had to change one spreadsheet, that cascades down. I'd have to change five, six, seven, eight of them. I found myself doing that time after time after time, going back and having to make sure that I covered all bases and brought everything up to date."

My, my, but a shill's work is never done. For $140,000, though, it wouldn't have hurt him to sweat a little.

"If Russell Allen had been using a computer program like Dairy Comm 305," Meckstroth said, "how would that have affected the amount of time it took you on this case?"

"Cut the time in half."

Was he trying to explain to WPS that the whopping fees he gouged from them were my fault? On the other hand, if I'd had the computer program he might only have squeezed out a measly 70K. I was using De Laval's latest program—again, they were throwing shit at the wall and watching to see what sticks.

Meckstroth, looking satisfied, walked up close to Mellenberger. "Doctor, what do you understand to be Mr. Allen's main contention in this case?"

"That milk production on his farm was adversely affected from 1976 through 1997."

What an answer. This Mellenberger had all sorts of slick responses. He worded everything to sound harmless. Whenever he said something like "adverse affects," I pictured dead, stiff-legged cows.

"And," he continued, "after the farm was isolated and the transformer moved in December of '97, that production increased substantially, indicating to him that voltage had been the main determinant to milk production for that length of time."

"Dr. Mellenberger, do you have an opinion as to whether Mr. Allen's herd has been adversely affected by stray voltage during any period of time since 1976?"

"Yes," he said emphatically. "I do."

"What is that opinion?"

"No, the farm was never affected by stray voltage or any electrical event."

The shock would've been if he'd said anything else. Of course, he was going to say it wasn't affected by current. According to the defense, if I hadn't been an inept farmer and manager, the cows wouldn't have stopped producing, got sick and died.

But, then, if my aunt had testicles, she'd be my uncle.

Meckstroth paused for effect and I glanced at the jury. They had no reaction at all to this, at least none that I could read. They all kept doing whatever it was they did while sitting there all this time—doodling, playing with paper clips, looking around the room.

Meckstroth then launched into a long Q and A with Mellenberger about farm management, milk/herd production, nutrition, reproduction and health. He asked for his analysis of production rhythms, with its seasonal ebbs and flows, all its variables, both generically and in relation to my farm. Mellenberger complained about my barn's slatted floors, my dry-cow facility and my breeding methods. Then he said that stray current would be the last issue of his concern. He would consider it if, and only if, it reached an "objectionable" level, above his "level of concern." *His* level of concern, that was, not the animals'. Which, to hear him tell it, certainly, definitely and absolutely had never been reached on my farm. Ever.

"Dr. Mellenberger," Kurt Meckstroth said, going over and picking up an exhibit and taking it to the witness. "I'm handing you Exhibit 448. Who are some of the authors and what is that document?"

Mellenberger leaned forward and placed his lips close to the microphone. "In the early 1990s, Allan Lefcourt of the USDA published the Red Book, *The Effects of Electrical Voltage and Current on Farm Animals.*"

It was also known as *Agriculture Handbook 696.*

"Do you consider the authors—Drs. Aneshansley, Gorewit, Gustafson and Stetson, among others, to be authoritative in the subject of stray voltage?"

"Yes, I do."

"Are they authors of portions of Exhibit 448?" Meckstroth asked.

"Yes, they are."

With their wrong-headed, destructive theories, this unholy quartet of Aneshansley, Gorewit, Gustafson and Stetson was responsible for much of the harm done to animals and farms throughout the dairy states, so many of them ravaged and ruined by electric current polluting their good earth. Gustafson was on our side early on in his career until the utility money started rolling in; over the years, the more he testified for the utilities, the more current the cows could handle. That really went for most of those university people making large sums of money. These university men, their studies underwritten and funded by the big utilities, had established a ridiculously high "level of concern" to suit their corporate benefactors and free them of guilt. According to the professors—false professors, to my way of thinking—any current beneath their artificial, inflated level had no effect and could be ignored.

Meckstroth went on to show the witness several more exhibits containing research papers written by still more university experts "proving" that current posed no problem if it was beneath their prescribed level. He was trying, I thought, to prove his contention with a quantity of material rather than its quality.

Satisfied that his expert witness had now blessed research favorable to WPS, Meckstroth returned to asking Mellenberger about all the deficiencies and mismanagement he claimed he'd seen on my farm.

"In making your assessment of Mr. Allen's farm, what period of time have you looked at?"

"1976 to 2002."

"What is Exhibit 1781?"

"The Allen annual milk production year by year. I needed to see what happened over time as part of my analysis."

"What conclusions have you drawn?"

"Well," Mellenberger said slowly, "as you can see, there's years it goes up and down. With the death of Russell's sister Geri Lynn in 1991—who had a lot of responsibility for herd management, specifically lactating cows—somebody else had to assume responsibility. Therefore, we see a drop when that occurs.

"And then from '91 until '98 or '99, there's a steady increase in milk production."

"Dr. Mellenberger, do you have an opinion as to whether the annual milk production is indicative of a herd affected by stray voltage?"

"If we take," Mellenberger said, "just going from 1976, that increase is a little over 8,000 pounds, especially when we get to 1996 for the almost 8,000 pound increase—That's more than a doubling of milk production in those 20 years. And I wouldn't expect that if there had been objectionable current or voltage accessing these cows."

Meckstroth went back over to the jury box and stood near the end of the rail. "What about the period after 1996? Do you have an opinion to a reasonable degree of certainty as to whether production during that time is indicative of a herd affected by stray voltage?"

"It has not been affected by stray current after 1996 either."

So simple for him to assert this when my animals had died. It took no effort at all.

"Have you looked at milk production for the period 1996 to 2002?"

"Yes, I have."

"Dr. Mellenberger, I'm showing you a blowup of Exhibit 1782." Meckstroth put it up on the monitor. "Is that the graph you prepared that relates to the period 1996 to 2002?"

"Yes. It goes from January '96 through December 2002."

"Why did it do that?"

"The main reason is that Mr. Allen started using BST on a continuous basis."

Those letters could've stood for *bullshit talk* coming out of the witness chair. Meckstroth turned slightly to better see the jurors and asked the witness, "What did milk production do in 1997?"

"Decreased."

"Why?"

"Essentially," Mellenberger said, "he stopped using BST. Also in '97 there was a major crop failure, a winter kill in '95 to '96 which affected alfalfa on a lot of dairy farms including Mr. Allen's."

Still standing by the jury, Meckstroth folded his arms across his chest and leaned back against the rail. "Let's turn to 1998. What did the production do on Mr. Allen's farm?"

"It increased substantially."

"How much?"

"Twenty-three pounds of milk per cow per day."

"Why?"

"Several reasons. If we allow for BST, about six pounds of increase is due to it. I'm talking about how much of the cows that received BST contributed to the whole herd. In early '98, Cow Group 3 started milking three times a day. Group 2 was already on three times. So we're going to get an increase. That's four pounds approximately.

"There was a change in ration in April, an improvement, that brought the high cows up six or seven pounds more. The forages Mr. Allen harvested from his spring seeding were absolutely excellent. Really good material."

We had the 30 pound increase before we harvested the 1998 crop, we could have gotten them on that one.

"Were there any other factors that affected production in 1998?"

"Some factors are a little harder to quantify."

I wanted to shout up to him, *Like a dramatic lessening of stray current due to the isolation transformer?* I stopped myself, but it wasn't easy.

"The dry-cow facility," he went on, "was renovated at the end of '96 and we've had a full cycle of cows going through the new setup. Four groups of lactating cows in the east barn, and the dry cows and prefresh maternity over in the old barn and in the shed. They made a lot of renovations there having a potential impact by 1998. I think we have other intangibles, harder to measure, but allowing the cows to perform better."

"In your opinion," Meckstroth said, turning and looking straight at the jury, "was stray voltage a factor in the milk production that occurred on Mr. Allen's farm in 1998?"

"No," Mellenberger said flatly. "It was not."

Meckstroth paused. Then he said, “Let’s turn to 1999. What did Mr. Allen’s herd produce in 1999?”

“By the fall of 1998 it increased on a per- cow basis to 68 pounds of milk, the highest ever. And we see another increase in ’99 over ’98, from 19,942 pounds per cow up to 23,005 annually. On a daily basis, 55 pounds in 1998 and 63 pounds per cow in 1999. Also, they started milking Group 4 three times a day. Therefore, we’re getting wider improvement.”

Kurt Meckstroth nodded. “In your opinion is the production increase in the spring of 1998 indicative of a stray voltage fix on the Allen farm?”

With a sly smile, Mellenberger said, “No. The current had nothing to do with it.”

I was anxious to hear just how he would skate around the questions posed to him by Barry Hammarback.

Meckstroth again paused. “Now, Dr. Mellenberger, were you provided production information surrounding the dates of December 31, 2001? And with a document marked as an exhibit at Dr. Johnson’s deposition pertaining to the December 31, 2001 date?”

“Yes.”

“Did you determine whether there was a production change after December 31?”

“Yes. I plotted the total shipment that went out of the tank. There wasn’t any difference.”

“What was your understanding of what occurred on December 31, 2001?” Meckstroth asked.

“My understanding is that Mr. Allen cut some grounds around his farm.”

Yes. And a Happy New Year it was.

“In your opinion, did cutting those grounds have any effect on the herd’s milk production?”

“None!” Mellenberger exclaimed.

No, not at the instant I cut them. But once the current lessened, and the animals had a chance to heal, production increased, but it took time.

“Dr. Mellenberger, what is a turnover rate? Have you calculated any for the Allen farm?”

“Turnover rate,” he said, “would be all cows that leave the herd in a particular year for whatever reason divided by the average number of cows in the herd for that year. If, like Mr. Allen, purchasing a lot of animals from the outside, it’s not unusual to have it go up to 40 or 45 percent.”

"What is DHI?" Meckstroth asked.

Mellenberger sipped some water, set the glass down and said, "Dairy Herd Improvement. Production testing. The farmer would test milk from his cows every month and DHI would do calculations, maintain records and send them back. It's done two ways. Farmers can take their own samples and send it into DHI with all the information on the herd: When cows calve, dried off, left the herd. It's called 'owner sampler.'

"Supervised testing is where they send somebody to the farm on a monthly basis, if the farmer wants. The supervisor takes the milk sample and they get their records from the farmer and put it on the barn sheet."

"What type of testing did Mr. Allen use?"

"Supervised, starting in May of 1989. The last test date was for January of 1996. So he had a seven-year time period that he used DHI."

"Did Mr. Allen name his cows?"

"They were numbered. To me that's a name. Starting in 1996, we don't have DHI. With a straight face he continued, "I don't have cow numbers. Patti Bowers kept a notebook where she entered cows that died or were sold, when they calved, got vaccinated. So I've got Patti's information. I've got some DHI. I've got sales and accounting records, when cows left the herd after July of '97. I had an individual computer sheet off the herd master system that said here is the cow. Here's her background, here's when she left the farm. Now I have individual pages. Milker communication sheets had cows on it once in a while that left the herd."

Meckstroth straightened up, buttoned his suit jacket and walked toward the witness stand. "Was the information from these sources consistent?" he asked.

"No." Again, he spoke with a straight face. "It was not. At times cows leaving the herd varied as much as 60 cows or more. It was very difficult. That's why I had to collect all information, because I kept getting different numbers from every source."

What did he want, tea and sympathy or a medal? He'd already collected his megabucks, so let that be enough.

"What conclusions have you reached with respect to turnover rate?"

"Like I said, Mr. Allen had been purchasing a lot of heifers or cows. It doesn't surprise me when the culling rate gets up to 40 percent or greater."

Did it surprise him that I was forced to cull pitiful, emaciated animals wasted by electricity pumped in by his employer? Probably not. But he had to make believe.

"Did the turnover rates for the Allen farm indicate the farm had been adversely affected by stray voltage?"

"Voltage currents had nothing to do with it."

No, of course not.

"Dr. Mellenberger," the attorney said, going over and picking up an exhibit and presenting it to the witness. "I'm handing you Exhibit 1790. What is it?"

"It's a summary of health events in 1997 to 2002, from two sources. One is from the Allen records, from their herd master computer records, and Patti also kept a handwritten record of individual cows.

"So I made a history. The top part is from Allen records. The bottom events, Dr. Homan, the vet, started in fall 1997. That part is from his notebook."

Meckstroth looked up at the exhibit, which was displayed on the screen. "If you look from 1999," he said, "it shows there are 22 pneumonias according to Patti Bowers' records. How much pneumonia does Dr. Homan's records reflect for 1999?"

"He lists 37."

"Do you have any explanation for why the numbers are different?"

Mellenberger shrugged. "Just didn't get entered on the herd master records."

"What's a DA?" Meckstroth asked.

"Displaced abomasum," Mellenberger explained. "It's where the fourth compartment of a cow's stomach flips either to the left and twists or it can flip to the right. Most are to the left. And, simply, it kinks off the intestinal track."

"How many DAs did you count from Patti Bowers's records for 2001?"

"Only four." Again, with a straight face.

"And from Dr. Homan's notebook?"

"Fifty-nine."

Kurt Meckstroth replaced the exhibit he was holding and walked slowly back toward the jurors. "Dr. Mellenberger, how would you rate Mr. Allen's management decisions as reflected in the material you've reviewed?"

"I would say average. Some good, some bad."

"What decisions would you characterize as bad?"

Mellenberger rubbed his chin and licked his lips. "Overcrowding the dry cow facilities, not adding more waterers to the east free-stall barn, not changing the stalls in the east barn."

He sure is something else. Each group had 10 acres of dry lot for exercise and to lie down—what more can you ask for? Everyone is trying to duplicate that kind of cow comfort. It was his story.

"He could disagree—I do not like slatted-floor barns. It's hard on the feet and legs of cows. Unfortunately, the slats are there, so it's tough to take care of them. On the feed, I didn't like the feed that was put up in '97. It wasn't very good feed."

Isn't that something. It's the same feed that we got the 30 pound increase.

"In 1995, Mr. Allen decided he did not have the facilities to raise his own heifers. All of his replacements in '96, '97, '98, '99 and 2000 are purchased from outside. In 1998, Mr. Allen makes a decision he's going to start raising heifers again. They make a decision to save heifers from some of the better cows, which Mr. Allen did not have the ability to do because he did not have the records and you can't do that. You don't know what a heifer is going to do until she calves. Then you will know how good she is.

"Then we have bred heifers coming back to the west barn. Those are housed with the dry cows. The day of our visit, June 27, 2002, there were at least 110 animals in there of bred heifers and dry cows. It was so crowded we couldn't even count."

He was really rolling now, a runaway train. All groups had ten acres of dry lot to lie down and exercise. Every pound that every cow produced went to the computer and was automatically recorded. Selected high-producing cows with good type were bred with the best proven bulls, and those were the offspring we were keeping for replacements. When I had WPS' downgrounds cut, some of the cows were producing 150 lbs per day. Bob got down to treating 3 cows during that period; that was a real joy to watch that I will never forget.

"And then," he went on, "we have a mixture of sick and fresh cows. I don't like bedded packs because of bacterial exposure where everybody's using the same area. So the dry cows' management has become a problem. It was okay in '98 and '99 because they didn't have the overcrowding, didn't have bred heifers in there. That's one of the reasons production is coming down these last few years."

This made me so angry I wanted to run up to the witness stand and start screaming, *None of this mattered before your bosses reconfigured the electrical distribution system—we thrived until then, "bad" management and all. We never had bedded packs.*

Meckstroth asked, "How would you rate Mr. Allen's supervisory skills?"

Mellenberger was silent for several moments. He stuck out his lower lip. "Things in the milker communication sheets concerned me," he said finally. "Seeing manure pits and slatted floors overflowing. Sometimes that comment occurs for a week, sometimes two. There's comments that some of the waterers sit empty for a week or two, or are overflowing. Seeing cows getting hung up on post railings or breaking through the slatted floors. I saw things that should have been taken care of.

"Whose responsibility is that? Ultimately, Russ Allen. He's the last man standing."

That was perfectly true even though my farm was not perfect, nor was I. Nor was anybody else's farm. There were always improvements needed and they were my responsibility. That was just life as we know it on a dairy farm. However, the point was that all of these things were manageable and wouldn't have been big issues if we hadn't become overwhelmed with sick and dying animals as a result of WPS' actions.

Next, Meckstroth played videotapes that Mellenberger had made in June of '02 at my farm. The witness pointed out what he considered defects in the free-stalls, their dividers, the neck-rail pipes, and cow housing and milking routines.

"Could you please explain to the jury," Meckstroth said, "what you saw the cows doing during that last milking?"

Mellenberger nodded happily. "On the milking routine, they take a foamer—this has a brush to add iodine to clean the teats off. This is where the teats are first touched and the cow is what you would label as 'kicking'. I could call it 'steppy,' 'dancing,' whatever. But this occurred essentially on every cow they milked either midday or night."

"What is the cows' lifting of feet when their teats are striped?" Meckstroth asked. "What does that indicate to you?"

Mellenberger sighed loudly and dramatically. Maybe he was auditioning for a part in a Broadway play.

"Pain!" he said. "Painful. It's the bottom part of the teat. It's like the first joint on your finger. If the vacuum equipment is not functioning right, too high, too low, every cow shows the same reaction."

"In your opinion," Meckstroth said, "has stray voltage contributed in any way to the behavior that we're seeing on this tape?"

Mellenberger shook his head furiously. "Absolutely none!"

Now, that was what I called earning one's pay.

"Dr. Mellenberger, in your opinion what are the milk production limits on Mr. Allen's farm?"

Mellenberger sighed, again loudly and theatrically. "I guess at the top of the list I would put cow comfort. I think a lot of the increase in diseases Mr.

Allen has seen is directly related to management of that barn. I would also list the heifer program. The Davister farm where calves were being raised, there were two different farms. Mr. Allen's were on another farm, but looked fairly sick to me. They had extreme diarrhea. Yet at one point they get put into a general building. And there's quite a bit of coughing in the barn.

"Maintenance of milking equipment is a concern. I would like to see more water. Right now, as a disease problem, I see mastitis, metabolic disorders. Diseases in the fresh cows are of real concern."

Still standing at the jury box, Meckstroth said, "Dr. Homan testified about records on the farm. Do you believe that records are limiting the Allen production?"

"Well," Mellenberger replied, "it's hard to say they are limiting production. But they are certainly limiting Mr. Allen's ability to make daily decisions on cows. Dr. Homan got frustrated because he couldn't get information that could help Mr. Allen make management decisions."

"Dr. Mellenberger, Dr. Behr testified that he believes Mr. Allen would produce at 50 percent or more above the Wisconsin average if not for stray voltage. Do you agree with that?"

Mellenberger shook his head. "No. I disagree with that."

Naturally, he did.

"Why do you disagree?"

"Because of the facilities and management decisions made or not made. That just won't allow Mr. Allen to be at 50 percent above the Wisconsin average." According to Mellenberger I can't do anything right, but when I reduce the current to the cows, I reach 85 lbs per cow per day. If I would have been able to keep the current away from my cows or if WPS would have fixed their system, I would have had the best herd in Brown County, and someday I will.

Meckstroth ended his direct questioning at this point and court recessed. I didn't think Mellenberger had hurt me badly yet. True, by offering an alternative theory for the cows' unruliness he'd scored some points, but there was no knockdown. And I thought it had been a mistake for Meckstroth to remind the jury of Dr. Behr's testimony about how stray voltage impacted my production, especially there at the end where it could linger in juror's minds. But only time and the jurors would tell if that was right.

When the court session resumed, it was time for Mellenberger's cross, and I was hopeful that we could wound him. Although he was a heavyweight witness with a big reputation in academic circles, there were indications that he might be vulnerable. It was for Barry Hammarback to find weaknesses and exploit them.

"Was it your testimony," Barry said, going up near the witness stand, "that half the time you spent so far in this case was sorting the farm records?"

"No."

"What percent would you say was allocated to sorting?"

"Probably 10 to 15 percent."

At his rate, that meant he'd charged WPS $15,000 dollars for shuffling papers. The WPS comptroller must've been thrilled.

"And the biggest thing was that it wasn't simply all nicely laid out on a spreadsheet?" Barry said disdainfully.

"Well, not that information, because you have accounting records and everything else."

Barry chuckled. "You said it would be nice to have records so you could analyze what you're doing, but I don't think that's a limiting factor, right?"

"That's correct."

"When you go through the list of what you consider limiting factors and get to the bottom," Barry said, "you have stray voltage down there as one of the things you need to rule out when there's a production change, right?"

"Correct."

"There's no doubt," Barry said, "that from the first part of 1998 up to 2000, the production in the Allen herd increased dramatically?"

"That's right," Mellenberger said. "It went up 28 pounds from January of 1998 to December of 1999, 28 pounds per cow per day."

Barry stuck his hands in his pockets, nodded, and looked down at the floor. "Would you agree that nothing like that increase had ever happened before in the history of the farm?" He looked up at the witness.

"The only time," Mellenberger told him, "was 1976 or 1977. The increase you're talking about was 22 percent."

This was just before WPS reconfigured its distribution system. What a surprise. This, I thought, was a telling point, and I watched the jury to see if they had a reaction. The lady who wore the floral print dresses and skirts was nodding and writing on a pad. I took it as a good sign.

"Now," Barry said, "you'd agree with me that actually putting together a ration is a bit of an art?"

"It is."

"You can't just look at one item or another and come up with any specific conclusions about it by itself, can you?"

"No. We have to set parameters to go by to make the ration."

"And you have never seen any, for example, length of cut because you haven't seen any haylage on the farm until you were there, right?"

"That's correct."

"You don't have any criticism of Mr. Root's ration, do you?"

"No, I don't," Mellenberger told him.

"I want you to assume," Barry said, "that Mr. Root testified that the ration did not in his opinion cause any production increase after 1998. Are you in any position to disagree with him?"

"I'd say the forages were good, but I can't disagree with him."

Another telling point. It wasn't the ration that increased production; it was the lessening of current, pure and simple. And no amount of verbal gymnastics could change that fact.

"The next thing," Barry said, "is that in order for there to be an increase in the protein going through the herd, you need to purchase more protein. And the purchased protein actually goes down from 1997 to the present, doesn't it?"

"That's because the forages got a lot better."

"Just answer the question if you would, Dr. Mellenberger. Isn't that true?"

"Yes," he said quietly. "It went down."

"Now," Barry said, "you made an analysis of all the feed purchased on that farm. And what the content was of all the rations, that is the feed tests. But in all the time you spent in this case, you've never done a spreadsheet that summarized changes in the ration, have you?"

Mellenberger's eyes narrowed. "That was not my role in this case."

Barry nodded. "And so if someone looked at the rations, they should be able to determine whether or not the content changed enough to change production. Wouldn't you agree?"

"That's correct."

"And you have not done that?"

"No," Mellenberger admitted.

We'd scored. Barry was on his game. I leaned forward so as not to miss a word.

Barry left the area of the witness stand and moved across the courtroom until he stood halfway between McKay's bench and the jury box. "Let's talk about stray voltage for a moment," he said to Mellenberger. "Your opinion is that stray voltage could not have had an effect on the Allen farm because you don't know of any measurements of 3 milliamps or more in a cow contact?"

"That's correct."

"If," Barry said, "I told you to assume that in fact there was evidence in this case that there were four milliamps or more measured at a cow contact area in this farm, you would have to take it into account, would you not?"

"The behavior only."

Barry grinned. "Behavior can be a serious problem for dairy animals if it deters them from drinking, can't it?"

"That's not a behavior," Mellenberger snapped in his best professorial, smart-ass tone. "That's avoidance. The behavior is, does the cow react to something versus avoiding something?"

I failed to see the difference; this was just more verbal gymnastics. But then, I was only a farmer, not a professional witness, one who manipulates words and ideas like puppeteers manipulate strings and marionettes.

Mellenberger continued in his hectoring, lecturing tone, "So that's why I make the distinction that once you get past 6 milliamps, then you'll get a range of cows that will avoid something like water, like feed. But at 4 milliamps, you've just got cows knowing there's something there, but they're not avoiding it. And so they're not going to avoid water. Not going to avoid feed, and so production is not affected."

It was truly miraculous how this man could read bovine minds. He was like The Amazing Kreskin. How the hell did he know what the cows thought and felt? Talk about arrogance.

"Your statement is based on the studies you've read, right?"

Mellenberger nodded several times. "All the way from the 1960s to the present."

"And that includes Exhibit 82 here, which is studies showing the effects of neutral to earth current on behavior, production and water intake in dairy cattle?"

"I've read that, yes."

"Okay," Barry said. "If we look at Exhibit 82, what it says on the front is that no significant difference in water consumption, feed intake, milk production or concentration of fat or protein in the milk was found. But it says drinking behavior, number of drinks per day and time of drink did change significantly. Isn't that right?"

"Yes."

Barry crossed his arms in front of his chest. "In fact what this study shows is that even at half a volt, drinking behavior and drinks per day changes significantly?"

"There are some cows that will respond to that. But they still go to the water," Mellenberger told him.

"In fact," Barry said, "this study shows that actually voltages as low as half a volt can reduce the cow's production, doesn't it?"

"Not that I recall," Mellenberger said sourly.

Barry kept up the pressure. "If you look at page 6 for animals receiving .5 volt treatment, there's a significant decrease in the number of drinks during the time the animals were receiving voltage, right?"

"That will happen with the most sensitive cows."

That will happen with *all* cows. What made Mellenberger and those like him think a little bit of electrocution was a harmless thing? They *torture* people with the stuff.

"If we look at the conclusions of the study," Barry said, "some animals will not drink from an electrified water bowl as evidenced by the heifer and cow that refused to drink for 36 hours during the 4-volt treatment. Right?"

"That could happen on sensitive cows, right."

I groaned loudly when I heard this B.S. and shook my head in disgust. Several people turned and looked at me, including the bailiff, who put his finger to his lips. But what did I care? Sensitive cows, my butt. I'd be half dead if I didn't drink for 36 hours, and I weigh 170. What does lack of water do to creatures weighing 1,400 pounds? All animals, not just "sensitive" ones, are made of tissue and nerves and cells. What the hell was this guy on the stand blabbering about?

"Item number 5 of the study," Barry told him, "says there is a trend toward more animals having significant reduction and there being larger reductions at increased voltage. So, clearly the trend was that it would affect production. Right?"

"In some of the cows, yes," Mellenberger said softly. I could see he didn't like having to give this answer.

"Common sense would tell you, wouldn't it, Dr. Mellenberger, that the higher the voltage accessing the animal, the greater the effect on milk production?"

Common sense? What was that? If common sense had been at work during any of my ordeal, I couldn't have found myself in this courtroom.

"When it gets high enough," Mellenberger replied, "that's right."

Barry nodded and slowly made his way toward the jury box. "Is it fair to say that if your definition of stray voltage is high enough before it has an effect, you're simply never going to be able to find stray voltage as being a problem?"

Not only was it fair, it was dead center accurate. It was at the heart of the case. Here was the shell game the utility always played.

"Not true," Mellenberger said testily.

"It has to be 3 milliamps?" Barry said.

"If I want a response I can see, I've got to be between 3 and 6 milliamps. Above that, I've got extreme or moderate behavior changes, plus, I'm going to get some effect on cows' health and production."

Barry stared at him. "Let me ask: If you go on a farm and see reduced production, erratic animal behavior in a parlor, strange drinking behavior, are you going to investigate stray voltage?"

"I have in the past, yes. Many times."

But did he ever find it after all his investigations? I think not.

"And if you go out and make a measurement and the voltage levels are less than 3 milliamps, are you just going to dismiss it?"

For certain that's what he would do, as per WPS standard practice. Equally certain, he would deny it on the witness stand.

"Not necessarily, no," he said. "Because you would like to think that on any farm as low as you can go reasonably. And yeah, we'd make that assessment all the time. I've done that many times."

What the hell did that mean? But I had to hand it to him. While avoiding Barry's question and saying absolutely nothing, Mellenberger managed to sound authoritative and convincing with such gibberish. I just prayed that the jury saw through him.

"Would you agree," Barry said, "that you may see behavioral reactions as low as half a volt, based on these studies?"

Now Roger Mellenberger hesitated before answering. "In the most sensitive animal, that's a probability. Because if you look at the Red Book, you have perception at those levels on some cows, only five percent of the cows. At half a volt, there is a perception."

Here again was The Amazing Kreskin reading the bovine mind. He continued, "Whether it changes into visual behavior, that's a question. You would have to be there on that farm at that time."

Inching closer toward the jury box, Barry paused. "Every farm is unique, isn't it?"

"Every farm is unique," Mellenberger agreed.

"Are there," Barry asked, "any studies at all on traditional or nontraditional stray voltage as it relates to a free-stall barn?"

"No, only laboratory research."

Which bears as much relation to the numbers and material conditions of animals on the farm as a pebble does to a planet.

"Cows have only their social behavior, don't they?"

"Yes, they do. If they're moved from group to group, they have to re-establish where they stand in the pecking order."

Barry began slowly to pace. "And so a laboratory condition where five or 10 animals may be cooped up in a pen is going to be a lot different than 100 animals in a free-stall situation?"

"That's true. But it's more controlled, so I can measure what I would like to know."

"You've got two choices," Barry told him. "You can either control all the variables like in the lab, or use a whole lot more animals, on farms."

Mellenberger agreed with this.

"And so," Barry said, "from a lab standpoint there aren't any stray voltage researches ever done in a milking parlor, are there?"

"That's probably correct."

"Would you agree," Barry asked, "that there's only one study that's ever been done on the effect of isolation on production? The University of Minnesota study?"

"It was never a controlled field study. They looked at the data."

"What they did," Barry said for the jury, "is go out and take 84 farms in Minnesota, right? Averaged about 50 to 70 cows a farm, I think, 4,000 or 5,000 animals. What they did is check the conditions before isolation for one year, and the conditions after for one year. Right?"

"Not true," Mellenberger contradicted. "All they looked at was the production records from DHI."

"All right," Barry smiled. "So they compared the records and you think those are good, don't you?"

"Yes."

"And they did that for a year before and after isolation. And there wasn't a single cow contact measurement made?"

"Never any measurements made."

Barry went over to the exhibits. "What they found," he said, "if we look at Exhibit 1062 here, is that several of the herds had a rapid change in production after isolation, for the better. Right?"

"It went about 50/50," Mellenberger said, grudgingly.

"Okay," Barry said. "I think 47 of those herds showed a substantial increase after isolation?"

"Yes," Mellenberger said. "About half."

"Okay. At the bottom of 1062 here, we see the graph of all the herds compared to the normal DHI trend line, right? And then at the top, we see the reaction of those high response herds?"

"Correct."

"Their angle of increase in production is surprisingly similar to the Allens', isn't it?"

Mellenberger waved his arm. "That's a stretch, sir. You don't have the same conditions. They never measured the voltage. Didn't even know what they were isolating. And so I've always criticized this study because they never had any controls. They never studied the farms so I could say, here's the situation before, here's after. Nothing changed except the voltage, and we don't even have the voltage levels."

The point was this: The farms had isolated and production shot up; end of story. No academic mumbo-jumbo could hide it.

"What you *do* know," Barry said, "is that the voltage did change on those farms that were isolated. Right?"

"Potentially."

What an answer! You had to love this guy.

"As a matter of physics," Barry said, "when that neutral is separated, the current going onto the farm will be reduced, shouldn't it?"

"Should go down, that's true."

"And so the only relationship from an electrical point of view to those high-response isolated farms is that the current was lowered?" "Yes. You could make that conclusion."

No other was possible.

Barry now came back over toward the witness stand. "You've testified on, probably, over 100 of these cases, haven't you?" he asked Mellenberger.

Mellenberger smiled sheepishly. "No, no. You've got me way too high there. Probably the max is 30 cases."

"It's over 100 you've looked at, then?"

"No. High '60s, '70s."

Suddenly, Barry whacked him, clocked him with a hard shot. "On a stray voltage case for the utilities, you've *never* had the opinion that stray voltage caused a problem. Have you?"

Truer words were never spoken.

"Yes," Mellenberger rushed to say. "At this point in time."

"You've got a new one now?" Barry asked skeptically.

He said he did. Mellenberger was the name, paid testimony the game.

"What kind of voltages are the cow contact there?"

"Running as high as five volts in the milking barn."

"No doubt that caused a problem?"

"They didn't have it bonded all the time, so at least milking in that situation the farmer didn't have a problem."

Barry studied him for a moment. Then he began to pace back and forth in front of the witness stand. "Back in 1989 or 1990 you were still at Michigan State," he said. "And in 1989, you billed 1,100 hours at least on stray voltage cases, didn't you?"

"I don't keep track of that," Mellenberger actually said with a straight face.

"Dr. Mellenberger, do you think that if you had a lower level of concern it would affect your employability with utilities?"

Did the sun rise in the morning?

"That doesn't factor into anything," he said, suddenly testy again.

And why not? Barry was zeroing in on his cozy, cushy little arrangement.

"Six milliamps," Barry said to him, "is the same current level it takes to trip a circuit breaker in the bathroom, isn't it?"

In other words, it was one hell of a jolt. Sadly, I thought of it constantly hurting my poor cows.

"I don't know what they're set at."

"Let's talk about the use of BST," Barry said. "Actually, the use of BST in 1996 didn't do anything for the Allen farm. Did it?"

"Yes, it did," Mellenberger insisted.

Barry quit pacing and looked hard at him. "You said they went on BST in May or June?"

"June 5."

"And they were at 51 pounds?"

"That's correct."

"And," Barry said sarcastically, "they went through that period of time and managed to get *all* the way up to 52 pounds. Didn't they?"

"Yes. In the space of a normal summer drop."

"And," Barry said quickly, "if there's stress on cows, they don't do well with BST. Do they?" He was working Mellenberger over, with quick left-right combinations to the body.

Mellenberger threw up his hands, exasperated. Barry was definitely getting under his skin. "Depends!" he almost yelled. "Heat stress is the only one that's been studied. There's some indication it could have a negative effect, but it's not as great as if they weren't on BST. BST moderates that effect."

"So," Barry said, "for whatever reason, BST did not gain them a pound in 1996, did it?"

"Yes," Mellenberger stubbornly insisted. "It did."

Walking toward the jury, Barry stopped in mid-stride, wheeled around and looked at the witness with his eyes wide. "One pound?" he said, with a curious smirk.

"No, about four."

Still smiling, Barry said, "So it's your testimony here today that the reason they stayed at 47 to 50 pounds or more is simply because they were using BST?"

"That's correct. The Group 3 went from 45 pounds to 51 pounds."

"All right," Barry said impatiently. "So how many pounds is that?"

"Six"

Barry grinned and shook his head. "Okay. If you take that six pounds herd-wise, what you get is about three, don't you?"

"No!" Mellenberger snorted. "We've got more than Group 3 involved. And they're not going to be the group potentially responding the most."

"But BST, we've heard here—I would have to ask you to assume that BST is only used on about 240 days or so?"

"Yes."

"So," Barry said to him, "there's a period of time where the cows are not on BST and it wouldn't be an increase in production?"

"The first 60 days or so. That's Group 1 in this case."

"So, if they're producing six pounds once you take that out through the cow's whole lactation, it turns out to be three or four. Doesn't it?"

"That would be true," Mellenberger had to concede.

No, it wasn't the BST at all; it was the effects of reduced current due to isolation, and nothing else. And though he would never admit it, Roger Mellenberger knew this as well as I did.

Barry nodded with satisfaction. He'd made his point. "Isn't it true," he said now, "that the longer a cow produces milk, the more overall milk she's going to put in the tank?"

"For a short term."

"So if her dry period is short, that lengthens the last part of lactation and allows more milk to be produced?"

"And cheats the next one."

"Let's take a look at the average," Barry said. "If you look at average dry days from 1989 through 1995, there's nothing that should change production very much, is there?"

"You can't look at the average," Mellenberger replied.

"Okay. If you look at 1996 through 2002, the one that seems to stand out is 1998?"

"That's right."

"And 80 days is still within an acceptable period for days dry, isn't it?"

"You're talking about 80 as an average for the herd? Not acceptable. That's too high."

"Okay," Barry said, coming away from the jurors and toward the stand. "Sixty-three as an average is okay? Basically, you don't want anything over 70, 75?"

"I don't want anything over the 60s here."

Barry paused. "Will you agree," he asked, "that if current were high enough and it was removed—let me ask you to assume that current was high enough so it affects the cows' feed and water consumption. Removing that voltage could result in a 21-pound gain in production, could it not?"

This was a no-brainer, but Mellenberger was certain to twist the answer into a hundred knots of confusion.

"Sure," he said, stunning me by his honesty. "And quickly."

Was he, I wondered, experiencing a conversion? Nah! He just must've figured there was no way around this one, so why bother trying? Whatever his reasoning, I was delighted with his answer.

"There's been," Barry said, "no substantial evidence of hoof injuries on the Allen farm caused by the slatted floors, has there?"

"I think the records will say yes, there has been, historically."

Barry again went over to the jury box. "Whatever the form of the slatted floor, would you agree that that would not cause an increase of production of 20-plus pounds of milk?"

"No, that would not cause an increase."

"That lid you put on production because of overcrowding, that would apply to their production today, right?" Barry asked.

Roger Mellenberger turned in his chair and took a swig from his water glass. "I'm not sure I put a big lid on it. I'm saying it's one of the points that should be corrected so they could go up in production."

"You can't put specific poundage on what they may or may not do?"

"No, sir."

"All right. When we go from milliamps to volts, we have to use resistance to do that?"

"That's correct."

"So, for example," Barry said, "if I've got 40,000 volts of resistance at 3 milliamps, that's 120 volts?"

"I'll accept your calculation."

"And six milliamps would be 240 volts? Ohms Law. So .006 times 40,000 is 240 volts?"

What Mellenberger is trying to tell this jury, the court, and myself, is that cows can handle 6 milliamps of current when 3 milliamps would be like plugging a cord into the wall, cutting off the other end and hanging on to both wires in the cord. Is this guy insane? I really think the attorney general of our state needs to take a look at these guys, how criminal.

"I'll accept your calculation."

"Mr. Lawrence did it, so I know it's right," Barry grinned. "Two-hundred-fifty ohms times three milliamps would be three-quarters of a volt?"

"Right."

"And so," Barry said, "if there was .75 volts or higher on a cow's contact point—say, the floor—you would expect to see behavioral changes in the cows?"

"Five percent or less and just perceptible," he said with a straight face. "So, nothing you could see visually as the farmer."

The Amazing Kreskin was at it again.

"By the way," Barry said. "On the day you were in the barn, there were a lot of new people milling around. And so in the free-stall barn, it didn't surprise you that the cows were up and moving around, did it?"

"I actually filmed Groups 3 and 4. I was the only one in there for a long time. So the cows you saw standing half in and half out of the stalls had not been disturbed."

"There hadn't been anybody through there taking cow contact-measurements?"

"Not by the time I took my videotape, no. They did afterwards."

"In the milking parlor," Barry said, "you were the stranger in the parlor that day?"

"Yes, I was."

"When that milking machine goes on, it puts about 14-and-a-half pounds of pressure per square inch on the teats, right?"

"No," Mellenberger corrected him. "The actual squeeze force, given the low vacuum levels they had many times on the Allen farm, you have to take out the resistance of the rubber and the inflation."

"How does that relate to pounds?"

"Inches of mercury. It takes a lot more pounds of water pressure to equal inches of mercury, because mercury is heavier. So everything on milking systems is in inches of mercury."

Barry put his hand in his pocket and jingled some coins. "If the teats are sore, their soreness is not simply going away when the milking machine is put on?"

"Yes, it does," Mellenberger replied. "That's what's interesting. Once it starts going through the massage and milking process, you have milk on the inside to buffer the squeezing action. If the machine stays on too long, you will see some cows doing the steppy sort of business. Because there's no milk flowing and the machine is massaging an empty teat. That cow is going to feel pain."

"Would you agree," Barry said, "that there have never been any major problems discovered in the milking equipment on the farm?"

"No, I saw some pulsation failure when Dr. Johnson measured the pulsation grip. The Allen records talked about some of that in their communication sheets, they had units that weren't working: Vacuums were too low. There was always something going on with the equipment that wasn't up to snuff."

"There's probably always a pulsator in 24 units that's not up to snuff, isn't there?" Barry asked.

"That can happen, sure," Mellenberger agreed. "I have to provide enough maintenance to keep that to a minimum."

"Actually," Barry said, "it's kind of normal. "They require maintenance, don't they?"

"On a routine basis."

"If the actual feed consumption of the cows went down after April of 1998, that wouldn't cause production to go up, would it?"

"Well, you would have to look at the total ration and the quality of it."

"Do you expect any change in milk production based on a 2.0 change in protein?"

"You should," Mellenberger told him. "And some of the protein could be used for energy, so it supplements for NEL that's not there. The more protein, I might see a production kick-up at the end of 1998."

"Okay," Barry said. "You've never analyzed what the dry matter intake was from the herd after April of 1998 and on, have you?"

"You can't," Mellenberger said.

"Nothing you can use to do that?"

"No."

Barry turned and went toward his counsel table. "That's all the questions I have, Your Honor," he said over his shoulder to McKay.

Kurt Meckstroth now got up to have a final word with this most important witness.

"Dr. Mellenberger," he said, walking up quickly to the witness stand. "That Minnesota study you were asked about. As part of the study, was any effort made to assess the level of stray voltage present on the farms prior to the time the farms were isolated?"

"There was not. It had to be one volt for them to get isolation. Beyond that there's no information if it was one volt or two or three or four. So it had to be one to get the isolation."

Meckstroth continued, "In reading the article, is there any way to tell how many farms, for example, were at four volts?"

"No, there was not."

"Is there any way to tell from the article what contribution to stray voltage came from any on-farm sources?"

"No."

"And did the article address whether any on-farm electrical contributions to stray voltage had been corrected in conjunction with the isolation?"

"No."

Meckstroth nodded. "Dr. Mellenberger, I wanted to go back to the days-dry distribution chart that you prepared. I think it's Exhibit 1399. It's marked as a separate exhibit. Mr. Hammarback asked you questions about the average days dry. Of what significance is the average days-dry figure?"

Mellenberger leaned forward to get closer to the microphone. "If I'm trying to analyze the herd, the effect of days dry on milk production is very little. I don't use the average at all."

"There's a reference in the exhibit to there being for 1997 in the 30 or less category, 22, and then in parentheses you've got nine. What do those numbers mean?"

"The 22 means there's 22 cows that had 30 days or less dry. And the 'nine' is nine percent of the numbers I have. So, nine percent of the 238."

"Of what significance are those numbers?"

"The 22? Extremely significant because six of those were zero. For the most part, any cows that have zero dry days are 30 percent or more, their next expected lactation is depressed by 30 percent or greater. So it's a very significant number. The 22 cows for their next lactation aren't going to do very well."

"So," Kurt Meckstroth said, "when you say there are six cows at zero, does that mean they calved while they were still being milked?"

"Correct."

"How would you assess Mr. Allen's efforts with respect to the days dry management in comparison to 1998 and 1997?"

"Very good," Mellenberger said. "Extremely significant improvement."

Now the message they were trying to send was that my efforts in this area had raised production, not the effects of isolation. I was a good farmer when it suited them and a bad one when it suited them.

"Dr. Mellenberger, if electricity—stray voltage—was present on a farm at a level above your level of concern at six milliamps, and if that problem was corrected, what would you expect the milk production on that farm to do?"

Mellenberger shrugged. "I would expect the milk production to rise dramatically within two weeks. And I mean *dramatically.*" He said that with a straight face.

Here is the right answer to that question. You may get a little response within two weeks depending on where (waterers, milking parlor, eating, et cetera) and how long the exposure. You will not reach peak potential of a herd for a couple of years. Cows need a lot of time to heal or may be so injured that they may not fully recover if they recover at all. Cows that aborted several times need time to get back into their normal days in milk, or may need to be culled from the herd. What am I thinking? Cows will not survive 6 milliamps of current running through them for more than 10 seconds; they will die.

The shills continued.

"Why would it do that?" Meckstroth wanted to know.

"Because they're going to go back to the feed and water and get maximum intake and, therefore, that's milk. Input in, output out. If I improve the input, the output goes up."

"Would that be true if there was any level of stray voltage causing them to avoid water or feed?"

"Well, if they didn't go at all they wouldn't live very long."

I heard a few chuckles behind me when he said that.

"My question," Meckstroth said, "is if there is a level of stray voltage present on a farm, regardless of what the level is, if it's enough to affect the cows' dry matter and water intake, what would you expect that herd to do if stray voltage were removed?"

"After it's removed, production is going to go up."

Meckstroth said a mouth full: "regardless of any level." Mine did exactly that, but it took 6 months to see the 30 lb increase.

After asking several more of what I considered useless and pointless questions, Meckstroth sat down and Barry got up to take his last licks with the witness.

"When a farm is isolated," Barry said, walking up near the stand, "if there's a voltage problem on the farm it stays on the farm, right? And gets worse, doesn't it?"

"It could."

"And if in fact those farms were isolated and had an on-farm problem, their production would not have gone up, it would have gone down?"

"Depends," Mellenberger told him, "on whatever level was there. That's the problem. We have no numbers."

"If we go back to your days dry—explain that again, please."

Mellenberger adjusted the microphone and spoke with exaggerated slowness, as if he were addressing an idiot. "Twenty-two cows had 30 days, 30 or less days dry. The other side you need to include is that 31 of the 50 days dry has an impact. This means that 26 percent of 238 is 63."

"That's a high percentage?" Barry asked.

"Very high," Roger Mellenberger told him.

"If the cows have been subjected to 6 milliamps of current, unless they are before 60 days they aren't going to come up much on milk, are they?"

"That's right."

"Once the cows are over the peak and are in the downhill slope of their lactation, they're not going to come back up in production, are they?"

"Yes, they are."

"No further questions," Barry said.

If truth be told, Barry's last questions seemed as useless and pointless to me as Kurt Meckstroth's. I thought we left Mellenberger off the hook a little. Now, unbelievably, Meckstroth got up to do a re-redirect, if such a word even exists. Talk about beating dead horses. Or dead cows.

"Why would you not expect the milk production of 1997 to go up because of the days-dry distribution exhibit we have on the screen?" Meckstroth asked.

"Because," Mellenberger told him, "these cows are going to produce less milk. I've got less milk going in the tank."

"Would that affect the average of the herd's production?"

"It's going to go down," Mellenberger answered, and then, finally, mercifully, Meckstroth let him go.

As Mellenberger lumbered down from the stand, I was feeling very encouraged about my chances of winning this case. Roger Mellenberger was a key defense witness whose testimony WPS was depending on to shoot big holes in my case. Although I wasn't a lawyer and knew very little about legal affairs, I knew enough about what had happened with stray current on my farm and all the measures we'd taken to try to remedy the situation to realize that Mellenberger's testimony, as shrewd and knowledgeable as it was, hadn't been so damaging. I remembered Scott Lawrence telling me that expert witnesses tended to cancel each other out, and that the case would likely hinge on my own credibility and testimony.

I would soon have the opportunity of testing Scott's theory, because they were getting ready to call me up to the witness stand.

Chapter Fifteen

IS THERE A DOCTOR IN THE HOUSE?

When Dr. Michael Lane, a veterinarian retained by WPS, entered the courtroom and stood before the bailiff to take his oath, it was all I could do to keep from laughing out loud. The guy was so disheveled—his cheap-looking, Kmart-style polyester pants so wrinkled—that I thought he'd slept on the straw of a milk barn surrounded by a dozen Holstein cows.

Sixtyish, short, balding, lean and fit-looking, Lane taught in the Agriculture Department at the University of California and specialized in treating dairy cattle which, for some reason, he insisted on referring to as "bovine cattle." Sort of like saying "canine dogs" or "feline cats," I guess.

Trevor Will was handling the direct testimony, and I was glad of it; I couldn't take another minute of Kurt Meckstroth's conceit. Dr. Lane at first talked about his work at my farm, which included reviewing depositions, analyzing farm data, touring the facilities and observing the operation. He'd spent about 10 hours on-site and watched the milking routine.

Will wasted no time in getting to the point he wanted the witness to make. "Doctor," he said, "based on the work you've done in this case, do you have an opinion to a reasonable degree of veterinarian certainty as to whether there is a stray voltage problem on the Allen farm?"

I loved that phrase, *veterinarian certainty.* Of course he had an opinion, and everyone in the room, including the mice in their hidey-holes, knew what it was before he gave it.

"Yes, I do."

"And what's your opinion?" Trevor Will asked.

"There is no stray voltage currently and never has been a stray voltage problem on the farm."

While dead wrong, this, at least, was a clear answer, not larded with the usual big-sounding words. Those would come later.

But for the life of me, I couldn't take my eyes off of Lane's baggy, saggy, rumpled and wrinkled pants that you would be embarrassed to donate to the Salvation Army. I couldn't believe he'd actually wear junk like that in public, let alone in a court of law. I wondered what Trevor Will, always dressed so elegantly in his thousand-dollar suits, thought about it.

Having gotten across his key point, Will next asked Lane to describe his background. A Penn State graduate with an engineering degree, he'd gone on for his MA while working at General Electric, and obtained a degree in veterinary medicine from the University of Florida in 1980. He then worked as a teacher at the University of Idaho. He was their in-house clinician, using local farm animals to help train students. Working with local communities to design herd health programs, he spent the majority of his time analyzing milking machines and establishing disease control processes.

After four years in private veterinary practice, where he did vaccinations, surgeries, dehornings and pregnancy testing, he took a position with the University of California. Lane went on to list his various board certifications.

"What do you charge for your consulting services?" Will asked him.

"One hundred thirty-five dollars an hour. On top of that will be expenses such as travel, lodging, food, car rental, lab fees, et cetera. which can be extensive."

None of these people came cheap, that was for sure.

"Is there any one client you've done a great deal of consulting for?" Trevor Will asked.

"The bulk of my consulting," Lane said, taking a quick sip of water, "has been one client in Michigan, Consumers Energy."

He was a utility man, and they had no end of stray voltage denial work for him. Over a 10 year period he'd worked on more than 200 cases for Consumer's Energy—now tell me there is not a problem with the multi-grounded neutral distribution system, using the earth to return their current back to the substation.

"Have you," Trevor Will asked him, "ever seen stray voltage on a farm?"

"Yes, I have," Lane said, surprising me.

"How important is nutrition in terms of milk production?"

"Nutrition is the cornerstone of milk production. If we think of the cow as a biological factory that makes milk, the components have to come out of the feed. So if the feed is not right, the cows are limited in milk production."

Trevor Will walked slowly over toward the jury, his three-piece, black pinstriped suit jacket neatly buttoned. "What role does water play in milk production?" he asked Dr. Lane.

"If cows do not intake the water they need, they would not be able to eat as much dry matter. Dry matter is what makes milk, contains all the building blocks of milk."

Will nodded. "Why doesn't water make milk?"

"Well, when the components of milk are taken out of the blood and put into the mammary gland, the particles that are in the mammary gland are what draw the water into the milk. So milk is 12 percent solids. And as those solids go into the mammary, it sucks water out of the blood stream so that 88 percent water surrounds those 12 percent solids."

Remaining beside the jury, Will asked, "Do you have an opinion as to whether nutrition played a role in the production changes on the Allen farm?"

Lane nodded vigorously. "Nutrition played a very big role. There was a bad feed component, low quality, through the last half of '97 and the first two or three months of '98. As that feed diminished, production went up."

"What feed components," Will asked, "can affect milk production?"

"The major thing is fiber levels. And the forage is the cornerstone of our nutrition program. Nutrition is the cornerstone of milk production. And forages are the cornerstone of the feed program.

"The cow," Lane continued, "is designed to eat forage—grass and hay. However, we can add grains and make the ration more high-energy. But we can't add too much or the cows will get sick. So we constantly monitor the fiber in the ration to make sure it's adequate, but there's not so much that we affect the turnover of the feed stuff."

"What's turnover?" Will asked.

"You've all heard the term 'a meal that stuck to your ribs'? That means you don't need to eat very often that day if you've had a meal that sticks to your ribs. That's not what we want with dairy cattle. We don't want a meal that sticks to their ribs. We want a meal that's highly digestible so they're hungry again and eating again and taking in more of the building components of milk."

"Why," Trevor Will wanted to know, "is turnover important?"

Dr. Lane pulled the microphone closer. "Looking at milk per cow per day, you can translate that to paychecks, or dollars per month—dollars that come from milk. We're interested in maximizing milk per cow per day.

"Say our cows are making 72 pounds per day, and that's a 24-hour day. She needs to make three pounds every hour. So in one hour the cow must digest all the feed it takes to make three pounds of milk. And she has to have that feed out of her system to make room for feed for the next three pounds of milk. And she has to do that within an hour.

"If our fiber levels are not right, if we have too much NDF—neutral detergent fibers—in the ration, it slows the digestion rate. Since the cow can't digest the feed fast enough, she can't make milk fast enough. If it takes her two hours more to digest the feed it takes to make three pounds of milk, she won't make 72 pounds a day. She'll make half that."

Trevor Will paused to let the jury absorb these facts. Then he asked, "Is there any kind of tool for measuring the quality of forage in terms of the turnover?"

Lane nodded. "There's a term," he said, "called 'relative feed value' which takes into account the types of fiber in the forage. And it makes a prediction on how fast this material will be digested."

"Doctor," Will said, "do you have an opinion as to whether all rations that are balanced to the same point will produce the same amount of milk in a cow?"

"Yeah. Introduction of the nutrients is the first step. Then we have to look at whether there's any forages impeding digestion of those nutrients."

"We've heard in this case about balancing the ration to 80 or 85 pounds. Can you explain that?"

"Yes. Usually high-producing cows, their ration is balanced to support 80 to 85 pounds of production. Since we have a certain volume in the rumen, and we can put a certain amount of protein, energy and so on in there, it only holds so much. We can only put enough feed in the rumen to produce about 85 pounds of milk.

"Now, many cows produce 150 pounds of milk, some only 40 pounds. If you feed them the same ration, what will happen is the cows that want to peak at 150, they're going to have to take fat off their back. And you will see high-producing cows in peak lactation lose 400 pounds in a matter of weeks because they can't get enough feed in to produce milk they're genetically programmed to produce.

"But if we spice up this ration to produce more than 85 pounds, it would make them sick. Too much energy, they would get an upset stomach."

"How," Trevor Will asked, "does balancing 80 to 85 pounds affect cows that are producing lower than that level?"

"Well, if you were providing a feed that could produce 85 pounds, you're taking a ceiling off the cow's production. If she wants to come up by eating more feed, she can get higher.

"For instance, if we had that ration balanced to produce 40 pounds and the cow was producing 35 and decided to make 45, she isn't going to 45 because she's only getting enough feed to make 40. At 85, that would remove the ceiling on her production."

"Thank you, doctor," Trevor Will told him. He walked up slowly toward the witness stand. "Have you done an analysis of the feed value on the Allen farm in 1997 and 1998?"

"I used all of the forages analyses of feeds they're feeding on a daily basis. Periodically, they take a sample and mail it for analysis. It comes back with nutrient analysis, tells us how concentrated particular nutrients are."

Asking Mckay's permission, Trevor Will went over and put up Exhibit 1804. "Can you tell us what this is, Dr. Lane?"

"Yes. I summarized forage analyses, by year. From 1996 to 2000. This shows there's a low forage quality on the farm on the average for 1996, '97; '98 we show a big raise. '99 and 2000 are consistent with 1998."

This was another way of WPS trying to show that my production increases after the isolation transformer was installed had nothing to do with isolation, nothing to do with electric current.

"Doctor, what impact, if any, would you expect to see from the inclusion of peas and oats in his feed?"

"Well," Lane said, "I would expect this feed to hang in a rumen too long and not be digested rapidly, interfere with new feed coming in, and depress overall production."

"Dr. Lane," Trevor Will asked, turning around and going back over to the jurors. "You also looked at the use of protein on the Allen farm in 1997 and 1998. What conclusions did you reach?"

"What that does," Lane said, "is increase the concentration of nutrients in the feed. The rations are becoming richer. The cows would improve milk production due to the higher concentrations of protein."

"Doctor, do you have an opinion as to whether the rations were kept constant between 1997 and 1998?"

"The rations varied widely," Lane told him. "They attempted to keep the nutrient density fairly constant. But the rations had wide variations."

"What impact would those variations have on milk production?"

Dr. Lane spread his hands out and shrugged. "Production would be impaired."

As impaired as from constant electroshock?

Will turned, glanced at the jurors, nodded, and looked back at the witness. "In addition to some things you've talked about, does palatability impact a cow's production?"

"Palatability," Lane said, "is another issue with poor quality feeds. As forages become stiffer, cows don't like the taste and do not intake as much feed."

"From your review of records, do you have an opinion as to whether palatability has ever been an issue on the Allen farm?"

"Yes. I believe that the University of Wisconsin's visit also identified a potential palatability problem while they were there."

Will nodded with satisfaction and remained silent for several moments. "Assume for the purposes of my next question," he said, "that Mr. Allen began using BST on his herd approximately March 18 of 1998. Do you have an opinion as to whether March milk production would be affected by resuming BST usage?"

The idea of production increases after isolation was a big problem for them. They were hauling out every other possible explanation under the sun to account for it.

"Yes," Lane said emphatically. "BST should have an effect on the March pickups."

"What impact would you expect to see on production after the second shot of BST?"

"Cows still increasing in milk. The second shot would give you an additional boost."

Still near the jury, Trevor Will asked, "What happens after the third shot in the BST series?"

"By the third shot the hormones have come into equilibrium. From that point on, your shots just maintain the extra milk produced. They don't continue to increase it."

Will went over toward the front of the courtroom to the exhibits and introduced a new one. "Dr. Lane, I'm showing you what we've marked as Exhibit 1807. Can you please tell us what that is?"

Dr. Lane squinted over at it. "Yes. This is a graph that shows what effect the season of the year has on milk production."

"Is that specific to the Allen farm?"

"Well, the data is, but you typically see this pattern in cows in the northern part of the United States."

"Why is this pattern typical?"

"Because cows are still seasonal breeders. They like to have their calves in the spring when grass is growing, so they have enough milk to feed the calf. So you will see milk production go up in the spring." This guy was in rare form and he could say this stuff with a straight face, too. This was the most asinine thing I ever heard of. The only people that paid attention to seasonal breeding were the rotational grazers. Other dairy farmers tried to get cows confirmed with calf by 100 days in milk, no matter what the season. The jurors knew no different.

"What sort of seasonal increase does this graph show on the Allen farm?"

Lane chuckled. "It shows that cows have a different calendar year than we do. But there is a six-pound increase from January to June of every year, and you should expect that."

"Now," Trevor Will said, "the graph shows that in the fall you're losing six pounds on the average. Did that happen in 1998?"

"No, it did not."

"Why not?"

"Well there were several changes in 1998 that occurred that would mask the curve line. I should say that I believe it did occur in 1998, but other management changes raised the entire production."

"What were some of those other management changes?"

"Nineteen-ninety-eight," Lane told him, "was a big year for management changes. There were improvements in the ration: Peas and oats came out. A veterinarian came onboard who intensified the vet program and improved overall cows' health and foot health on the farm.

"There were fans put in the barn, so ventilation was better. Cow mats were put in. There was use of BST. Three times a day milking occurred."

"You mentioned, Doctor," Trevor Will said, folding his arms, "a more intense vet program in 1998. What's your basis for saying that?"

"Dr. Homan's deposition. He's there more often, checking cows for pregnancy: increased veterinarian intensity."

Trevor Will stood silent for a long moment. Then he asked Lane about peer review and about articles the doctor had published.

"There are," Lane explained, "journals that accept articles from vets and scientists that are directed toward the farmer, and some are directed toward the scientific community. You submit the article anonymously. It's then sent to reviewers you do not know."

True, but that doesn't matter. Reviewers have their instructions and know the line they're expected to toe. Anonymity makes the conformity that much easier.

"And," Lane went on, "they look at the scientific merit of the piece to make sure it has proper statistical methods. If it passes the test, it's recommended for publication."

But the key test, of course, is whether or not the article makes the utility subsidizing the effort look good. And Lane was in the habit of making his paymasters look good.

Trevor Will suddenly changed the subject. "Based on your analysis of Allen farm vet expenses doubling from 1997 to 1998, what effect would you expect that to have on milk production on the farm?"

Lane drew a deep breath. "I would expect that to increase production. When you think of vet expenses, which is your dog getting sick, it's not a good thing that your veterinarian expenses go up each year. But on a dairy farm where we do routine service like vaccination and preventative medicine, this doesn't mean there's more disease or illness. It means there is more money spent on preventative programs."

Will nodded. "Doctor, did you look at the overall disease situation on the Allen farm? The issue of cow turnover?"

"Yes. Some years were higher than average, some lower. The turnover rate was relatively normal."

"Is that unusual on a dairy farm?"

"No," Lane shook his head. "In some years more cows leave the herd than others."

"From your review of the death data," Trevor Will said from the jury rail, "did you form opinions about what happened on the Allen farm?"

Dr. Lane shifted in his seat. "The data showed that years 1995, 1996 and 1997 were elevated and coincided with problems they had with retained placentas and displaced abomasums. Other years were low. The average death rate nationally is somewhere around four to six percent per year."

"Does electricity have anything to do with causing those diseases?" Will asked.

"It has," Lane told him, "nothing to do with those diseases."

This was a pure technicality, a lawyer's word game. Of course electric shock didn't of itself cause disease. Germs, bacteria, viruses, infections caused disease. Electric shocks delivered over time weakened the animals' immune systems to the point where their bodies couldn't fight off disease and they got sick, grew weak, and I was forced to cull or shoot them.

"Do you have an opinion as to whether electricity causes displaced abomasum?"

"It does not," Lane replied. "This is a surgical disease. A more intense, hot-feed feeding pattern is what caused these."

"Did you," Will said to him, "look at the issue of mastitis on the Allen farm?"

"Yes."

"What is a somatic cell count?"

"Somatic cell count is a white blood cell that migrates from the bloodstream into the mammary gland when the mammary has an infection."

"Looking at your chart of somatic cell count at the Allen farm, what conclusions do you draw?"

"That it was fairly low," Lane said. "Ideally, you would like to be down somewhere in the 100,000 area. But that's a goal all dairymen have. The 200,000, 300,000 range is common and readily acceptable in the industry."

Trevor Will unbuttoned his jacket, showing his vest. "Is there anything about this chart that suggests there's an electricity problem on the Allen farm?"

Lane shrugged. "Well, again, somatic cell counts have been blamed for electricity though no scientific research shows there's any relationship at all. This is very unusual for a stray voltage claim, in that most claim that electricity caused cows to get mastitis. This shows there is no mastitis problem on the Allen farm."

No, but then I'd gotten rid of scores of cows that did have it.

"On the graph," Trevor Will said, "you have something marked as 'legal limit.' What is that?"

"The federal government," the doctor said, "actually has a limit to how many white blood cells can be in the milk. In 1994, that limit was changed from a million cells to 750,000. The Allen farm at no time approached legal limits."

"Your chart covers what time period?" Will asked.

"From January of 1986 to January of 2001."

"Why not go back before January 1986?"

"I suppose I didn't have the records."

Will paused. "Did you," he asked, "look at the issue of pneumonia on the Allen farm?"

Dr. Lane had looked at it.

"Can you tell us if you'd expect to find pneumonia on a Holstein dairy farm?"

"Pneumonia is rare in Holstein cows," Lane replied. "It's a virus, and when you see it, you will see an epidemic. Pneumonia on the Allen farm is different. It's sporadic. I attribute that to the gases coming from the manure pit under the slatted floor."

"What's the basis for attributing pneumonia to gases from the pit?" Will asked, still remaining beside the jury box, with his vest showing and his arms folded across his chest.

Lane grinned. "As the manure decays, it releases ammonia. Ammonia can then interfere with the mechanism by which an animal keeps its trachea clear of bacteria. And so by breathing ammonia, cows can develop pneumonia."

"Have you ever seen that in other situations?"

"Only in slatted barns."

"Doctor," Trevor Will said, walking down the length of the jury box to the opposite end, "overall, how would you assess animal health on the Allen farm?"

Lane considered for a moment. "Like any dairy, there are episodes of disease. And that, I think, is why both veterinarians in their deposition felt that disease levels were not extreme but fairly normal."

Easy for him to say; he was alive and well, but most of my cows were sick or dead.

"Are you familiar with the concept of differential diagnosis, and is it something you use?"

"Yes," Lane told him. "Every day."

"If I were to ask you to apply your differential diagnosis methodology to claims Mr. Allen made about his herd, could you do that?"

"Yes."

"Mr. Allen," Will said now, "has complained that the cows were jumpy in the parlor. He complained about a high death rate. Assume he's complained about a high disease rate. Assume complaints about mastitis. Assume complaints about lapping at water. Assume complaints about reproductive problems. Assume there has been low milk production. If you had those complaints, could you go about analyzing them under the differential diagnosis method?"

Maybe Will should've asked him to assume I'd complained about global warming, loss of the ozone layer and depletion of the human gene pool. I was ready for anything at this point, but thought I knew where Will was headed: Back to the idea of my "mismanagement." That would be beating a dead horse—or should I say a dead cow?—but it didn't seem he had anyplace else to go. And even though I expected it, it still made me mad as hell, because I *knew* how hard I'd worked to protect the animals and the farm.

"Yes," Lane said. "Let's start with number one, being jumpy in the parlor. On our visit to evaluate milking, there was absolutely no movement of cows in the parlor. But when their teat ends were touched, it was obvious they were swollen and painful. The cows were completely comfortable till then. So, I would translate that into a milk machine or milking technique problem."

"What about a high death rate?"

"Other than years we had known causes of a high death rate—displaced abomasa and retained placentas—we do not have a high death rate that's a problem in this herd."

That just depends on your definition of high.

"And," Lane continued, "there's no reason to believe that stray voltage, if it existed, causes a high death rate. I would take 'high death rate' off Mr. Allen's list of complaints."

Actually, there were lots of reasons to believe it, as Dr. Johnson and Don Zipse had repeatedly explained in detail. I wondered what Lane would say about stray voltage, *if it existed*, since all of WPS' experts readily conceded that it did exist. I glanced over at the jury to see if they were reacting to this but couldn't tell much from their expressions. They were still fidgeting, or doodling, or looking around the courtroom. I leaned back against the bench and crossed my legs, which were stiff.

Lane went on, "Mastitis should have been lumped in with high disease rates. I've looked at somatic cell count, looked at a number of treatments. Those looked reasonable and I don't think there's a mastitis problem. Lapping at the water, you see in the literature that's a sign of stray voltage. I would say stray voltage or normal behavior, because that's normal boredom behavior of cattle."

Ah, so they acted up out of having nothing to do except hang around the farm all day. Maybe I should've given them decks of cards or some checkerboards to keep them busy.

Will moved a few steps toward the witness box. "Let me ask about that. Lapping at water is associated with stray voltage?"

Lane nodded. "Yes. In the early works of the literature."

"Have you observed a cow drinking water when voltage was passed through?"

"I have," Lane said.

Will suddenly went back to the jury. "On that basis, can you tell what mechanism was used to provide water in instances where this is associated?"

"In the stray voltage literature, lapping at waterers was noted in tie-stall housing where each cow has a water cup next to her. The water cup is made of metal. The cow sticks its nose into the cup and depresses the lever to get water. In that instance, lapping would be unusual behavior. But lapping at a freestanding trough is not unusual behavior."

"What type of system is in Mr. Allen's barn?"

"All concrete waterers."

Will paused. "You mentioned your experience. Could you tell us a bit more about what you observed with a cow at a waterer?"

"Yes. On one stray voltage investigation in Michigan, we had a waterer that had seven volts measured on it. We videotaped the cows coming to the seven volts, and they all just came up and drank the water. There was no lapping, no hesitation. These were Holstein cows. Seven volts did not impede water consumption."

This demonstrated nothing, proved nothing. What would really matter was the presence of that voltage over time. For all anyone knew, those seven volts might've just shown up that very day.

"Finally," Dr. Lane said, "we have low milk production. The first thing I think of is, of course, nutrition. Because without good nutrition there can be nothing but low production.

"So my complaint list has been whittled down, and I then I come up with a differential diagnosis for several of the complaints."

"What type or style of farmer is Mr. Allen?" Will asked.

"Mr. Allen feeds all his forages that he grows to his cattle. That means there are going to be years, such as 1997, where the quality was poor."

"Did you see anything abnormal about his production?"

"No," Lane told him. "Not really."

"Based on the information you had, do you believe stray voltage was a cause of low milk production on the Allen farm?"

Dr. Lane answered quickly. "I believe it was not associated."

"If," Trevor Will said, "stray voltage had been a problem and it was removed, what impact would you expect to see on production?"

"I would expect an immediate response."

"Why?"

"Because," Lane told him, "dry matter intake would be limited by water intake, and as soon as we took the electricity away from the cow, her water consumption would go back up and she would go back to maximum dry-matter intake for the feed she was given." If she is still alive.

Trevor Will turned from the jurors and went up to the witness. "Doctor, did you form an opinion as to what caused the production change in 2001?"

"Yes. There was a severe butter-fat reduction with some butter-fat inversions. A butter-fat inversion is caused by cows that eat too much grain. The bacteria in the rumen breaks down the corn so rapidly that it produces acid. And the aciditate, a building block of butter fat, is reduced. Butter fat inversion is where a normal cow has a 3.5, 3.6, 3.7 percent butter fat and 3.1 percent protein, so there's always this relationship. And when they get acidic the butter fat drops and the protein rises. That's the diagnostic of acidosis."

"How does that inversion affect production?"

"Well, there were times during 2001 where production did take some dips."

Will nodded and slowly crossed back over to the jury box railing. "Did you, Doctor," he said, "also investigate a claim by Mr. Allen that after he buried the ground ring in spring of 1999 that it caused the production, which had dropped, to rebound?"

"Yes, I looked at that."

"And did you form an opinion as to whether or not that claim was accurate?"

Dr. Lane straightened up in his seat. "It was not accurate."

No, of course not. Nothing I said was ever accurate. According to these people, even my lies were lies.

"And why is that?" Will asked.

"Because the claim was that production dropped 11 pounds and rebounded within a couple of weeks after the electrical fix was instituted. There was no change of that magnitude, no change greater than three pounds during that time period."

"Doctor," Trevor Will said, "based on your review of scientific literature relating to electricity and harm to Holstein dairy cattle, do you have an opinion as to the amount of current flowing through the cow that is required to affect milk production?"

"Yes—somewhere about six milliamps. Some people say as much as eight milliamps still has no effect."

Not surprisingly, the people who make these ridiculous claims are university professors affiliated with or directly employed by utilities. No farmer dealing with this problem in the real world would ever say such an outrageous thing.

"Why," Will said, "do you believe you need at least six milliamps to affect milk production?"

"Well, basically from the Red Book."

Ah, the Red Book again, 696, their holy writ. Lane sounded like a *mullah* citing the Koran.

"Is there any scientific literature in the peer-reviewed press that says earth currents below 1 milliamp could reduce milk production in cows?"

"No."

Why would there be? The entire peer review process was a stacked deck.

"Dr. Lane," Trevor Will asked, "are you familiar with the concept of a limiting factor? From your review of the Allen farm, what do you believe are the limiting factors out there today?"

"Feed, periodically," Lane said. "And consistently bad cow comfort."

I still couldn't fathom his reasoning. If this was true, then why had production been so high when the current was reduced, with the identical housing and "consistently bad cow comfort"? Something didn't jibe, but no WPS witness was going to go near the truth about that disconnect. The comfort of a cow lying out in the field was trying to be duplicated by many, and my cows had their choice.

Will nodded. "Do you have any observations on the records maintained by the Allen farm?"

"That would be another limiting condition. There's no user-friendly record system. DHIA has been discontinued. The other thing lost with DHIA was monthly cow milk weights and somatic cells. So we have no way of monitoring mastitis, don't know which cows to take out of the herd and treat at any one time because we have no somatic cell counts."

Funny, but I never had any trouble identifying animals with problems. My trouble was identifying healthy ones, because there were so damned few of them.

"Dr. Lane, do you believe the lack of adequate records has any impact on production?"

"Yes."

Trevor Will began slowly pacing. "Given the situation on the Allen farm today, in your opinion would he be able to produce 27,000 pounds per year?"

"No."

Still pacing, Will said, "Based on what you've seen in Wisconsin, Michigan, Idaho, California and everywhere else, where would you rank Mr. Allen as a farmer? How would you classify him?"

Lane shifted in his seat and held onto the microphone with both hands as he answered. "Most of the herds in Wisconsin are very small, where Mr. Allen has a 400-cow herd. Most people with a 400-cow herd do not just market their milk as—do not market their forage as milk. They use high-quality feeds to maintain good cash flow.

"I think Mr. Allen is a unique farmer in that he has a lot of debt, yet is willing to just feed forages and take whatever Mother Nature gives him. So in his peer group, as a farmer he would be way below average."

Anger rushed through me at this. I felt like running up there and slapping Lane. How dare he belittle me! He knew nothing about me, my farm, my life, my animals. My father had left me a thriving business and I'd worked day and night to improve it. If WPS hadn't come calling with its electrical poisons, I'd have had one of the best dairy farms in the nation. And I still planned to have one of the best.

Will ended his direct on this last note and Barry got up to start the cross-examination. I hoped he would do something quick to refute Lane's last remark, which was sticking in my craw like a chicken bone.

Barry walked briskly up near the witness stand, unbuttoned his suit jacket and stuck his hands in his pants pockets. "One of the ways to check your differential diagnosis on stray electricity is simply to reduce the level of electricity and see what happens, right?"

"Yes, if you were trying to prove that something was caused by electricity. You would have to measure an effect—measure that the cow was receiving electricity. You would have to see what her production level was then. You would remove the electricity and see her production level. Then you would put it back and see if it had an effect."

"Out in the real world," Barry said, "no farmer is going to let you do it, right?"

"No. That's why we do controlled research at universities."

Which bears no relation to actual farm conditions, and is designed to reach a predetermined conclusion favorable to the utilities underwriting the research. I could say that the universities and the utilities scratch each others' backs, but a more honest way of putting it is that they're in bed together.

"Okay," Barry said. "But there's never been any study of somebody putting current on waterers in a free-stall barn to see what would happen to the cow, right?"

"Your answer is no."

"There haven't been any studies in a milking parlor of the effects of electricity on milk production of cows either, have there?"

"I don't believe so," Dr. Lane told him.

"And as far as you know there have been no studies of ground current and effects it may have on cows in a free-stall, have there?"

Lane scratched his chin. "In a free-stall? I'd have to review Dr. Reinemann's report where he studied ground currents, but I believe he did that in a controlled environment."

Controlled? Reinemann's environment was downright rigged.

"And as far as you know," Barry said, "nobody ever put a clamp around a cow at the Allen farm to figure out how much current is going through?"

"No."

"And as far as you know cows aren't good conversationalists. So we can't just ask her if she's getting shocked, right?"

"That's right."

"You have to see what her behavior is?"

"Yes."

Barry nodded and went over to the jurors. "Would you agree," he asked, "that if there was electricity of a high enough level to affect intake of either feed or water on a dairy farm, that would also affect how well BST worked?"

"I agree with that."

"Okay. Now, if you've put stray electricity back into the equation, then, by removal of stray electricity on this farm, would you agree that there's about 17 pounds of milk increase that could be attributable to the removal of electricity?"

"Well," Lane said, "if you assume it was there in the first place and you removed it, there would be no way to get a response if the feed had not changed. So there are other things that have to be changed at the same time to get a milk response."

Barry folded his hands in front of him. "The rations presented to the Allen cows in January, February and March of 1998 would allow a cow to produce 85 pounds of milk—in the high group—right?"

"I wouldn't say that," Lane replied. "They were balanced to have enough nutrients for cows with a full rumen to produce 80 pounds of milk if the turnover was not impaired."

"Okay," Barry said. "And for the rations at the Allen farm, would you agree that through 1998 the high group was balanced at least 80, 85 pounds or higher? I think 85."

Dr. Lane nodded. "I would agree with that."

"You made the statement that you didn't think electricity was causing the increase in somatic cell count. Are you assuming there is no electricity affecting the cows now?"

"I'm assuming there never was any electricity affecting the cows."

That was a hell of an assumption, considering the consistently high readings on the voltage meters, among other things.

"On the affects of the Allen farm's increased production when electricity was reduced, the slatted floors would not cause the increase, would they?" Barry asked.

"No."

"The ammonia would not cause an increase in production?"

Dr. Lane smiled. "That's correct."

"Okay," Barry said. "The records or lack thereof would not cause an increase in production?"

"That's true."

"Now, if voltage was removed from the Allen farm about the end of December when the ground was moved, would you agree that could be a cause of increased production?"

"That would be a possibility if the voltage was on the waterers and it was decreasing dry matter intake. Otherwise, the increase has to be due to ration changes."

Barry moved up closer to Dr. Lane. "Actually," he said, "you've never really ruled out stray voltage as being the cause of the production change on the Allen farm. Have you?"

Lane bristled at this. "That wasn't my job to measure stray voltage on the farm!"

"Dr. Lane, you testified that it takes six volts to affect production?"

"I think I said six milliamps: Three volts would be what I meant."

"You said you wouldn't have a substantial effect until you had six volts?"

"I think I misstated," Dr. Lane told him. "I might rather have said 'eight volts.'"

"You should have said 'eight volts'?" Barry quickly asked, pouncing on this.

"Well," Lane said hesitantly, "the Red Book, which I rely on, has production losses that begin at six volts—excuse me, six milliamps. And then there's some papers that have values at eight milliamps that still don't show any production changing or any decreases in water consumption. So it's somewhere between six and eight milliamps."

"If," Barry said, "you assume that 10 percent of the cows have a resistance of 244 ohms or less, that would mean in a volt you could have four milliamps. Right?"

"I think," Lane said, "the work that was done on the 200 ohm cow was done from leg to leg with electrodes where they measured the resistance of the cows and did not take into account the hooves."

Barry went up even closer to the stand. "I'll ask you to assume that we've got a study in evidence here that indicates that 10 percent of cows have a resistance of 244 ohms or less from the mouth to the four hooves, okay?"

Lane took a deep breath and sipped some water from his glass. "I've read many articles and I know some of them have measured cow resistances without taking into account hooves and with EKG electrodes on the legs and just measured through the cow, and did not take into account the resistance of the hooves. And those numbers are somewhere around 200 to 250 ohms. So I'm not sure whether the 200 ohm cow you're talking about was measured in that way or through the hooves also."

"Would you agree that cows definitely have a variation in their resistance?"

"Yes."

"And the research you were talking about actually says that the cow's least sensitivity to current is about the same as human sensitivity?"

"Yeah, I've read that," Dr. Lane said.

"Well, if the current levels are the same for people and cows, then the difference in voltage is because of resistance difference, right?"

"The source resistance for people—the contact resistance for people who wear shoes—would be quite different."

Barry now turned away and went back over to the jury. "Dr. Surbrook, in an article, has said that people have 40,000 ohms of resistance on dry skin. So compare 40,000 ohms resistance to 250 or even 500 in a cow and it takes a lot less voltage to affect the cow than a person?"

"I don't know that to be true," Lane grunted.

Barry smiled thinly. Then he spoke more loudly than he had before. "In your business, where you go out on the farm and evaluate, isn't common sense important?"

"Absolutely," Lane responded.

"So, assuming, as Dr. Surbrook indicated, that cows have the same sensitivity to current as people, what would be the voltage you're talking about for a person that would do the same milliamps for cows?"

Dr. Lane was confused by the question, so Barry rephrased it. "I have 40,000 ohms of resistance, and I've got 3 milliamps; .003 amps."

"Okay."

"How many volts is that?"

"I have 120."

"And if I go to 6 milliamps—twice as much—how many volts would that be?"

"Well," Lane told him, "it would be 240."

Barry nodded. "Now, if I had an extension cord and had you hang on to both ends, are you telling us that the level of current that would flow through you, assuming you've got dry skin, would be such that it wouldn't affect you?"

Dr. Lane paused for several long moments before answering. "Well, I...don't know. I'm not sure about the 40,000 ohms. But I would not be interested in picking up a 120 or 240 line, if that's what you're getting at."

"Because you would get a shock?"

Lane again hesitated. Then he said, "Yes."

This was good stuff from Barry. And, as he'd mentioned, it was all common sense. Living tissue, whether animal or human, could be hurt by electricity and feel pain, regardless of this gibberish about being "beneath the level of concern." That was just a cheap trick.

"As an expert," Barry said, "when you were out to dairies in Idaho and someone had a concern about stray electricity, what did you tell them they should stay under for voltage?"

"I...don't think I ever had that discussion. I measured dairies in Idaho, and have not found anything that needed to be corrected."

How convenient for him, and for WPS. Convenient, but not very believable. Barry apparently thought the same thing, because he then said, "In your work as an expert, the 200 cases or so that you've been on have all involved stray voltage, haven't they?"

"The 200, yes."

"And since 1991, all of them have been jobs for utilities?"

"Of the stray voltage cases, yes," Lane replied.

"And you've never found in *any* of those couple of hundred cases that a utility was causing a problem that affected cows?" Barry sounded highly skeptical, and rightly so.

"That the utility was causing? No, that was never found."

Surprise, surprise!

"Just for Consumers Energy utility, your fees were over a million dollars for work you did?"

"Over a 10-year period, yes."

Let's see, that averages out to 100K per year. Just for that one utility. And he'd done plenty of work—shilling—for others. Clearly he wasn't going to find them liable for anything—that was as certain as night following day.

And on that perfect note, perfect at least from my viewpoint, Barry ended his cross. I heard some murmuring among the jurors at the mention of the million-dollar figure, and watched happily as Trevor Will nearly fell over his counsel table in his rush to run up there for redirect. Barry's last series of questions had hurt Lane, and Will knew it.

"Regarding the last point," Will said quickly to the doctor. "What was included in the million dollars?"

"That would be air fares, lab fees, car rentals and so on."

This sounded strange because he'd testified earlier that his expenses were paid *on top* of his fees. So, which was it? Probably he was too embarrassed to admit the truth. And yet even if he actually did have to pay for coffee and donuts and carfare out of that one mil, so what? For sure he wasn't sweating next month's rent.

Apparently realizing that nothing good could come from further discussion of Lane's fees, Will quickly changed the subject. "In terms of the turnover rate of feed on the Allen farm," he said, "what is the significance of the trend you pointed out earlier in the forage?"

"Well, these feeds would be expected to turn over faster. The worry would be whether they are going to be healthy to the rumen or whether they cause acidosis."

Will nodded and then quickly said, "What is the basis for your belief that electricity has not affected the Allen herd?"

Lane leaned forward, seemed anxious to answer. WPS was forking over a ton of money to him.

"Well, I think we can more than explain all the increases in production from changes in the rations that occur during the first part of 1998 through the middle of 1998. Rations, BST, lots of other things. But overall, the rations must change or the increase in milk production could not have happened."

What, no mention of the isolation transformer? That was the 800-pound gorilla in the room. I hoped the jury could see it.

"You were asked questions," Trevor Will said, "about whether you could test if electricity is causing a problem with milk production by removing the electricity and seeing what happens to production. If you were to design such a test, what would you do to make sure you were measuring the change in electricity?"

"Well," Lane said, "like all good research done at universities, you have to control extraneous variables. So you would not start BST. You would not make any ration changes. You would not add mats to the free-stall barn. You would not add fans. You would keep your veterinarian programs the same. Then, if all those things were constant, you may be able to see an effect of removing electricity from cows."

Will went over to the jury box and rested his elbow on the rail, facing Lane. "What would you do in terms of measuring the electricity?" he asked.

"At the very least you would have to make sure it was there, make sure it was affecting the cows. And then make sure it had been removed from the cows when you made the electrical changes."

"Have you personally done any sponsored research?"

"Yes."

"Tell us," Will said, "what is the process to get a grant to do independent research?"

I couldn't see what this had to with the price of beans, but they were Will's questions and he was entitled to them.

"Yes. Different organizations have grant money available for universities or private research foundations. We'll send a request for proposals and they specify what kind of projects they're willing to lend money on.

"We write our proposal, apply for money, specify how data will be collected, specify how statistics will be used. Then the package goes into the research foundation. And then grants are awarded as people qualify."

"And in your experience, what control does the person funding the research have over the results?"

Dr. Lane shrugged his shoulders. "My experience at the university has been none."

This was amazing in that all the research as well as peer review always accrued to the benefit of the utilities doing the funding. I doubted that WPS did anything as obvious as taking a red pen to the writing or telling professors how to conduct experiments; they didn't have to. Self-censorship and experimenters' knowing which side their bread was buttered on was enough to produce the desired results.

"Last question," Trevor Will said. "Do you know what the resistance is of dry human skin?"

"No," Lane said. "I don't."

Will sat down and Barry got up to take another crack at the good doctor.

"So, Dr. Lane," Barry said cheerfully, walking up near the stand. "You would be critical of any study that actually changed the feeding program prior to doing any electrical experimentation?"

"I think any feeding program could be changed prior to as long as the cows were given time to come into equilibrium with the program so that they were on a stable diet through the test period."

"If the feeding went from a Total Mixed Ration to a TMR with a computer feeder, that would be a significant change?"

"Yes. Those changes can be compensated if there is a control group involved."

"You've never gone back and looked at changes in rations before 1998 as they may have affected production, have you?"

"Yes, in 1997," Dr. Lane said.

"But nothing before that?"

"No."

"So you have no idea what the rations were at any time prior to 1996?"

"True," Lane told him. "You have no data on that."

"That's all," Barry said, and I was surprised to see Will get up yet again. He must've been deeply worried. And that made me very, very happy.

"Dr. Lane," Trevor Will said. "You looked at the rations in 1997, correct? Did you form an opinion as to what caused production to decrease in 1997?"

"Yes."

"And what was that?"

Lane held up two fingers. "Two things. One was the removal of BST in 1996 from the herd so the production per cow per day started to diminish. And the second thing was as of July 1997, the peas and oats came into the ration and were fed in increasing amounts towards the end of the year, so production took a real dive in 1997."

Here, finally, Trevor Will ended. So did Barry. They were both done with Lane, who left the stand looking exactly as he had when he'd taken it: Wrinkled, disheveled and a complete mess in his cheap polyesters. Doctor or not—and he looked like anything but—I thought Barry had done a great job on him, punched holes in his theories and testimony. But my mind didn't linger on Lane for very long. Because next up was one Russell Allen, and he was going to have a lot to say, if only he could get over that case of the heebie-jeebies he'd just suddenly come down with.

As Judge McKay told Barry Hammarback to call the next witness—me—I felt the hot gaze of those 26 eyes in the jury box beating on the back of my neck like a couple of dozen flaming suns.

Chapter Sixteen
FIGHTING FOR EDEN

As I approached the witness stand, I was so nervous I thought my heart might beat right through my chest. I was breathing like I'd run a marathon. Which, in a sense, I had. Right now, everything that had happened in the course of this marathon trial—the expert opinions, lawyers' arguments and objections, the exhibits, the rebuttals, the rebuttals to the rebuttals—meant next to nothing, because the case had come down to what it had always been about: *me*.

And about my family, alive and dead. They knew that if I did not get a fix, the dairy business would have to be sold. The weight of their expectations hung heavy as a butcher block around my neck. And now I was alone in this courtroom, alone in a crowd. Robinson Crusoe shipwrecked on his island couldn't have felt more solitary.

After they swore me in I plunked myself down in the witness chair and picked up the glass of water beside me. My hand shook so badly that the ice cubes rattled; they sounded as loud as a window shattering. Embarrassed, I managed to lift the glass to my lips and sip a little water without spilling any and, thank the Lord, was able to set the glass back down without dropping it.

The day was another in a string of perfect spring days. Bright sunshine filtered in the courtroom windows and spread across the spectators' gallery, which was crammed full. But I didn't notice faces. I didn't even notice the jurors' faces although I felt the force of their gaze: 24 eyes boring into me all at once.

I looked at the gold chandeliers, turned and glanced up at white-haired old Judge McKay perched behind his oak bench, saw the uniformed bailiff standing near the door, watched the stenographer, Karen Nagorny, pecking away on her machine, noticed the mahogany counsel tables, the jury box with its glossy railing, and the plush carpet with the state seal covering the floor. I felt like I was in a cathedral.

Heart still pounding—I was trying to hide my nerves—I folded my hands on my lap and waited. As Barry got up from his counsel table and walked up close to me, he shot a quick wink and flashed a smile.

"So," he said. "You're *the* Russ Allen, the one we've all been talking about?"

"I guess so," I said, feeling sheepish.

"Welcome. Give us a little history first, about your background. Basically, your folks were who?"

"Alvin and Alice. Allen."

"When was it they moved to this area?"

I had to stop and think. "1937."

"And they moved where?"

"5206 Little Apple Road."

"Is that the very same farm we've been talking about for weeks?"

"Yes," I said.

"When they came to the farm, had they started a family yet?"

"No. They purchased the farm in 1937, got married in '38, and then had their first child. Larry was born a year after they married."

Barry nodded. "Okay. After that, how many brothers and sisters?"

"I got six brothers and four sisters. Eleven kids altogether." I thought achingly of the ones who were gone.

"How many were involved on the farm?"

I took another sip of water and started to calm down. "Larry went to college, Jim went to college. The rest of us were all on the farm."

"Larry and Jim today are where?"

"Larry's in East St. Louis. Jim's in Seattle."

Barry came a step closer. "As you were growing up," he said, "tell us how you learned to take care of dairy cows."

I realized what Barry was doing—showing the jury that I was flesh and blood, with a family and a history, a person just like them; not just some docket number.

"Well, from 5, 6 years old, we started helping with all the chores and milking. We had buckets and the bucket was held on with a strap and we changed straps and one of the older kids would put the unit on and take the unit off when the cow was done."

"How," Barry asked, "would the milk get from the bucket to the bulk tank?"

"We would dump it in pails and we'd have to carry it to the milk house."

"And there wasn't a milk tank then. What was there?"

"Milk cans," I said.

"How many cows?"

I glanced at the jurors. "We started with 32, and when I was 7 or 8 years old we rented another farm and were milking the 32 plus another 27."

"As time went on, did the family add to the number of cows?"

"Yeah, we did. After we rented that farm, we built a hundred-stall barn. Then we moved the 32 cows and the 27 cows in, and then we filled the barn the rest of the way with more cows. We kind of grew into it."

"Okay," Barry said. "When was that barn completed?"

"In 1962."

"How was that barn built?"

"We laid the block ourselves," I told him. "The lumber came out of our woods. We built the rafters ourselves. Then we had a raising bee."

"What's a raising bee?" Barry asked.

I smiled at the memory. I can still taste all the wonderful food from that day. I don't think I've ever eaten as good since. "When you get the neighbors together and raise the rafters and you do it in a day or two."

"Did you help with that? You were a kid then."

"Yep. I helped nail rafters together."

Noticing that I was no longer so nervous, Barry turned and walked over to the jury box. "How did you figure out how to put feed together for cows back in the early '60s?"

"From the University of Wisconsin. And we had a mix mill."

"Back at that time—we've all heard about total mixed rations now—was all the feed mixed together?"

"Back then, no. You would feed about half hay and half corn silage, and then vitamins and minerals and supplements would be weighed up."

"Who, in the '60s—was your dad still running the place?"

"Well," I said, "we were getting up there where we'd have a little say, too."

"Who, then, in the '60s was actually running the farm? The whole family?"

"Cletus went off to school for a while," I explained. "Les was in the marines, then came back. Cletus came back from school. We were all involved except for Larry and Jim."

"Once you graduated from high school, did you want to do anything other than work on the farm?"

"That's all I did." It was all I ever wanted to do. I loved it.

"What jobs did you have around the farm?"

"For the first couple of years we rented another farm, and we milked 120, 130 cows down there. Me and Bob would drive down there and milk. My other brothers were milking up here, and then we'd all crop together."

Still by the jury, Barry asked, "Who was doing the paperwork?"

"Mostly Pa," I told him. "The decisions, we'd all kind of get together, sit around and talk, but mostly my Pa was doing the paperwork."

"As you came to the middle of the 1970s, what decision was made regarding the dairy?"

"We just decided we were going to combine herds and build a bigger barn. It took us three years to build."

"What did you do to prepare?"

"Visited quite a few farms," I said. "The University of Wisconsin helped us with blueprints and stuff."

"Did you look into parlor setups?"

"We did. We even have blueprints of that. I think ours was only the seventh one in the world like it."

"What had you decided to do with electrical power at the new facility?"

"Well," I said, "my pa took care of that. We got three-phase running for the new barn."

"Okay," Barry said. "After the barn got going, how is the new facility different than the one you had before?"

"The new one had the parlor—the cows came to you."

"What did you do for manure control?"

"Built pits underneath the whole barn; 16 pits, individual, 10 feet deep with a 10-inch wall, and then a slatted floor on top of that. We bought our own cement truck. We mixed all the concrete for that barn ourselves. We had our own crane which moved the forms for the walls."

We'd done everything we could to improve the farm's quality and productivity.

"We've heard quite a bit about use of the slatted floors. Why did you go with that?"

"Well, you're not in there disturbing the cows scraping. And to this day, I don't think there's anything wrong with them." The hell with WPS' experts. "The biggest thing nowadays is economics—there's cheaper ways but I don't think there's anything wrong with slatted floors."

Barry stepped away from the jury rail and moved toward the partition in front of the spectator benches. "How did you decide on the ventilation you used back in '76?"

"I don't remember who designed it, but there's 12 4x4 fans on the roof and the air would come in on the eaves and go out the peak. It was an insulated ceiling, had ducts going up to those fans."

"In the new parlor," Barry said, "what was the setup for 1976?"

"It was called a 'polygon parlor', which is herringbone. You milked from the side, six, six, six and six. You didn't have to wait for 12 cows to be milked out. You could let six out at a time, let in another six."

I shot a look at the jury and took a drink as Barry walked toward me, stopping a few feet in front of the stand.

"Records-wise," he said, "back in '76 what were you using to keep track of cow births and treatments, things like that?"

"We had separate books for dry, fresh and treated cows."

Barry was laying the foundation of my case as well as humanizing me. I felt pretty good. Cross-examination was a long way off.

"Could you explain to us how you ended up being the one running the farm?"

"Well, first off, my brother Les passed away. Shortly after that, Cletus left and went farming on his own. My brother, Kevin, ended up farming on his own. And Bob—I ended up purchasing his share—he's working for me now."

"Your sisters were involved too?" Barry asked.

"Yeah, Geri Lynn was and Josie too."

"What year was it you took over operation of the whole thing?"

"Approximately '89," I said.

"At that point was anybody helping you with the records?"

"My pa was for a bit."

"And the finances, were you doing that?"

I nodded. "Shortly after I was sole owner."

"As of 1989, then," Barry said, "who were the people working on the farm?"

I had to think about that for a moment or two. "Me, Geri Lynn, my brother Bob, his three boys and I don't remember exactly when Patti Bowers came into the picture."

"For the next two or three years, who was in charge of the milking?"

"Geri Lynn managed the parlor, and she would hire people. Mainly, that's what it was, Geri Lynn and hired help."

"And who was in charge of the cropping?"

"I did. I managed it. I had help from Bob and his boys, you know."

"Who organized the feeding?"

"I was doing the feeding until '91, when we got our first mixer."

"Who was taking care of sick cows?"

"Bob."

"Back in 1976," Barry said, "what type of business arrangement did you have with your brothers and father?"

"It was a partnership," I explained. "We'd all get so much a month and the rest would go to pay the bills."

"Did you have a revenue plan?"

"Well, we started with 250 cows. We planned over the next five, six years, to build a herd within the herd, shooting for 16, 17,000 pounds, which is not asking too much."

"Okay. So the total number of cows you were looking at would have been what?"

"Five hundred. Four hundred milking, one hundred dry."

"In terms of total milk, what kind of poundage was that?"

"About eight million a year."

Barry folded his arms and turned to the side so he could see the jury as well as see me. "As time went on with the partnership" he said, "how did you progress toward those goals?"

"Well, we did the normal things you do. Worked with the nutritionist, worked on the parlor, balanced the ration."

"How did production do compared to where you planned to be?"

I shook my head. "Seemed to stay stagnant, never could get moving."

"Did that affect the partnership?"

"Certainly."

"How?" Barry asked.

"Well," I grinned, "if money is not there, you can't get paid." That wasn't very profound.

Barry nodded, unfolded his arms, put his hands in his pockets and went over to the jury again. "Who was first to go out of the partnership?"

"Cletus. Went farming on his own. Next was Les. He passed."

"The next one?"

"That would be John Hermsem. He went to build homes."

"And the next one?"

"That would've been Bob."

"And the next?"

"Kevin."

"Where did he go?"

"To farm on his own."

"And that left you?"

"Yes."

Barry paused, and I sipped some more water. Then I pulled a hard candy from my pocket, unwrapped the cellophane and popped the candy in my mouth.

"How did the subject of stray voltage come up on your farm in 1988?"

I shrugged. "Cows kicking in the parlor, couldn't keep the units on. Pa and I read about it, talked about it. Thought it was a good idea to get it checked out."

"So, did you?"

Clearly I did.

"Who came to check?"

"Rob Gelhardt," I told him. "From WPS."

"What did he do?"

I laughed. "Monitored an electrical panel. Said there was no problem."

That was their story from first to last. Now they were stuck with it.

"Did you believe him? At the time?"

"Sure," I said. I had no reason not to.

"What did you do about production?"

"Kept plugging away, working and working on it. Bought our first mixer, tried to boost it like that."

"Did you think that would affect production?"

"We were hoping. Research says it."

"What affect did that mixer wagon have on production?"

I sighed. "I was disappointed—nothing."

"After the mixer wagon was purchased, did you do any upgrades to the parlor?"

"I thought a new parlor, a parallel parlor, was the answer to my kicking problems."

"Did you see how cows behaved in those parlors?"

I'd visited four farms in Michigan and observed them. "Yeah," I told him. "They would stand perfect. I thought it was the answer to my problems. I went back home, we put it together."

"How did you finance the new parlor?" Barry asked.

"A lease, through Farm Credit Services. It's a lending institution dealing mostly with farmers. I think it was completed in April of 1994."

Barry began slowly to pace. "Was the new parlor hooked up to a computer of some sort?"

"Correct," I told him. "It kept track of milk weights. There was a processor box that came with it. We could punch in numbers, cut out cows automatically, stuff like that."

"Would this system allow you to keep records of individual cows for milk production?"

"Yes, yes," I said.

"With this new system, did you need to be on DHIA anymore?"

No, we didn't.

"Okay," Barry said, walking back toward the witness stand and stopping next to me. "Once the new parlor was in, was an equipotential plane put in?"

"Yes."

"At the time, Mr. Allen, did you know what that was for?"

"I didn't know much about it, an equipotential plane. No."

"What was cow behavior like in the new parlor?"

I shook my head disgustedly as I remembered. "Not any better than the old parlor." And after I'd spent a hundred grand on it. "We had many kick-offs. I ended up putting a second person in because one person just couldn't handle it."

Barry paused. "What else did you consider as far as production?"

"Well," I said, "we bought a second mixer, we intensified labor and management. We were always trying different things."

"How do you know when something works or doesn't on dairy farm production?"

That was simple. "You change one thing and keep everything else the same."

Barry nodded. He was building up the case. "By 1996," he said, "after the computer equipment had been put in, what did you do about BST?"

"Gave it a shot," I told him. "Put available cows on it which means all cows 55 or 60 days after calving."

"Expectations concerning BST?"

"We hoped it would boost production."

"How did it work out?"

"We didn't see much at all. Hardly anything."

"Were you able to get the three pounds extra you needed to break even with BST?"

"No," I told him.

Barry turned away, lowered his head in thought for a moment, then turned back to me. "Toward the end of 1996, beginning of 1997, what was the financial condition of the farm?"

I sighed. "Heading for bankruptcy." For some reason when I said this I was overcome by emotion. Tears welled up and I could not stop them from running down my cheeks.

Barry waited. "Did you get hold of Mike Hoerth?" he said softly.

"I…I got tired of cows kicking in the parlor," I said, and wiped the tears with my index finger. "I told Mike, 'It's your parlor, your tit dip, your crowd gate, it's all your stuff. Get those cows to stand still!' He knew of a guy on another parlor testing for stray current and thought I should get it checked out. Because he'd gone over my parlor a zillion times and couldn't find anything wrong."

Poor Mike. He was an honest and conscientious man and deserved better than how I treated him.

"That's when Larry Neubauer came into the picture," I said, regaining a measure of control. I took out a hankie and blew my nose and then sipped some water. "He came and tested for a couple of days." And discovered the 800 milliamps of current flowing through the parlor.

"What did he do that stands out in your mind?" Barry asked.

"Shut off power to the farm so there was nothing incoming. The measurement then was 800 milliamps—I couldn't believe there was current there when the power was off to the whole farm."

Barry nodded. "Anything else stand out?"

I nodded back at him. "He disconnected the utility's neutral."

"What happened after that?"

"The current on the parlor steel was nearly gone."

It was one hell of a coincidence.

"When you got a report from Mr. Neubauer, how did you take it?"

"I didn't know if he was pulling my leg. I wasn't sure. I wanted a second opinion. So I had WPS come and test. Mike Moore came. He said there wasn't a problem, and that if I was counting on stray voltage to save my farm, it wasn't going to happen." He was a hardass, and nasty. A real company man.

"Let me ask you this," Barry said. "Have you ever restricted anybody from WPS on where they could test or what they could do on your farm?"

"No, I never did." Despite the lies of Vern Peterson, who claimed I coldly kicked him off the place and that his tender feelings were hurt.

"What was your understanding of what the Dairyland Isolator he offered you would do?"

"Get the current off the parlor. I didn't know where to turn or who to trust, so I called Larry. He said to try it."

Barry turned and went over to the jury box. "Had Mr. Neubauer told you the cost to put in an isolation system?"

"Yes. About $58,000."

Barry now walked me through the events dealing with the Dairyland Isolator and the moving of reference rods. "I was excited for the first couple of days after installation," I said. "I thought, man, this is just what I'm looking for. The cows settled down and production even came up a bit in the first couple days."

"After this improvement, then," Barry said, leaning one elbow on the box rail, "what happened next?"

I felt like screaming out the answer, but managed to control my voice.

"The cows were *terrible* that first weekend I had to milk. So terrible I called up Bob to come help me." My heart started to pound as I recalled my anxiety and fear. Then a lump formed in my throat and tears again began welling up. I tried to keep from crying but succeeded only partially.

"How did the behavior of the cows go for the next week or two?"

"Well," I said, my voice cracking, "I was keeping an eye on them because I had such a problem that first weekend. Couldn't understand how they could be so bad and then the next day they'd be perfect. Next weekend they predicted hot weather. WPS recorded the largest power usage ever. I told Larry to come and monitor what's going to go on."

Larry came and measured tremendous amounts of current flowing into the parlor, this despite the Isolator.

Frantic, I called WPS and Vern Peterson came back again and Larry showed him the current registering on his meter.

"What did Vern say?" Barry asked from across the courtroom.

I shook my head in disgust. "He told me it was on-farm problems."

When in doubt, blame the victim.

"After that," Barry asked, "what did you do?"

"After a couple more times seeing the cows standing perfect in the parlor and then seeing them unbearable, I came to the conclusion that it had to be the Dairyland Isolator opening and closing, just the way it is designed to do."

Barry waited and the courtroom was very quiet. Judge and jury were watching me.

"Did you come to any conclusions about whether that isolator was working?"

"It wasn't working for the cows. So I decided to—made up my mind we had to get an isolation transformer."

I recounted my struggles to obtain the financing.

"When," Barry asked, "was the isolation system put in by Mr. Neubauer?"

"January '97."

"Once that transformer pole was moved and the isolation system up and running, what observations did you make of the herd?"

"Production started to come up. And," I said with sarcasm, "one of the things I noticed is they quit dying."

That became obvious when there were no carcasses on the ground and no buzzards circling overhead.

"What other observations did you make?"

"General health improved. They didn't have rough coats. Things started rolling along."

Barry left the jury area and came back to me. I grabbed a quick drink and popped another hard candy in my mouth. A cherry-flavored one.

"When the isolation system went in, did Mr. Neubauer make any warranties about how much milk production you would get?"

"He told me 75 pounds."

"Was he right?"

"No, he lied," I said. "We got to 80. In less than six months."

Barry smiled. "At that point," he said, "what did you believe your herd should produce without the effects of electricity?"

"Well," I said, and took a deep breath. "To see the cows take off like that, you keep current away and you can't stop them cranking out the milk. There's no reason the herd can't be at 27, 28,000 pounds."

"And so when you were meeting with Dr. Homan and Steve Root and others, one of your goals was to get how many pounds at that point?"

"Ninety pounds."

"Okay. Dr. Homan made some suggestions, do you remember?"

I remembered. "Cow comfort. More waterers, basic stuff I already knew would get a herd over 30,000 pounds."

"Why hadn't you done these before?"

I shifted in my seat, crossed my legs, and leaned forward. "To do some of these things, you almost have to tear down the building and start over."

"In the dairy business, about how long does a farm facility last till it's outdated?"

"In general, 20 years."

"In your facility what should be done?"

"Well, everybody claims I should tear it down and start over, but, jeez, when you take the current away we can sure get milk out of the facility."

"Are there things you could modernize or improve?" Barry asked.

"Sure. In October '98 I put in 14 wall fans on the sides of the barn and created tunnel ventilation."

"What had production done at that point?"

"At times over 80 pounds."

"What was the reason for changing the lights in the barn?"

"My old ones were 110; Larry needed to make them 240."

"After the lights were changed, was it darker or lighter in the barn?"

"Well," I said, "it's still a dungeon."

Barry stuck his hands in his pockets and began pacing in front of Judge McKay's bench. "Now, by July '98, what was your feeling about whether electricity had been removed from your farm?"

"I thought finally this thing is behind me, in the past. Life can go on."

"About your research to what caused the problems, what was your conclusion?"

"No doubt. It had to be electricity."

Barry nodded. "In '98, did you again call WPS?"

"I called Jerry Held, to take a look at electricity again affecting cows—I wanted to be compensated for all the damage that had been done."

Barry stopped pacing. "What did he tell you?"

"He said, 'We just don't hand out money to anybody.' I knew if I was going to get what I needed, get compensated for what I'd lost, I was going to have to do something else."

"In early '99," Barry said, "when the cows again started twitching, what did you think?"

I stared straight at the jury foreperson. "Current," I said, loudly. "I got hold of Larry, had him test. He was picking up earth current."

That was when we'd dug the trench around the farm and laid in the copper ring. I explained all this to Barry.

"Where did you understand the current was coming from?" Barry asked.

"From WPS' down grounds."

"So what did you do?"

"Cut the wires," I said.

"How did you go about doing that?"

I shrugged. "Went in the shop, got a side cutter and cut them."

"Are any houses or people located near these down grounds?"

"No."

"Any buildings next to them?"

"No."

"Why did you cut them?" Barry asked.

"I saw how much the current affected my cows, saw how effective the buried ring was, and thought if I lowered the current on the ring, I'd be better able to keep the current away from my cows."

Barry went back to the jury and stood near the foreperson. "Did you have any additional contact with Jerry Held?"

"Yes." I knew where we were headed now, and not a moment too soon.

"What was that contact?"

"I saw him at a neighbor's house, in a WPS truck."

"On the basis of that," Barry asked, "did you decide to do something?"

I sure did. "I decided to sue," I told him.

"And why was that?"

"I learned that the way WPS was addressing their electrical problem was by digging up dirt on Russ Allen."

Barry paused to let the jury digest this.

"What kind of things were they trying to find out?"

I grinned. "Inquiring about me going on vacation. Inquiring about my boat."

Again Barry paused.

"At the present time," he said, "what's the debt load on your farm?"

"In the neighborhood of $1.6 million."

"With the present production of your farm, do you have cash flow?"

"No," I told him. The truth was that I was flat broke and falling behind 50 thousand dollars a month.

"If the farm was producing at the rate it should have but for the electricity, would there be cash flow?"

"Yes."

"Milk prices. Are they up or down?"

"Down."

"With prices down, would cash still be flowing if production was where it should've been?"

"Yes."

"Have you tried to do other things to keep WPS electricity from coming onto your property?"

"I tried to live with the down grounds connected. I used an insulated wire to try to bypass the current around my farm."

"After this lawsuit was started, were the grounds hooked back up?"

"Yes."

"Did you disconnect them again?"

"Yes."

"Why?"

I shook my head, hard, and threw up my hands. "Too many problems. The cows cannot handle that current."

Walking down the length of the jury box to the opposite end, Barry asked, "How did the cows do after that?"

"They responded positively," I told him. As all mammals would.

"What happened to production after they were hooked back up?"

"Cows got sluggish." As all mammals would.

"And production did what?"

"Fell."

"Has the electricity on your farm affected your enjoyment of that farm as a dairy?"

"You bet!" I told him.

"In what ways?"

Again I threw up my hands. But this time I blinked back tears. "With the cows kicking in the parlor, no cash flow, things not going right...nobody on the farm is happy. It's just...a struggle."

Barry waited before asking the next question. "Did a representative from WPS indicate to you what your options were regarding electricity in the ground on your property?"

"Yes. My options were to put up with the current coming through my property affecting my cows, or have no electricity at all."

"Is there," Barry asked, "another power company you can hook up to?"

"No." WPS was a virtual monopoly.

"What is it you want from WPS?" Barry asked.

What did I want? That was simple. I wanted justice. "I want compensation for everything they caused," I said. "Not a penny more or a penny less. I want the current off my property so it's not affecting my cows."

"That's all the questions I have, Your Honor," Barry said to the judge.

Now I braced myself for the onslaught. I sat up straight and stiff in my chair as Trevor Will, thin, gray-haired and graceful as ever, rose from his seat and walked slowly toward me, stopping about 3 feet away and crossing his arms in front of his chest. I took a big gulp of water. As I emptied the glass, I felt sweat start trickling down my back.

"Good morning, Mr. Allen. How are you?" Trevor Will asked.

"Good," I mumbled without looking at him.

I knew Will a little. He deposed me for six days, billing WPS by the hour, I suppose. It was like having a tooth drilled. Today figured to be more like a root canal. Or an extraction without anesthetic.

"You're probably the person who knows the most about this farm, right?"

"Correct."

"And not only do you know all aspects of the farm, you were involved in various electrical experiments?"

"Correct."

"And you understand that the reason we did deposition is because WPS wanted to find out the basis of your claim?"

"Correct."

"And I can't just call you up and talk to you on the telephone when you're represented by lawyers, you understand that?"

I wondered what he was setting me up for.

"Correct," I muttered.

"You understood I would be relying on the answers you gave?"

Now it was clear. He planned to impeach my deposition testimony. *So what?* I thought. *Let him try.*

"That's correct," I said.

Will smiled, the way a cat smiles at a mouse pinned in a corner. "Now that we've got that out of the way, let's talk a little about electricity on your farm. It's true that you believe that before electricity can harm a cow it must flow through the cow, correct?"

"I don't believe that anymore," I said.

"Well, you believed it at the time of your deposition. Didn't you?"

"I'd never thought about it and I answered 'yes'. But I don't believe it anymore."

"You're saying your belief has changed?"

"Correct."

Will shook his head as if confused. "Well, Mr. Allen, let's be clear. In your subsequent deposition sessions you brought in lists of things you wanted to correct. Never once did you tell me you wanted to change this piece of testimony. Did you?"

"I couldn't remember everything that was asked the day before, and the day before that."

Will nodded, turned and went over near the jurors and started walking up and down in front of them, practically tiptoeing. "Mr. Allen, let's focus on the question. When you showed up for the fourth session of your deposition, you brought a list of things you wanted to correct, didn't you? Things you had observed from reading the transcript?"

"Probably."

"This wasn't on the list. Was it?"

I shrugged. "Maybe not. I don't know."

Let Will score this point, who cared?

"So," Will said, "I walked in here thinking that was your testimony. Didn't I?"

Poor Trevor, how I had deceived him!

Here Barry got up and objected, acidly noting that the thoughts of Trevor Will were irrelevant to the proceedings. McKay sustained him.

"So, Mr. Allen," Will said. "You think that current is flowing from WPS' system into the ground? And up out of the ground and through your cows? And you think that takes place everywhere on your farm?"

I had to admit Will was a good actor. He sounded so skeptical that for a moment I almost didn't believe myself. "I think uncontrolled," I told him. "I don't know where it all is. I think it's all over."

"You think it's 24 hours a day?"

"Not when WPS has a power outage," I said, and the whole courtroom laughed.

Trevor Will didn't find it funny. "Any time the power is on, you think it's flowing?" he frowned. "If you want to know whether ground conductors are affecting animals, the first thing is to measure how much current is flowing through the earth, right?"

I looked at him. "I don't know," I said. "You're starting to talk on electricity and I'm not an expert on it."

Stopping dead in his tracks, Will spit out the next question. "Well, you offered a number of opinions this morning about whether electricity was harming your cows. Didn't you?"

"Yes," I said without blinking. I was ready to rumble.

Will wasn't—not yet, anyway. He said amiably, "If we want to know whether ground currents are affecting animals, the first thing we need to do is measure how much current is flowing through the earth, right?"

"Or remove it and see what the cows do," I shot back.

Will smiled. "One thing at a time. First, you want to know if there's any. So you'd measure to see how much was coming out of the earth and getting to the cows?"

I disagreed. "You'd measure how much is on the rings and how much is in the earth. And then you *remove* it."

Now Will grew a little more combative. "Well, Mr. Allen, isn't it true that you take a measurement to see how much current was flowing out of the earth and affecting the cows?"

Again I disagreed. "The best way is to remove it and see what the cows tell you."

Will didn't like that answer. "Mr. Allen, I'm referring to page 892 of your deposition. Line 19: 'Yesterday we were talking about earth currents affecting cows. You said that if you wanted to know to what extent currents affected cows, you'd want to know how much current was in the earth, correct?'

"Answer: 'Yes.'

"Did you give those answers, sir?"

"Yes."

"All right," Will said, looking pleased. "And if you wanted to know how much current is actually reaching the cow out of the earth, you'd take a step potential measurement. Correct?"

"I told you several times through deposition that I'm not an electrical engineer or expert."

"Well, does that mean you've never heard the term?"

"I've heard it several times, but don't know—"

He cut me off. "To your understanding it measures earth current that can flow through a cow, right?"

"Yeah."

"A couple of weeks ago," Will said, "had Mr. Neubauer told you he had taken step potential measurements and the numbers were so small they were insignificant?"

"I've heard it here."

"Mr. Allen," he said, slowly walking the length of the jury box, "you've been running around taking measurements on these ground rings yourself, for something like three years. And you've been taking measurements from WPS' system. And measurements on some ditch wires you ran. And you set the instruments up, put the data in your computer, right?"

"Correct," I said.

"And did you ever ask them to teach you how to take a step potential measurement?"

"No."

"Well, you agree, don't you, that the measurements you took on the down ground wires don't tell how much electricity was actually flowing through your cows?"

Where else did he think it was flowing but into my cows? Down the river, to the sea? "I believe the measurements on the ring don't capture all the current in the earth."

This was another of my answers that Will disliked. "My question is different. The measurements you take don't tell how much current is flowing in the cows, do they?"

"No."

"Okay. Isn't it true, Mr. Allen, that that you could've had your experts come to the farm and take step potential measurements any time you wanted?"

"Sure."

"Sure," Will said, nodding. "And if you wanted a measurement somewhere you thought there was a problem, you could tell them, 'Please take a measurement in the parlor, the cows are acting up'?"

"My understanding was that we had all the measurements we needed."

Will quit walking but remained near the jurors. "You've heard of the Public Service Commission's level of concern, haven't you?"

Only until it was coming out of my ears. So, now he was going to fall back on voodoo. "Yes."

"And you understand the level of concern is now one volt. That that translates into two milliamps? And you're not aware of any measurement that exceeds the level of concern, are you?"

I leaned back and crossed my left leg over my right. "I guess I don't know that."

"Well," he enthused, practically rubbing his hands together as he went over to his table, "let's get your deposition out." He picked it up. "You said, 'I don't think I've ever seen a level higher than the Public Service Commission.' Did you give me that answer?"

"Yes," I told him. "I did."

"All right," Will said, replacing the transcript and walking back toward the jury. "It's also true, isn't it, that you think you could live with one volt or two milliamps of current flowing through the cow and it wouldn't affect your milk production?"

So he was admitting there was stray current on my property and in my animals, caused by WPS? The only thing open for debate was how much? "I don't know," I replied.

"Well," he said and strode quickly over to his table, "let's get your deposition out. Question: 'Do you have an opinion as to what the threshold level is?' Answer: 'I'm thinking if there was 1 or 2 milliamps I could live with it.' Did you give me that answer?" I misunderstood the question in deposition; I meant on the buried ring.

"Yes."

"Okay." This time he kept the transcript with him. Maybe to slap me with later. "Mr. Allen, you've never seen a measurement that suggests there are more than 1 or 2 milliamps flowing through your cows, have you?"

I was getting angry. It showed, too, because I raised my voice, and snottily. "The only measurement I do is when I remove the current and see what the cow is doing."

"My question is different," Will snapped back. "You've never seen a measurement that indicates as much as 1 or 2 milliamps flowing through cows, either step potential or cow contact, have you?"

"No," I said.

"You said a moment ago, Mr. Allen, that when current was removed you saw production take off. If you didn't take measurements, how do you know current was removed?"

I shook my head. "That was an error; it was reduced."

"But if you didn't take any measurements, how do you know it was reduced?"

"Well, I reduced the current on the ring."

"I'm talking about what's flowing through the cow," Will snapped again. Both of us getting under the other's skin. "How do you know there was a change in flow through the cow if you don't take measurements?"

I didn't answer right away. He stared at me and blinked several times. Finally I said, "The only thing I go by is current reduced on the ring."

"So," he said, "if it turns out that's induced and not earth current, you don't have any basis for saying the electrical changes had any impact at all. Do you?"

Exasperated, I struggled to answer civilly. "All I can tell you is what I did. And what the cows did."

Will grinned. "Right. Now, about the electrical changes that took place since 1995, Mr. Neubauer has been responsible for those?"

I wondered if he was driving toward the idea that Larry had conspired with me to get rich by bilking WPS. If he was, I'd be ready.

"Yes."

"He was trying to remove what he described as loop current? And charged $9,800 to do that?"

"I believe so."

"And he told you the money you had spent on farm rewiring didn't solve the problem. And then in July WPS put up the Dairyland Isolator, right? And within a month, Mr. Neubauer told you it wasn't working. Right?"

"He told me and told Vern Peterson."

"Okay, yeah. He then began to suggest you needed to isolate, correct? Said that would be a $50,000, $60,000 project. Said you needed to isolate in order to fix the WPS problems. And he prepared quotations for you to take to various lenders to try to get financing for this. Correct?"

"That could be."

"Okay. Mr. Neubauer didn't tell you that you had an on-farm electrical problem in August of 1997, did he?"

"No."

"And that when the rewiring work that he did was finally inspected, it had 22 code violations in it, didn't it?"

"I guess you're correct."

Will patted the deposition transcript against his thigh. "Isn't it possible, Mr. Allen, that some of the work Mr. Neubauer did might've been responsible for what you say was the electrical problem with your cows?"

"I don't believe so," I told him.

"And isn't it true that he told you WPS didn't have your best interests at heart?"

My best interests? The only interests WPS ever had at heart were its own. "With regards to the isolator that wasn't working, you may be correct."

Will continued pacing. "Isn't it true that even before he installed the Dairyland Isolator he told you in a letter that WPS, to solve a problem, always designs around their system's capabilities, and in that message lies their interests—stockholders, legal costs, et cetera.? He told you that in a letter?"

Larry sure had. And he sure was right. But it didn't take a rocket scientist to know this. Anybody with an ounce of sense knew it. WPS, like all powerful corporations, existed to protect and perpetuate itself and its profits.

"Isn't it also true," Will said, "that he told you they were incompetent because of where they put the reference rod, whether the Dairyland Isolator was functioning properly?"

"He may have."

"And isn't it true that Mr. Neubauer told you WPS would not move your transformer pole because that would 'fix' your farm?"

"Yes."

"Okay. And isn't it true that Mr. Neubauer began taking you to *meetings* where *stray voltage* was discussed?"

Will's dark tone implied that I was a conspirator, as if mixed up in terrorist cells or going to KKK meetings. "I think I took myself," I told him.

Will didn't see anything funny about it. He frowned too. "Well, didn't he invite you to the meeting at Helzer's farm in summer of 1997?"

He had me there. "I was invited, yes."

"And it was by Mr. Neubauer, wasn't it? In fact, Mr. Neubauer helped sponsor that one, didn't he?"

Grinning, I said, "I really don't know."

"Mr. Neubauer also got you to a meeting in 1997 at Vande Hay's to discuss stray voltage, right? And told you that farmers with stray voltage problems could bring lawsuits and recover damages?"

A regular Lenin, this Larry Neubauer was, forever stirring up the peasants and fomenting revolution.

I shook my head. "I think a farmer would know that."

"And you knew that?"

"Well, any time somebody is causing damage…"

Will stopped pacing and pointed the transcript toward me like an accusatory finger. "Isn't it true you discussed with Mr. Neubauer bringing a lawsuit for stray voltage?"

I thought: *So what if we had? People are free to discuss anything they like concerning their business.* "I don't think Larry is the issue with a lawsuit," I growled.

"No—my question was different," Trevor Will said, lowering the transcript. "I just want to know if you talked with him. Isn't it true that by January of 1998 you discussed bringing suit against WPS?"

"I don't know," I said.

"I want to go back a little, to July of 1997. Isn't it true that when you were talking with Vern Peterson you were not a happy camper?"

That was putting it mildly. "Probably."

"And isn't it true that you might have said something to the WPS people that they might have misinterpreted as an instruction to get off your farm?"

I did a double take. Were we really back to this again? Did he believe that harping on this nonsense would influence the jury against me? If he was counting on that to beat me, I was in better shape than I thought.

"Mr. Peterson had said, 'I want to go into the basement of the house and start disconnecting things.' Right?"

"Correct."

"And you said, 'I don't want you to do it.' Correct?"

"Yes."

"Isn't it true at that time WPS left your premises and didn't complete the testing?"

"I didn't know it needed retesting."

"Well, they left your premises. And you recognize you said something which in your view they misinterpreted as an instruction not to come back. Right?"

"I don't buy it," I told him.

I supposed Will's idea was to try showing that WPS could've fixed things if only I hadn't been so miserable to their nice, helpful employees.

"Now," Trevor Will said, "I want to run through the chronology of isolation. Isn't it true that his isolation work didn't have any immediate impact on production?"

"When you say 'immediate'—"

"How about a month? Didn't have an impact on production the first month, did it?"

I refilled my water glass from the pitcher and took a long drink before answering. "The cows quit dying," I told him.

"I'm talking about production."

"Okay. It took a little bit. A month and a half, two months."

"So the first couple of months there wasn't any change?" He came toward me and stood about ten feet away, transcript still in his hand.

"It took a while for the cows to heal."

Will fell silent. Then he said, "Isn't it true that you were talking with Mr. Neubauer about filing a stray voltage lawsuit even though you hadn't seen an increase in production?"

"That never happened," I said.

"Isn't it true that in February of 1998 you were telling Mr. Neubauer that you got hold of Jack Kevorkian's attorney, over in Michigan?"

"Yeah."

"Okay. So by this time, then, you were already talking about a lawsuit, weren't you?"

"Could be."

"Now you knew that to succeed in your lawsuit you have to show an increase in milk production after making electrical changes?"

"Right. I would've had to see a response before I did anything. You don't have a case if you don't have an increase."

"Right," Will said, nodding vigorously. "There was a meeting in November 1997 with you and your veterinarian and nutritionist and electrician and also the county ag agent, right? At that meeting you discussed whether there were problems with your management style, right?"

"Correct."

Still nodding, as though his head were attached to an elastic band, Will, nervous as ever with all his facial tics, began walking back and forth in front of the witness stand. "Now, isn't it true, Mr. Allen, that there are probably 20 to 25 factors that can influence milk production? For example, balancing rations, quality of feed, use of BST, times milked per day, weather, cow comfort, lighting, vet care, handling by milkers?"

"Sure."

"And isn't it true that a large number of things can be controlled by the farm owner?"

"Sure."

Will next went through a laundry list of everything I'd done to improve productivity. "By May 8," he said, "you were reporting to Mr. Neubauer that you had production up to 74 pounds per cow. And you called him back the next day and said it was up to 75. That was his guarantee, right?"

"Yes."

"I mean, you would call his office and leave messages when he was out, right?"

"I was pretty excited when we were climbing."

"Sure," Trevor Will said. "One of the reasons you were so excited was because this would help with the lawsuit. Right?"

"I got to say no," I told him.

"Now, we were going to talk about the Public Service Commission." He went over to the counsel table, put down the transcript and picked up some papers. "This, Your Honor, is from Exhibit 669, page 523. This indicates Steve Sauer is on vacation. Do you know Steve?"

"He was an attorney, at one time with the Public Service Commission of Wisconsin. He spoke at the Vande Hay stray voltage thing."

"The meeting," Will said. "Then you said to Larry, 'Not placed to file complaint.' Doesn't this suggest you were calling Mr. Neubauer for advice on whether to file a complaint with the Public Service Commission?"

"I've got a difficult time figuring out what it says."

"To your knowledge Mr. Neubauer is not a lawyer, is he?"

"No."

"You never contacted the Public Service Commission about your situation, did you? The Public Service Commission had something called a Stray Voltage Action Team, a S-V-A-T team? They would come out to your farm and inspect for stray voltage, right? And you never asked him, them, to come out to your farm. Did you?"

"No. Correct."

"All right," Will said. "Let's move on. In November of 1998 you were going to meet with a lawyer about this lawsuit and invited Larry Neubauer to come along. And by December of 1998, isn't it true that Mr. Neubauer put you in touch with four other farmers, and the five of you were going to join together and bring a lawsuit?"

Sure, we were five Klansmen suiting up in bed sheets and pillowcases to lynch helpless WPS. Neubauer was the Imperial Wizard. "I'm not sure Larry did that. I knew the other farmers."

"One of those farmers was a fellow by the name of Vern Manke?"

"Manke."

"And he's from Michigan. There was a problem with him joining the suit, wasn't there? He had made electrical changes but didn't have a production increase, correct?"

I shrugged. "I don't know the specifics with Vern Manke. I do know that at one time when Larry walked into his barn he saw a grown man crying."

WPS had wrecked his business too, and good.

"Mr. Allen, this note of Mr. Neubauer's regarding your call to him, 'Vern Manke at 50 pounds, got to get him cranking if he is one of the five,' is correct?"

"Correct."

"So you had decided to file suit by December 1998?"

Again, I shrugged. It didn't matter much when I'd decided. Fact was, we were here in court. "That's very possible," I said.

"Two weeks after the document about Manke, you called Larry Neubauer about a videotape taken of your cows, taken by Doug Sutter, the county ag agent, right?"

"Could be."

"You wanted to demonstrate the cows' hooves were bad. And that they were acting up in the parlor, right?"

"Right."

"And isn't it true Mr. Sutter told you he'd seen worse dancing on other farms?"

"Yes." But did that mean things were good with me, because they were worse in other places at a given moment? For certain WPS was an equal opportunity electrocuter.

Trevor Will turned and went over and stood beside the jury foreperson. "And isn't it true that Mr. Sutter says he gave the tape to you? And that the tape has disappeared?"

"That is correct," I said.

"And isn't it true that you refused to let the University of Wisconsin come out and take samples at your farm?"

"Yes."

"You were worried, weren't you," Will said, "that if they came they would conclude there was no stray voltage on your farm?"

I glanced over at Barry. He was scribbling away on a yellow pad. "Stray voltage had nothing to do with it," I said.

"Isn't it true that the reason you did not let them come is because you had issues with Dr. Douglas Reinemann?"

Issues? The guy was a flack, a shill, a mouthpiece, a propagandist and a paid puppet for WPS. He was an arrogant enemy of farmers. Why would I have an issue with him?

"That is correct," I replied.

Will folded his arms and started pacing along the jury box. "You understood that Dr. Reinemann does not believe earth currents can cause dairy cows harm?"

"Utilities are relying on Reinemann's research, and it's not coinciding with what's happening in the real world. *A lot of cattle and a lot of families are being harmed.*"

"And so you disagree with what you understood to be Dr. Reinemann's position?"

"I don't agree with the research," I told him. I crossed my legs and shifted my weight in the chair. My butt was getting stiff.

Will nodded. "All right. So you didn't want him on your farm?"

"It's sad that I shunned the whole university because of Reinemann's research. I was upset. Sad. Yes."

Again Trevor Will nodded. "In 1998 and 1999 Mr. Neubauer was a frequent visitor to your farm: You were always talking to him about stray voltage, right? And during this time period—December of 1997 until November of 2000 when you filed this lawsuit—you never asked WPS to come out and test or measure for stray voltage, did you?"

I tried to remember the dates. "No, I don't think so."

"And the reason," Will said, "is because you and Mr. Neubauer were working to put the lawsuit together, weren't you?"

Back to the conspiracy theory again. Oliver Stone would've been so proud.

"Well," I said, snickering, "if that's the case, I would have made sure he got more measurements and pictures and everything else."

"More measurements would've been helpful, wouldn't they?" Will said sarcastically.

"That," I said, snippy as a grounded teenager now, "is an incorrect statement you made, Mr. Will."

Trevor Will did not like hearing that. He scowled. "Well, isn't it true that in May 1999 you called Mr. Neubauer to talk about the lawsuit?"

"Sure."

"Do you remember the note of your conversation with Mr. Neubauer that was subpoenaed in this case and not produced? You said, 'after we win this lawsuit, on the back of my boat, like at the end of a letter, I will put instead of P.S. I love you, WPS I love you.' That's what you left for him that day, correct?"

"That is correct," I said.

"You own a 27-foot boat, correct?"

"Correct."

Will paused, then said, "In 1999 you sat in on a stray voltage trial, the Voss case, down in the Manitowoc area. And in 2000 you went over to Waupaca and spent four days sitting in there on the Hoffmann case. And that was all before you brought this lawsuit, correct?"

And before I brought this lawsuit I dated girls and ate in restaurants and rooted for the Packers. What did that have to do with the price of milk?

Trevor Will went away from the jurors and tiptoed to his counsel table. He looked at some papers and then came tiptoeing toward me. "When WPS people came to your farm on June 27 of 2002 to take measurements, one thing they did was investigate that buried ring, right? Excuse me, the surface ring."

"That's correct," I said.

"We've heard about loggers you've used in this case. What you do is place them at various locations around your farm and hook them up to wires you have. Then you take readings to see how much current you're measuring?"

"I download the loggers."

"Right. You load the data into your computer?"

"Correct."

"The reason you started logging was for purposes of this lawsuit?"

"The purpose of logging the down grounds is to see what the heck is going on."

Will stared up at McKay. "The person you initially discussed logging with was Gerald Bodman, correct? And he was an expert that you retained for purposes of this lawsuit?"

"Correct."

"And after Mr. Bodman suggested you do it, there was a conversation with your lawyer. At that point you decided to go forward, right?"

"Yes."

"You were trying to prove it was coming from WPS' system, right?"

"We wanted to figure out what was going on so we could redirect it so it ain't coming through to the cows."

"In April of 1998," Trevor Will said, "Dr. Homan prepared a letter for you. You said this morning that your milk goal was 90 pounds a day. Isn't it true that it was 80 pounds a day as set out in Dr. Homan's letter?"

"We set a new goal by then."

"Fair to say that he lists a number of things in order to reach 80 pounds, a number of them that you did not do?"

I shrugged. "We didn't change a thing."

"Now, Mr. Allen," he said, "I want to shift gears and talk a little about crops. A crop year is crops you grow in a calendar year, what you're planting this spring and harvesting this fall, right?"

"There's some also you're going to harvest through the summer."

Will moved backward a few steps. "As a general proposition, you generally feed the older foodstuffs before starting with the newer?"

"You would like to, but in general that ain't what happens."

"Mr. Allen, you now have crop insurance, correct? And you've had it as long as you've had the farm?"

"Yes," I said.

"Do you know what level you're insured at?"

For a moment I drew a blank. "Sixty-five percent, maybe," I said finally.

"Now in 1996 you had a major crop failure. Both your alfalfa and corn were essentially wiped out that year. Isn't it true, Mr. Allen, that you received over $200,000 in disaster relief because of crop failures in 1996?"

"That could be true, yes."

"One of the things you did in '96 because of that crop failure was you planted peas and oats as a cover crop for the new alfalfa. You will typically harvest them and put them in the silo as well, won't you?"

"Yes."

"Isn't it true that once peas and oats are removed from the ration you would expect cows to produce more milk?"

"Not necessarily."

"But it may be true?"

"It may not be true."

"Will you agree it depends on the situation?"

"If you keep the ration balanced," I said, "they should produce the same."

"Do you remember that in 2000 you asked to have your loan payments restructured because of a crop problem in 1999?"

"I don't remember."

"Do you remember restructuring at all?"

I clucked my tongue. "Seems like I was always restructuring."

"Mr. Allen," Will said, "as part of your job as a farm manager, you subscribe to various dairy publications. In one of them, *Country Today*, there was an article about you. And in your deposition you told me that Larry Neubauer was the first person to tell you stray voltage electricity could be a problem for cows. But that's not really true, is it?"

"I've read about it."

"And as time went on, you continued to read and hear about stray voltage. Didn't you?"

"Run into it in the papers now and then."

"Isn't it true that before November of 1994 you had seen articles about stray voltage?"

I had no idea where he was headed with this and didn't much care. For all his probes and lines of attack, I honestly felt Trevor Will wasn't making much headway. My confidence in my testimony and my case was growing.

"If they were in there and I read them," I told him, "I'll agree."

Will inched closer toward me and put both hands in his pockets. "You give the cows their BST shots, right? Bob and Patti give other medications to the cows?"

That was true.

"Fair to say you know exactly what medication they give?"

"I know some of it."

"Not all?"

"There's always new stuff coming out."

"You said that at the end of '96 you were heading for bankruptcy?"

I nodded. "I was in tough shape middle of '97, but the end of '96, sure, it was not a pretty picture."

"Didn't you indicate at your deposition that toward the end of 1997 you had $100,000 in unpaid bills that you needed to deal with? That was when it was worst?"

"The checkbook was empty," I said. "I didn't know where to turn to pay bills."

"You never missed a payroll? Never laid anybody off?"

"I asked people, during the winter, to not work as many hours."

"And you never missed a loan payment? Never sold farm assets to raise cash? Never sold any cows you'd otherwise keep just to raise cash?"

"Correct."

"Isn't it true you were never required to go to a lender and try to renegotiate a loan at the end of '97 because of your financial condition? And your utilities were never shut off? And you never got sued by anyone for nonpayment?"

"That's correct," I told him.

"Nobody started foreclosure proceedings against you? And you were able to get the M & I Bank to loan you an additional $58,000?"

"Correct."

Will nodded, turned, and again tip toed over to the jury area. "Now, you believe, Mr. Allen, that stray voltage was the only thing limiting production on your farm from 1976 to today—the only reason the cows didn't produce more was stray voltage, right?"

"Yes, yes."

"You don't believe there's anything within your control that would have improved production, right?"

"Correct," I said.

"And you also think that in the last three years, all changes in production, behavior and health in your cows have been due to earth current flowing through them, right?"

"Yes."

Will looked back and forth among various jurors. "And there have been ground rods and grounding wires on the distribution line since 1976. And yet Dr. Behr's graph shows that in 1978 your rolling herd average was 9,800 pounds, production 6,600 pounds on a monthly annualized rate. By January of 1997 you'd been able to double your rolling herd to 18,600 pounds. You managed to do all that while current was being put into the ground, didn't you?"

"Yes."

"And you've continued to make changes since," Trevor Will said. "I want to talk about some of those changes. As you explained, in the '70s and early '80s decisions were made by committee, by partners of the group. Each partner was responsible for a particular area of farm operations. Within that area, they tended to make their own decisions, unless it was a big investment or something that required the group. Correct?"

"Yes."

Will nodded. "Today you're the guy who makes decisions for all areas of the farm, aren't you? Yours is the final voice?"

"Yes."

"A, you're the boss. And B, you've got the checkbook. Right?"

"I got an empty one," I told him. Several jurors smiled as I said this.

"Do you," Will said, "agree that feeding or being a nutritionist is an art?"

"When you fine-tune and get into that 30,000-pound range, sure."

"In fact, you have parted company with some nutritionists because you didn't think they were getting results for you, right?"

I shook my head. "I blamed a lot of people that I shouldn't have."

"Like WPS?" Will said quickly.

"*I think we've proved differently,*" I said equally quickly. I thought I saw a juror nod ever so slightly, as if in agreement. But maybe it was just my imagination.

Ignoring the remark, Will continued, "Now, there have been a number of different rations prepared over time. In 1997, you were feeding three rations to the lactating cows. Then you went to four. Now you're back to just two for the herd. Those things have been changes that may affect production. Right?"

"Not necessarily," I said. And with that, we broke for the day.

That night, I barely slept. Tossing and turning, I tried to prepare myself for another heavy dose of Trevor Will in the morning. I even dreamt of him: In my dream I was seated on the stand testifying and whenever Will didn't like an answer, he ran up and poked me in the neck with an electric cattle prod that Judge McKay handed him, the pair of them laughing and joking as the sparks flew and my skin smoked.

I sure hoped tomorrow wouldn't be quite that bad.

Chapter Seventeen
DOWN THE STRETCH

Both Barry and Scott were pleased with my testimony thus far. The worst of the grilling, they whispered as we filed into the courtroom, was over. Whether that was true or whether they were simply trying to give me confidence remained to be seen. But I thought I'd held my own fairly well yesterday. Today, though, was a new day. And the end was in sight, whatever it might hold. We were galloping down the home stretch.

The image of Trevor Will zapping me with that cattle prod was still stuck in my head; I couldn't get rid of it. Will, wearing a light gray suit, white shirt with French cuffs and silver cuff links, and sporting a burgundy tie, picked up exactly where he'd left off: Badgering me about my management. Or lack thereof.

After briefly studying his notes, he came up near the stand and stood with his arms folded, close enough that I could smell his cologne.

"When we broke yesterday," he said as soon as I was sworn in and seated, "we were talking about changes on the farm, nutrition. If you go back through your ration sheets, will you see that over time the ration has been balanced at different amounts?"

"Very possible," I said.

"You rotate your crops year to year?"

I told him we did.

"And crop quality varies?"

"Yeah."

"You now have Dr. Homan as your vet, preceded by Dr. Foust, right?"

"Correct."

"I don't want to say, 'fired', but you elected to stop using Dr. Foust because he was not open to stray voltage?"

"He was a good vet," I said. "We needed answers and it just happened that way. Terry Homan did not believe in stray voltage and ended up being our regular vet."

"You culled 32 cows at the end of 1997. And you recognize that by culling cows, that alone would increase production in the following month? Culling low producers."

I shook my head. "That had no effect."

"My question was different," Will said. "As a matter of mathematics: If you cull 32 low-producing cows, that will increase your average in the following month. Won't it?"

Again I shook my head. "If you go two or three months beyond that, it had no difference."

"Again, my question was different," Will snapped.

"Maybe you might see it for a month or so," I told him. "That would be it."

Will looked across at the jury and then went over to them. He turned and faced me obliquely. "You mentioned that in July 1997 WPS hit its highest system peak?"

"If I'm not mistaken."

"Isn't it true, Mr. Allen, that the disruption in production on your farm in July 1997 was caused by heat?"

It was a wonder he didn't blame it on little green men from Mars. "I don't believe so," I said.

"Isn't that what you told Larry Neubauer at the time?"

"Could be."

"All right," Will said. "I want to talk about disease and health on the farm. You never calculated disease rates?"

"No."

"And you never tried to correlate incidence of disease with measurements on various wires?"

I shook my head. "I go by how many cows were in our treated group."

Will frowned at me. "Isn't it true, Mr. Allen, that you've been told your animals get pneumonia because of the slats, living over manure pits?"

Now it was my turn to frown. "That's Mr. Mellenberger's opinion." WPS' top gun.

Trevor Will squinched up his eyes. "Do you agree, Mr. Allen, with the principle that if you can't measure it, you can't manage it?"

"I agree," I told him.

"And isn't it true that you don't have a computer system that allows you to get information on BST usage? You don't have the capacity to do that with your records?"

"Ninety to 95% of the dairy farmers in the country can't do that." What did he think, that we were IBM?

"Right, and you can't, can you?" Will said, sounding like a cranky kid.

"That's correct."

"And so Dr. Mellenberger had to take hours and hours and hours to go through boxes and boxes of records to come up with that kind of data. Didn't he?"

Poor, misused, downtrodden Mellenberger, having to sift and sort through all those dusty boxes to earn his million-dollar pittance.

Earlier, they'd accused me of having too few records, and then suddenly Mellenberger had to sift through tons and tons of record boxes. Which was it—did I have records or didn't I? They were sending mixed signals when it was convenient for them. I didn't have any records, and when they were trying to justify Mellenberger's pay it was boxes and boxes and boxes of records; they were trying to have it both ways. And if you matched my records against any records they had showing maintenance on their system, mine would've stacked up a thousand times higher. Full of split bolts, their system near my farm was of 1935 vintage. WPS possessed no record of having ever inspected it.

Barry stood up here and objected to the last question and McKay sustained him.

Will backed off and delved into something a little different. "You have told me," he said, coming toward me, "that you think earth currents have been there from WPS' distribution system since 1976 to the present, correct?"

"That is," I said.

"And yet during that time your production has almost doubled. Hasn't it?"

"That's probably correct," I said.

Here was the refutation of the idea of my bad management. The farm was able to grow despite the constant electrical contamination; my methods lessened it to some extent. Which proved that under normal conditions we should've increased, probably, ten times or better during those years. I felt a deep regret over what might've been. Over what *should've* been.

"So earth currents aren't preventing you from dramatically increasing, are they?"

I grinned. "I'm doing my best to keep the current away."

Will shook his head as if confused. "The first thing you ever did directly aimed at earth current was the ring in the spring of 1999, correct?"

I knew what he was doing here and tried to head him off. "Moving the pole away from the feed room had some effect with earth current," I said.

"Wouldn't you conclude, Mr. Allen, that earth current had absolutely nothing to do with any change in production from 1976 through 1996?"

"I don't *think* so."

"Mr. Allen, did you ever consider the possibility that the east barn is holding your farm back?"

"I firmly believe that facility is capable of at least 27,000 pounds of milk," I told him.

Will folded his hands together in front of his waist. "Mr. Allen, I want to ask a couple of questions about cutting the grounds. You first cut WPS ground wires in spring of '99. And you were asked by WPS in the fall of 2001 not to cut them again, correct?"

That was very true.

"And yet you went ahead and cut them again, correct?"

That was also very true.

"Now, you told us you thought there were improvements after the groundings were cut in December of 2001, right?"

"Yes."

"And yet you know there was no production change outside the range of normal fluctuation?"

"Ours steadily climbed," I said forcefully, for the jury.

"Is it true," Will said, "that when you looked at averages for December of 2001 and January of 2002 after the grounds were cut that they just showed normal variation?"

I snickered. "It don't happen that quick, Mr. Will."

"Do you understand it's your testimony that it takes three or four months for cutting the grounds to show up in production?"

Understand it? I was the one who told *him*.

"After I cut the grounds, it was a steady climb," I said.

"Isn't it true that your herd always makes a steady climb in the spring?"

"Not like we were seeing."

"Mr. Allen," Trevor Will said, "I want to change topics a little. "Do you know what a normal death rate is in a herd of 500 cows? One that's completely unaffected by electricity?"

"I think three or four percent."

"Well, let me get your deposition out," Will said, going to his table and picking it up. He thumbed through the pages and then held it out, a priest flaunting a Bible. "Page 830: 'What type of death rate would you expect in a

500-head herd not affected by electricity?' Answer: 'I don't even know.' Did you give me that answer?"

"Correct," I said.

"You don't know what it is over the last seven years, do you?"

"I know an awful lot of cows died," I snorted.

Will made a sour face. "I don't want to argue with you," he said, "but could you confirm for me that you don't know the death rate?"

"That's correct."

"Isn't it true," Will said, turning belligerently, "you didn't even know what the cutoff was from Land O' Lakes as to the somatic cell count? Didn't know if it was 300 or 400,000. Did you?"

"I had an idea."

"You talked yesterday about the new parlor in the east barn. You didn't know when I took your deposition whether that was leased or purchased, did you?"

I started to get angry. I didn't mind him trying to make me out as incompetent, but now he was implying that I was stupid as well.

"That was a trick question," I told him, sounding as belligerent as I could.

"It wasn't a trick question! You didn't really know if the lease was up and you purchased the parlor, did you?"

Here Barry objected that Will was being argumentative. Which he clearly was. McKay, on one of his rare occasions, sustained him.

So Will tried another tactic: Making me seem oblivious. "Isn't it true you don't know when your electronic ID system was installed?"

"If I'm not mistaken, when the RFs were installed."

Will marched up near the stand. "Let's look at page 759 of your deposition. Question: 'You had in the barn an electronic identification system for the cows. When was that installed?' Answer: 'I don't know exactly.' Was that your answer?"

"Yes."

"Okay. Now, there was a time when you turned that system off to see if it affected the cows, correct? And you don't even remember when that took place. Do you?"

I couldn't decide if Trevor Will was trying to make me out to be a fool or an Alzheimer's patient.

"Not exactly."

Will tapped the deposition against his thigh. "Now, Mr. Allen, you rely on herdmaster records instead of using DHIA. If an animal is not entered into the herdmaster system, you won't have a good record, right?"

"Not necessarily. If we have a card on that cow, we still have pretty good records."

"But in terms of the computerized milk system, you won't have a good record if you don't have the animal in herdmaster. You don't even know, Mr. Allen, at what point animals are even entered into your computer system. Do you?"

"It's entered when it goes across into the milking herd."

"No," Will said shaking his head. "I mean the first time. You don't know at what point a springing heifer or a purchased cow is entered in as being one of your animals. Do you?"

I took my sweet time answering, sipped some water first, shifted position. Let Will wait. "After they have their calf and they're in the west barn a couple of days, when they go to the east barn they get a transponder put on. That's when they're entered into the computer."

Will kept patting the sheaf of papers in his hand against his leg. "Isn't it true you don't know how to read a great deal of data on herdmaster records?"

In addition to being incompetent, ignorant and oblivious, now I was illiterate.

"There was some confusing data there," I told him.

Will recited a litany of the information and cow data I'd either forgotten or did not fully understand, until Judge McKay wearied of the subject and warned him to move on.

"Okay," Will said to the judge. Then to me, "You told me you didn't want the three-phase line reconstructed because that would mean more grounding wires and rods. You're not in favor because you recognize that if you have a multigrounded system, you're going to have some current that, as a matter of physics, goes in the ground. Right?"

Now, that was right as rain. "That current is harming my cows!"

"You know that with a multigrounded system, there's no way to avoid putting some current into the earth. Correct?"

There it was. His case in a nutshell. Current was there all right, they admitted that, but according to WPS and its cohorts on the Public Service Commission and in the universities, the amount was within acceptable levels. They were tiptoeing a fine line, doing a delicate balancing act; they were going to have trouble bringing in their verdict. I felt it in my bones.

So I tried hammering a big nail into their coffin lid. "It's a bad system," I said, leaning forward and looking past Trevor Will and at the jurors.

"And what you want, Mr. Allen, is a system where there is zero current going down ground wires and into the earth?"

That was it, exactly. That was what I wanted and needed, just as I needed air to breathe that was free of poisons. Every person and creature in the state of Wisconsin and in the United States of America deserved no less.

"That's correct," I said.

"Yet you don't know of any system anywhere in the world that has those attributes you want, do you?"

Man, but I felt like I was on a roll. Will's logic was ridiculous and I was confident the jury would see it. Why should citizens be expected to instruct corporations on how not to destroy their property and the environment? It was for those causing the harm to cease and desist. They knew full well what they were doing. All people, including employees of WPS, including lawyers for the utility, including its spokespeople and flacks, deserved to live on clean, unpolluted American soil.

But here I gave a bad answer; I should've shouted to the rafters what I just wrote. You always come up with the best stuff to say after the fact. I said lamely, "I'm sure there are safe systems to use."

"In fact," Will said, "you publicly stated that you think an ungrounded delta would be the system Wisconsin ought to adopt. Right?"

"Whatever is safe," I said. There was the essence of things.

Will went to the table and dropped off the deposition papers and then tip toed again toward the jury box.

"Now, Mr. Allen, you've never asked your neighbors if they wanted their power shut off, have you?"

"If they realized how much current was going through their property, maybe they would want it off." A few jurors chuckled.

At this point Will decided he had no more questions; I'd see him again though, since Barry planned a redirect. I figured Will would hog the last word and try to go out with a bang.

After a recess, during which I ate an apple and some cashew nuts, I went back to the stand and got ready for Barry. I was in good spirits. Barry and Scott were hopeful of our chances.

"Mr. Allen," Barry said, coming near the stand, "what's your understanding as to whether all cows are the same with resistance or not?"

"Some are more affected than others."

"Was there current in your parlor?"

"Yes."

"What level?"

"Over 800 milliamps."

"In the real world, what would you do to test if your cows are affected by electricity?"

"Take away the current."

"And what would you expect to happen?"

"Well, if they go up in production, that's whats causing the problem."

"On your farm," Barry said, "what's the best indicator of the current affecting your cows?"

"Behavior," I answered. "Production."

"In your opinion, what's the best measurement device for the amount of current affecting your cows?"

"The buried ring."

"Okay. And do you have a maximum level in mind that you think will cause trouble to your animals?"

"One to two milliamps," I said.

"If it goes above that, what do you believe?"

"It's going to be a problem."

"Do you," Barry asked, "realistically believe that all current in the ground can be removed?"

"Darn close," I told him.

Barry paused. "What did isolation do concerning current on your farm?"

"Removed the majority off the parlor."

"What measure did you use to see if it had an effect on your cattle?"

We were pounding away, both of us, on the key points. "Production. Cow behavior."

"Does current have to pass through a cow to affect it?"

"I don't think so. If you have an electric fence and a cow gets a shock off of it, it's never going near it again. Same thing with the waterer, the feeder, any of that stuff."

Barry stepped over toward the jurors. "How many times does it actually have to go through a cow before it would have an effect?"

"If she's just exposed one time, it could have a lasting effect."

"After July of 1997," Barry said, "did you do more research into stray electricity?"

"Sure," I told him. "I heard about other people having problems with it—Al Mazna, Dave and Dan Vande Hay, Jim and Vicki Helzer—and I talked to them about it."

"What was the reason for talking to those folks?"

"Well, things aren't going well at the farm, you are left out there by yourself." I was too embarrassed to say that misery loved company.

"Did these conversations play any role in your decision to put in an isolation system?"

I nodded. "Sure."

"What did you find out as to the effect of isolation on their farms?"

"After current got removed, the cows at times reached 85 pounds a cow."

"Did you," Barry asked "consult with the Brown County ag Agent as to effects of isolation on farms?"

"Yes, I sure did." That was Doug Sutter. "He told me that in places, it was working well."

"Did you come to any resolution in the event your experiment showed electricity from WPS was harming your cows? What were you going to do?"

I raised both my arms. "I wanted to be compensated for it."

Still with the jury, an elbow against the box rail, Barry said, "How were you going to go about this? Have you ever sued anyone before in your life?"

I shook my head. "No."

"Can you tell us what you knew about stray voltage lawsuits at that point?"

"I know that when you go up against a major corporation that has an open checkbook—"

Trevor Will vaulted to his feet, almost apoplectic to shut me up. This was the last thing he ever wanted the jury to hear.

"Your Honor, I move to strike this!" he hollered.

McKay nodded. "Motion granted. Jury is to disregard that answer as nonresponsive."

Personally, I thought it was very responsive and I liked it a lot. But even McKay couldn't put that particular genie back in the bottle. And it was good for the jury to hear the truth. Because they were footing the bill as well. WPS could have an open checkbook because whatever they paid out was recouped in the form of rate hikes, which the Public Service Commission granted them at the drop of a hat. In a real sense, I was helping WPS to fund the case against me. The law should be changed, it should come out of the utilities profits, then things would get fixed in a hurray.

Barry grinned. "Did you find out what stray voltage lawsuits were all about?"

"Sure. I sat in on a couple."

"When did you actually retain a lawyer to help you?"

"October 2000," I said.

"And who was that?"

"Hammarback Law Office."

Barry nodded. "So, other than talking to lawyers up through that time, you never hired one?"

I had not.

"Did you," Barry asked me, "go to trials after January 1998?"

"Sure. You want to see how it goes. How it's done." I reached for my glass and sipped from it.

"You found out that hiring experts for one of these cases can be pretty expensive? That's why you talked to these other folks about maybe trying lawsuits together, to save costs?"

"Correct. It just didn't work out."

"What conclusion did you eventually arrive at as to whether there was a prospect of settling your case with WPS?"

"I felt it wasn't going to happen. Jerry Held said, 'We just don't hand out money to anybody.'"

"Let me clarify something," Barry said, still with the jurors. "Do you actually live on the farm?"

"For the last four or five years," I explained, "my mother was confined to a wheelchair. I bought the farm from her on a land contract. She had the right to stay in the home until she passed. We would take turns staying with her. I stayed two nights a week for the last four years."

"Did you have a place of your own?"

"Yes."

"And did you actually live on your boat a couple of summers? Isn't December in Green Bay on a boat a little tough?"

"Yes," I grinned.

Barry started toward me, then thought better of it and returned to the jurors. "While we're on the subject of the boat, did you talk to Larry Neubauer about changing the name of your boat to something relating to WPS? How did that come up?"

He was referring to my "WPS, I love you" comment.

"Wisconsin Public Service made my boat an issue in this lawsuit. Someone there provoked me to respond the way I did. They addressed their electrical issues by—digging up dirt on Russ Allen." Or trying to. There simply wasn't any. My life was pretty bland.

"So when you gave Larry Neubauer that message, were you kidding or not?"

"Well, yeah."

Now Barry walked away from the jury and toward me. "On limiting factors on your place, do you think your barn would be?"

"Sure," I told him. "I think we can get 27,000 pounds easy. If you want to get in the higher range, though, you start looking at other things."

"At the level you're now at," Barry said, "what do you believe is the limiting factor at the 27,000-pound level?"

I sat upright in my chair. "The only thing limiting is that current."

Barry paused.

"Can you," he asked, "ever think of a time when there were rations fed below the level the herd was producing at?"

"Never."

Barry started shooting the questions rapid-fire. "There were changes in vets over the years, in nutrition, in milking frequency, hoof trimming, changes in personnel. Based on your experience as a dairyman, were any changes significant enough to cause a milk production of over 30 pounds?"

"Those changes don't cause stuff like that," I said.

"Do you know of anything of a sudden nature on your farm that could cause that big an increase from 1997 to 1999 if we exclude electricity?"

I drew a deep breath and sighed. "No."

"Have there been any significant changes in capacity of the farm before and after December 1997? Population density of the animals? From when the isolator system was put in until afterwards?"

"No."

"As a backup to your computer records on individual cows, you still keep cards?"

"We do."

"Right up to today?"

"Yes."

Barry was a machine gun of questioning, and I fired back my answers. He asked everything without looking at any notes. I was impressed.

"You were asked about electrical systems. You're not an electrical engineer?"

"Nope."

"In terms of what it would take to prevent electricity from getting through on your farm, do you know?"

"I've heard of a uniwire system, where the grounds are not hooked to the neutral and pumping current in the dirt. They have a separate wire for each. I'm not an expert."

"That's an engineer question?"

It surely was.

"To clarify the question Mr. Will asked—whether or not you wanted the electricity on the whole feeder off, or just your electricity—if electricity to your farm was turned completely off, would there still be ground currents on your farm?"

"We demonstrated that, with the loggers," I told him. "Even if I disconnected from WPS right now and ran on a generator, currents are still going to go through my farm. All that current is *still* flowing through my property."

Barry nodded rapidly. Then he said to McKay, "That's all the questions I have, Your Honor."

I glanced up and noticed the judge fidgeting. "Any recross?" he asked impatiently. It was clear McKay wanted to take a break. Maybe he needed the men's room. I knew I did.

But it was the bottom of the ninth inning, two outs, and Trevor Will was not about to put off his last at-bat if he could help it. "Just a bit, Your Honor," he said. "I'll keep it brief."

He rose from behind his table and came forward. "In terms of the .657-volt reading from Mr. Neubauer, that was taken in June of 2002, after the lawsuit had started? And after Mr. Neubauer had done about $230,000 worth of electrical changes to the farm?"

"Yes."

"It was taken on a system that did not look at all like the system in place in December of 1997?"

"You're asking tough questions about the electrical thing. I don't know a whole lot about it."

"There had been modifications made?"

I nodded. "It was changed."

"Right," Will said. "Now, you were asked about the .914 measurement. Would it be fair to say, since you were not there to see Mr. Neubauer taking that measurement, you don't know exactly how he did it?"

"I wasn't present when he took it," I said.

"Now," Will said, "you mentioned the 800 milliamps—that was current clamped on the parlor steel?"

"Correct."

"Again, Mr. Neubauer told you that?"

What did he want, to put Larry on trial? Larry had already testified. "Yes," I said.

"In terms of your contact with Jerry Held, it's true you called him about getting compensation?"

It certainly was.

"And you refused to provide information?"

I shook my head in disgust. These WPS people would fabricate anything. The truth was a stranger to them all. "I didn't refuse to provide it. He led me in a different direction. I knew what he was thinking."

"Well," Will said impatiently, "you *assumed* what he wanted to do. Correct?"

"I made an honest attempt," I grunted.

"Mr. Allen, is it fair to say you did not give him the information he requested? So there was no further discussion between the two of you on a settlement?"

"Correct."

Will paused. "You were asked whether current has to flow through cows to affect them, and you gave the example of an electric fence. And the first time a cow touches an electric fence they get a shock?"

"I don't necessarily know that."

"A shock is current flowing through somebody?"

"Well," I said, "if you or I touch the fence I don't know if that's true. But we won't go near it again."

"The second time, what happens, the cow avoids the shock it can perceive?"

"Yes," I said.

Will went over to the jury. "You talked with Mr. Hammarback about your boat. In fact, at your deposition we went through various places you'd lived. When you're living off the farm, it may take a while to get to it if something comes up?"

"Right now it would take 10 minutes." *How long,* I thought, *does it take for a response if you call 9-1-1?*

"And that can be important if an animal is in distress or sick?"

It was clear that Trevor Will had nothing worthwhile left to say. He was just talking to hear his own voice. The jury couldn't have been pleased. They were fidgety too. *All the better for me,* I thought.

"Sure," I said absently.

"You were asked a question about whether you knew of any one thing that could cause a rise in production of 30 pounds a cow. But isn't it true, Mr. Allen, that a combination of factors could produce the increase?"

"Not if you don't change it while they're climbing."

"Last question," Trevor Will said, glancing up at a clearly antsy McKay. "You have publicly advocated that Wisconsin adopt an ungrounded delta system for its electric distribution method, correct?"

"Yes," I said.

"Thank you, Mr. Allen," Trevor Will said. "That's all I have, Your Honor."

And with that, my torment came to an end. As I rose from the witness chair, I was weak and stiff in the legs. But that was nothing. I kept thinking of the thirteen people sitting in judgment across the way.

The trial portion was done and the lawyers were ready to make their closing arguments to those good folks in the morning. But I couldn't think about that now. That was for tomorrow. Right now the only thing that mattered to me was getting the hell out of that courthouse. It felt like a prison cell.

Chapter Eighteen
LONG DAY'S JOURNEY INTO NIGHT

And so the drama was nearly done. After a defense motion to dismiss was denied and several other legal questions ruled on by McKay, the jury trudged in and the lawyers prepared to make their final pitch. I wondered if anybody was persuadable at this point. Maybe the thirteen (one alternate) had already made up their minds. I had no idea which might be better or if it even mattered. All I knew was that I was more tense and anxious than at any time since the day the trial opened, three and a half weeks earlier.

McKay issued some instructions to the jury, requesting a finding of guilt or innocence on each question and a decision regarding monetary damages. Agreement by 11 or more jurors would constitute a verdict.

The rule was to put all juror names in a hat, pick one name out and that person would go home. I told Barry and Scott that all 13 sat there a whole month. It would be unfair to send one home. If we would send one home, we would have only needed 10 jurors for a verdict. It certainly would have lessened my odds, but I was willing to take the chance.

Barry, dressed handsomely in a dark-blue suit, powder-blue shirt and white-and-navy polka-dot tie, spoke first, and I prayed he'd brought his A game with him this morning. He stepped briskly over to the jury box and started by thanking everyone for their diligent attention throughout these long weeks.

Then he plunged into his presentation by recounting the origins of our farm. He placed its founding in historical context and spoke movingly of my mother and father, of either characters, their hopes and their dreams.

"Alvin and Alice Allen," he said, "learned about dairy farming not from universities, not from books, but from themselves, from each other, and from the dairy animals that they took care of."

Barry wasn't five minutes into the speech when I was overcome by emotion. As he talked about Pa and Ma living the American dream in the American heartland, about their love for the farm and for one another and for the good Wisconsin earth, I began to weep. I could not hold back my tears. Their dream was nearly dead; our land had been raped, and for what? Money. WPS profit. I *had* to keep them alive, both the farm and the dream. This was my land now. It was my inheritance, my patrimony, my birthright. I'd fought for it here in court and I was willing to die for it. Head in hands, elbows on my knees, I struggled to cry quietly, without disturbing anybody. I'm not sure I succeeded.

"They learned how to feed, care for, tend, mend and help these animals do the best job they could," I heard Barry saying. "It required a lot of time and careful attention to detail, required they make do with what they had. That's the way dairy farming is. As time went on 11 children came to work with them, to do things that a farmer, both men and women, do. And *that's* what this case is about—the Allen farm and what WPS should have done to prevent harm, the harm that is *still* occurring today."

As I long as I lived, my dead parents and dead siblings would live, too, through me and through our farm. But only if I could save it.

Holding the rail with his right hand, Barry looked back and forth between jurors. He discussed McKay's instructions and the verdict forms. Then he defined the WPS liability.

"Negligence," he said to them. "Negligence means that all of us have a duty. A duty to exercise ordinary care in pursuit of the things we do. Electrical companies have other standards, statutes and regulations with which they must comply. A breach of those is a breach of duty. The duty for WPS is to prevent harmful current from getting to cows. There's some question about that? Because when I asked Mr. Steve Schott of WPS whether it was their duty to prevent harm to dairy cows, the response was: 'By whose standard?'"

He shook his head in sorrow. "I suggest, folks, that it's by the *cows'* standards. We have to look at what those Allen animals did as to whether current affected them. The evidence in this case is overwhelming—it did. How it got there is another matter. The evidence is strong that there was no maintenance on the line. There was no inspection of split-bolt connectors, no tests of return current. In fact their own tests showed that 70 percent of their return current was coming through the earth near or at the Allen farm. That is a violation of law."

He walked to the far end of the jury box, near the spectator benches. As he did so, I straightened up and wiped my cheeks with a handkerchief, tried to look nonchalant. I didn't want Barry to see me crying.

"Cause," he said. "How do you show that something caused something else? Use a substantial factor. There's no question that electricity at a high enough level can cause injury to an animal, affect their behavior. Behavior for a dairy animal is highly significant. A cow is not going to eat or drink—it affects her whole life, not just milk production but her reproduction, general health, cost to maintain her, everything. And so when we say it's *only* a behavioral response, that's a very serious thing."

Now Barry walked slowly up the length of the box and stopped at the opposite end. "What is a nuisance?" he asked, then answered. "Nuisance is unreasonable interference with use and enjoyment of property. Who gave WPS the right to put any level of electricity through Russell Allen's cows, any level at all? What right—and how dare they come out and say, 'Oh, we can put a little bit of electric current into the cows at Russell Allen's farm.' The fact that WPS did not maintain their line and allowed 70 percent of their return current onto the property was without question a nuisance."

It was a whole lot more than just that.

Next he explained burden of proof. "As plaintiffs we have the burden of presenting a preponderance of evidence to any question we want answered 'yes.' Preponderance is more probable than not by clear and convincing evidence that what we're saying is correct."

He went on to explain circumstantial evidence by telling a story about Abe Lincoln. "Back in the 1800s there was a person in a bar fight and the allegation was that he bit off another guy's ear. Lincoln asked a witness, 'You didn't see him bite off Joe's ear, did you?' And the guy says, 'No, I didn't see it.' 'So,' Lincoln says, 'you didn't see it. How do you know?' 'Because,' the guy says, 'I saw him spit it out.' Very strong circumstantial evidence."

The first question for the jury asked if stray electricity from WPS had harmed my operation. And so Barry went through the evolution of the operation. "From 1937 to 1962," he said, coming and standing at the midpoint of the jury box, "changes were made in agriculture. In 1962 they had a nice new barn. They had a nice pipeline milking system and plugged along as a dairy farm. By 1972 they start talking about a new facility. They do research and say, "We can build a parlor that will take the whole family here—we can make this work.' Back in 1976, when it actually started, that was the premier farm in Brown County, a 500-cow facility. One thousand three hundred acres, it was one of only six in the world like it at that time. As Dr. Johnson stated, it was the gold standard.

"And what happens with production? In '76 they immediately had a couple-of-thousand-pound loss. Pretty serious today, but back then was a *fourth* of their production. Then production drops another couple thousand pounds and they're working very hard." Barry began pacing alongside the railing, walking from one end to the other and back. "All the partners were doing everything they could, with the best equipment and best feed to make the cows produce. But they didn't. And remember Bob Allen talking about being in the milking parlor and the cows were not calm. They were nervous and kicking, not like before. And he thought that was a problem.

"Nobody knew at the time that placing the transformer pole next to the feed room would cause WPS return current to come right into their dairy. They thought they were modernizing so they could get more milk, make more money and take care of their families. They didn't know, and WPS didn't tell them, it was putting electricity in the ground.

"So for all these years they struggled. There's been talk about nutrition changes, but let's face it, in 20 years there's got to be a year or two of good crops. Wouldn't they have one good crop? And they were balancing the ration, checking the milking equipment, their housing, milk quality, forages, disease, trying to work on labor and management. They had DHIA back then, so records couldn't have been a problem. But they did not look at electricity because nobody thought of it."

He discussed repercussions. "The partnership falls apart. One by one the partners leave and go farm someplace else. Because farming is a way of life—maybe a hard way of life, but it's the Allens' way of life. It's the way Russ Allen wants his life."

As he said this, tears again started welling up in my eyes and a lump formed in my throat. *Pa,* I thought, *I'm glad you didn't live to see how they tortured our poor animals that you loved so.*

"As a farmer," Barry continued, "he doesn't want to be interfered with by outside things. He just wants to milk his cows. Gradually the partners go out. Russ was left alone.

"In 1988 when Russ took over, things got a bit better. And so when Mike Hoerth came out, he said you could save money by putting in a new parallel type of system. And Patti's arms won't be black and blue anymore. In this new parlor the cows stand still in steel plates. They can't kick you as much." He stopped walking and made a kicking motion with his right leg. "They adjust the ration but get no response from the cows."

Barry stopped and faced the jury. "What happens in 1992? Russ says let's try a new system, maybe the cows will milk better. We'll spend $100,000 trying to make these cows produce to where everybody thought they should've been. It still didn't work. They were worse than ever."

He sighed. "Now they won't come into the parlor. They have to send a person, Patti, to shoo them in. Everybody in the dairy business said, hey, when a cow wants to be milked she's going to come right into a parlor. Sometimes they would, sometimes they wouldn't, and nobody could figure out the difference. *The cows were saying 'Something is not right.'*

"They started using some BST in 1994, a little more in 1995, quite a bit in 1996. We looked at Mr. Mellenberger's numbers that showed if you take the seasonal variation into account in 1996, they didn't get a single pound of improvement from BST. Why use it? Why spend the $10 a month or the $11 a month per cow if you're not getting any benefit? So he stopped." Barry frowned and stared down at his shoes.

Then he folded his arms over his chest and looked up. "So we get to 1997 and Russ Allen has had enough. He's spent $100,000 on this parlor that the cows are not standing still in. The cows are not milking out and they can't do production things that they should be doing with this new equipment. Mike Hoerth says, 'We've looked at that parlor a zillion times. You should check out stray voltage.'"

Barry shrugged and pushed out his lower lip. "Thank God for Larry Neubauer," he said. "He's a tough character, but your choice is to decide if he's a straight shooter. To me, if you hire a gun, he better be straight and use all the available information, not just pick and choose.

"And so because Mr. Neubauer is an industrial electrician, wasn't caught up in cow contact things that WPS promotes, he was able to make measurements and found amazing things: 800 milliamps on the parlor steel, almost an amp of current. When you disconnect the neutral, that went away." Barry let that linger a few moments.

"You have your life experience," he said to the jurors. "Common sense says if there's current flowing, you've got voltage someplace. So Larry finds this current flowing, and there's no doubt there's at least 8/10 of an amp flowing on the farm from WPS. When you turn the farm power off, it's still there.

"So they get a second opinion. They have WPS come out. WPS comes and finds no problem. They measure in the parlor which has this equipotential plane and get low voltages. They didn't make any current-flow measurements. The measurements they made were on the single phase transformer, way across the yard, instead of making the test where the cows were. But what they got should've turned on red lights. If you look at the current return, they had 70 percent by their own measurement returning into the ground. It was coming right back on the neutral. Even if it's across the yard, they've got almost 1.3 amps of current and 70 percent of it is in the ground."

Barry clasped his hands together and shook them. "The other really interesting thing is that farm resistance was very low. If I've got a resistance 100 times lower on the farm than the ground rods, where is the electricity going to go? The path of least resistance—the Allen farm.

"Now, what obligation did WPS have at that point? Their obligation was not to tell Russ Allen how to farm. Their obligation was not to tell Russ Allen how to wire his farm. Their obligation was to fix their line and make it operate so there was no objectionable current coming down those ground rods. Did they do it? No! They didn't."

Barry shook his head and resumed pacing. "And so," he said, "when you look at question number 1, 'Did stray electricity from Wisconsin Public Service Corporation's system harm the Allen dairy,' sure. There's no question. The path is there. The electricity is on the farm. WPS admits it's on the farm."

He stopped pacing and looked from one juror to another and touched the index finger of his left hand to the middle finger of his right. "Next question is, what did it do to the cows? Larry Neubauer measures .914 volts. Russ doesn't necessarily believe Larry. Larry says, get a second opinion. Russ has explained he's tried everything else. His vet, Dr. Foust at the time, thought, it can't be management but I can't rule out stray voltage.

"So WPS admits that the way to test is to isolate, because isolation reduces electricity that goes onto the farm. Russ Allen isolates, with their isolator. And then gets very strange behavior: Sometimes the cows are horrible, sometimes fine. There's no predictability.

"Russell Allen had a hard decision to make when the isolator didn't work. The voltage at some point had to be high enough to trip it and let current onto the farm. So they called WPS."

Barry laughed sourly. "One thing we learned from this case is that WPS does what it wants—it does not listen to its customer. Vern Peterson was going to tell them his line no matter what: This is what we can and can't do, take it or leave it. There wasn't any effort at all to help Mr. Allen."

Barry made eye contact with several jurors. "So, when Russ knew WPS wasn't going to do anything other than leave that useless isolator where it was, he finds out it's going to be $57,000 to change the system over and be rid of the problem. Was WPS going to help? No. They've done everything they would. Not *could,* but *would.*

"In December of 1997 they put in the isolation system. And lo and behold, what happens? There is a substantial and significant change. Production is up six pounds in a couple of months, 10 pounds in another month, and it goes right on up until it's 30 pounds at the end of five months. 30 pounds! The herd increased substantially and stayed there for a couple of years. This is not just one game with five touchdown passes—there were a couple of years with a whole lot of them."

Suddenly he wheeled around and squinted over at the defense table. "Look," he said to the jury. "Mr. Will is smiling."

Everyone in the courtroom turned and stared at Trevor Will, who appeared sheepish and embarrassed. His cheeks flushed red.

"He thinks that's funny," Barry said, stone-faced. "But the fact of the matter is that there's no other reasonable explanation for an increase that big. Slatted floors couldn't cause an increase in production, bad management practices do not, bad records do not, bad breeding does not. Russ Allen is only a bad farmer to WPS when production is going down. He's a good farmer apparently when production is going up."

This was one of the two or three really crucial points of this trial, one I'd been harping on endlessly to anyone who would listen. Barry had picked a perfect moment to highlight it for the jurors.

"All of a sudden," he said, snapping his fingers, "like Rip Van Winkle, he woke up and remembered how to farm?" The jury liked this; all 13 smiled and a few even nodded sympathetically. I glanced at Trevor Will to gauge his reaction but his head was down and he was scrawling notes on his yellow pad.

"Remember Steve Root, the man that actually did the rations, testifying that there was no way that any change in rations would account for that 35 pounds of increase? He'd ruled everything else out—it had to be electrical changes reducing current. Circumstantial evidence? Yes. Strong circumstantial evidence? Yes!"

Hands behind his back, Barry nodded and began to walk up and down the length of the jury box. "Did electricity affect production?" he asked, pacing. "Yes. There's much more than a preponderance of evidence.

"Question number 2: 'Was Wisconsin Public Service Corporation through its agents and/or employees negligent in its distribution of electricity?' The answer to that must also be 'yes.'"

He looked over at the monitor. "I've got Exhibit 1020 up," he said. "WPS is required by the code not to allow objectionable current in its grounding conductors." He read, "'If an objectionable flow of current occurs over a grounding conductor due to use of multigrounds, one or more of the following methods should be used: One, abandon one or more grounds, two, change location of grounds, or three, interrupt the continuity of the conductor between ground connectors.'

"Now. Russ Allen should not have cut the grounds on his farm. They're not his property. WPS should have. It was WPS' obligation under the electric code to cut and abandon those grounds. They had a duty to reduce that current. They never did. Never offered. It would be as cheap as having a man come out and cut those wires. Wouldn't have cost more than a couple of hours' time.

"I don't want to belabor it, but let's talk about the split bolt issue for a second. There's a code that says ground connections shall be periodically inspected. They just didn't do this. Another code violation. Now, Mr. English said that part of the reason the return wire wasn't working was it was old, 1935 wire. The reason it didn't fall down in an ice storm is because it's steel. It better not fall down!"

The jurors grinned. "We know there are bad connections on that line because Mr. Schott himself found one. So they do need to be inspected. What is WPS' response? We don't want to change split bolt connectors because it's going to cost us another $1.40. That's $56,000 a year. We don't want to spend the money. We're too cheap. We want to squeeze every cent we can out of that line that went in when Babe Ruth retired, so we're not going to do it."

Even the spectators laughed at that one.

Barry was holding the jury box rail with both hands. "When you look at the questions on the verdict form," he said, "the answer to the second question should also be 'yes'. Clearly they didn't maintain their line properly. And the third question—we've already talked about it somewhat—'Was such negligence a cause of damages to the Allen dairy operation?'—should also be answered 'yes.' They caused harmful levels of electricity to come onto the farm—that's it.

"Once you've answered those questions, the next one we deal with is: 'Was Russ Allen, his agents or employees negligent in the use of electricity on his farm?' Let's see the evidence."

Barry walked a few steps away from the rail and put his hands in his pockets. "Back in 1997, Larry Neubauer said, 'There's too many wiring things you should change on your farm. These need to be eliminated before we can go on.' Russ Allen did not hesitate. He said 'All right, let's go ahead and fix them.'

"He fixed everything, thousands of dollars it was. What happened to production? Nothing. The power outage in June showed there was no current on the ring of life. The farm was up and milking—the generator was running. The wiring system on the farm was working, so it rules out the farm as causing any of the current. Did Russ Allen do anything negligent regarding his wiring? The answer has got to be no.

"'Did Wisconsin Public Service Corporation distribute electricity to the Allen Dairy Farm in a manner that constituted a nuisance?' I'll say! There's no doubt. The current was there. There's no doubt it's objectionable. There's no doubt Mr. Allen didn't want it there. There's no doubt objectionable current is not allowed under the statute. So, clearly that needs to be answered 'yes.' I'll point out to you that you need to carefully look at these instructions, but you can only answer some if you've answered other questions yes. So take your time and see what you need to do with each one."

Now Barry went over the counsel table and glanced at some papers. Then he walked slowly back to the jury.

"'Did such nuisance cause unreasonable interference with Russ Allen's interest in his use and enjoyment of the farm?' Well, I'll say! You've got a partnership that broke up, all the self-doubt he had trying to figure out why in the world his cows wouldn't produce. All that has to go toward his use and enjoyment. So that answer should also be 'yes.'"

Barry stopped for a moment and stared at the jurors. "Eventually," he said, "if you answer 'yes' to all the questions so far, then you need to answer the one on comparative negligence. What that says is, if you find Russ has caused a problem and WPS has caused a problem, you have to differentiate between those two. But, how can it be more Russ' problem than WPS'? Electricity is coming down their down grounds, from their distribution system. He's not causing it—WPS is. Russ Allen can't be responsible because no evidence would support dividing it out.

"The last questions you look at are economic damages for annoyance, inconvenience, Loss of use and enjoyment of his property. Dr. Behr calculated damages: lost milk production resulting from this current, lost calves, credit back. The history is there.

"Dr. Johnson said they should produce at 26,000 pounds. Dr. Homan said they should produce at 26,000 pounds. They actually did get over it for a while. They maintained 23,500 pounds for a couple of years.

"So if you're satisfied that Mr. Allen with his slatted floors and all could produce at the 26,000 level, then Dr. Behr's report is sound and doesn't need to be adjusted. If you take the notebook which has calculated the same thing—

young stock, cow losses, cost of mitigations, capital loss, inflation, it comes to $13,000,870. That's a lot of money and it's spread over a long time."

Barry did not walk, did not pace. He stood dead center of the jury, hands on hips, as his presentation wound down.

"That's the summary of economic damages. The annoyance and inconvenience and loss of use and enjoyment of his property, I can't tell you what that number is. As folks in the community, you have to decide what the cost is for that. But I'd suggest some things to think about: As if you were a farmer who loved to dairy farm, and loved to dairy farm as a family. This certainly was instrumental in affecting their enjoyment about how they liked to go to the barn and watch those cows milk, how they liked to take care of them.

"There are some bits of testimony that stand out. Steve Root stood out as an unbiased person there simply to evaluate the feed. He never saw anything that would stop the Allens from producing where they should have been. Dr. Foust was critical of his management but couldn't rule out stray voltage. Another person who testified was Doug Sutter, the county extension agent. He has no axes to grind one way or the other. Simply made observations on the farm before and afterwards. And didn't see anything but that electrical change could have caused it."

Barry stopped and took a deep breath. "I suppose there are alternative explanations. I think Mr. Will gave 25 different excuses of why production went up. But you have to ask yourselves why in the world it didn't happen in the 20 years before. I mean, they couldn't have had bad feed every single year. The slats were slats. Nobody, none of the experts, liked them. Maybe when they get to the 27,000 pound range those slats would be important, but not at the range they're looking at now.

"WPS was actually taught by Dr. Johnson. He testified that if you're looking at stray voltage, half a volt is a good standard, *but you have to listen to what the cows say*. You just can't ignore everything there. What the cows say, *that's* the important lesson here.

"Mr. Seiltz, the hoof trimmer, said those hooves wouldn't heal well before, and after the electric change they were better.

"I want to talk about experts for a minute. Mr. Allen testified he tried to associate with other farmers regarding stray voltage lawsuits because he was worried about costs. It's a problem that needs to be looked at in a serious manner. There was not a single person called by WPS that wasn't either an expert or on their payroll. Not one."

Barry leaned against the rail as close to the jurors as he could get. He spoke in a low voice, sounded as if he was speaking confidentially, to close friends, about matters of utmost importance to them all. "Mr. Mellenberger, for example, I thought he was smooth and calculated. Never even looked at stray voltage as being an issue because that's impossible! It's below the level that can affect them, so I'm not looking at it. The $140,000 fee, I suggest, may have helped."

He fell silent, looked down. Then he said, "Dr. Homan outlined exactly what he saw. He started out a skeptic. He wanted to prove exactly what the problem is, find out, measure it. He checked on the management, housing, milking equipment. The only significant change was that the electricity on the farm was reduced."

He nodded at the jurors and smiled. "I thank you," he said, "for your kind attention. I thank you, and Russ Allen thanks you."

Here I grew wet-eyed again.

"I'm proud," Barry said, "to represent Russ Allen in this case. I know that we're getting a fair trial here. Whatever your decision, we'll be here with you, waiting. Whatever your verdict, let it speak the truth."

I could ask for no more.

Trevor Will, I knew, would have a much different slant on the truth of these matters. During recess, I prepared myself to get skewered. Personally, I didn't care what names he called me or what lies he told. Will was nobody. My concern was the impact his falsehoods would have on the jury. Will was slick, Will was smooth, Will was persuasive; he could've sold snow cones during the Green Bay Packers' Ice Bowl, and he was speaking last. I hoped Barry's words—and they'd been very strong—would be remembered by those 13 people in the jury room when they made the decision that would affect the rest of my life.

After the break, everyone in the courtroom having resumed their usual positions, Will rose from his chair and started out by flattering the jury and trying to ingratiate himself. Wearing a three-piece black pinstriped suit, pale gray shirt and maroon and gray striped tie, along with a matching pocket handkerchief, he said fawningly, "You've been a very attentive, intelligent jury judging by your responses in *voir dire*, and we all thank you. We are just very fortunate that we have citizens like you in this country that provide this obligation."

Having gotten that out of the way, Will cut to the chase. "The nub of the issue in this case is did stray voltage harm Russ Allen's cows? Number two, was Wisconsin Public Service Corporation negligent in distributing its electricity? Number three, I will touch on the other questions on the special verdict form. And the last thing I'm going to talk about is this question of damages.

"In order to answer the question of whether stray voltage harmed Russ Allen's cows, there are three issues we need to look at. One, was there current through the cow? We're not talking about current reaching the farm. There's a lot of electricity in this room, but none is flowing through us. We need current through the cow. That's the first issue. The second question is how much current on the Allen farm was flowing through the cow. And the third issue is how much current does it take to reduce production when it flows through the cow?"

It was clear to me that just like his expert witnesses, Will would play parlor tricks with numbers. Maybe he planned to confuse the jury, although he could offer little more than pseudo-science, a kind of numerology. I hoped the jurors wouldn't fall for his tricks at this late stage.

"Those, I submit, are the three critical issues." Referring to Barry's remarks, Will said, "I didn't hear anything about that in the last hour and fifteen minutes." Up on the balls of his feet, he tiptoed over toward the foreperson in a ballet movement that struck me as very odd. I could picture him in a tutu.

"What did the evidence show about these things?" he asked, looking at juror number one. "There wasn't a witness in this case who didn't agree that in order to affect a cow's production, electrical current must flow through the cow. Mr. Zipse, Dr. Johnson, Dr. Foust, Mr. Neubauer, everybody who got up there said that. There was one exception. Do you remember who that was? Do you remember the one person who said something different? It was Russ Allen."

A stubborn, conniving bastard, that Russ Allen.

"Other than that, it was unanimous. I don't think that's an issue. Let's go to number two: How do you find out how much current was flowing through Russ Allen's cows? Well, again, there was unanimity. You take measurements. If you want to know if it's flowing through the earth, you take step potentials. If you want to know if it's coming through metal work, you take what are called 'cow contact measurements.' Again, there wasn't a single witness who disagreed with that.

"So what measurements do we have in this case? Well, Wisconsin Public Service took measurements at four different times. They took a lot more than four—look at this chart over here to your right." He went over to it and tapped it with his knuckle. "Keep in mind the levels which show what the level was according to the United States Department of Agriculture. You begin to get a production effect at 2 volts and 4 milliamps according to the Public Service Commission of Wisconsin. That's 1 volt or 2 milliamps. According to the PSC, the utility has to take action at half a volt, or 1 milliamp. That's uncontradicted. It's also uncontradicted that Wisconsin Public Service is regulated by the Public Service Commission and we've got to do what they tell us."

If the PSC was so important to the oversight of WPS, then why didn't Mark Cook, the mouthpiece for the PSC, testify? Trevor Will told the jury at the very beginning of the trial that Cook would be here as a witness. He never showed up. Will never mentioned him again.

There was a good reason for this: Cook, by testifying truthfully, could only hurt their case, not help it. Scott Lawrence would've demolished him on the stand and they all knew it. As an example of what Will could expect, here is part of a most interesting Q and A Scott did with Cook at his deposition.

"Can you," Scott had asked, "describe your formal training with respect to utility distribution and electrical engineering subjects?"

"My training," Cook had answered sheepishly, "is on the secondary side of the meter."

This was a huge understatement.

"You do not," Scott had said, "hold a bachelor's degree in electrical engineering, correct?"

"No."

"You have no college education in electrical subjects—correct?"

"That's true," Cook had answered.

"Have you," Scott had wanted to know, "ever been responsible for the construction of utility distribution systems, where you are involved or in charge of putting one up?"

"No."

"Same thing for maintenance. Have you ever had personal responsibility for day-to-day maintenance of utility distribution systems?"

"No," Cook had said.

"Same with respect to operation?"

"Yes."

"You've never done that on a day-to-day basis as a utility employee or contractor?"

"Right."

"Do you," Scott had asked, "have any personal experience in inspecting or testing power distribution systems that are unigrounded or ungrounded?"

Cook had fidgeted and looked at the floor. "I've not done a unigrounded system and never seen an ungrounded distribution system."

"Is there anything about the Wisconsin modifications to the 1997 NESC that prohibits the use of a unigrounded distribution system?"

"I believe that the Wisconsin Administrative Code says that you have to have a multi-grounded neutral and it has to be grounded nine times per mile."

"But," Scott had said, "does it require the use of a multi-grounded neutral in the first place?"

"That's my interpretation," Cook had answered.

"So it appears that, under the Wisconsin code, single-grounded and specifically unigrounded or delta systems are available as options under the code. Is that correct?"

Cook had shaken his head. "Doesn't mean they're available. It means that they wouldn't be a violation if they are in place and it would be subject to interpretation."

"Where does the code or anything in Wisconsin law, to your knowledge, prohibit the installation of unigrounded or delta systems as the law exists now?"

"We used interpretation in the Hoffman case. I said that once you establish multi-grounded distribution systems, you stick with it. That's our interpretation. So a unigrounded system not grounded nine times per mile in a rural situation is a violation of our code as we interpret it."

Scott had grinned. "Can you describe for me the background of how this paper came to be?"

"Sure. One of our lead counsel people submitted an affidavit or a brief in Supreme Court Case Hoffman versus Wisconsin Electric. In that brief he spelled out our interpretation of the code on multi-grounded systems and how we see the multi-grounded system. He then asked us to put that in a document memorializing our interpretation of the code in Wisconsin as it relates to multi-grounded systems. So we did that."

"Is there," Scott had asked, "a generic name for the system depicted in the schematic again, when I've only shown the phase conductor serving the farm after taking the third phase conductor out of our drawing?"

Mark Cook had shrugged his shoulders. "It appears to be a wye-connected unigrounded system. But, like I said before, I don't have any experience with these systems."

"In your opinion, would a system configured in this manner, as depicted in the drawing, comply with the requirements of the National Electrical Safety Code 1997, the national code at the moment?"

"I don't think so," Cook had replied. "I think it's my interpretation that, regardless of transformation, the primary has to continue with its Public Service Commission interpretation of that."

"So it is your interpretation that a unigrounded system is prohibited in Wisconsin?"

Cook had begun to stammer. "I think…if you were going to install it today…it would be prohibited based on our interpretation. You've got to have flexibility to interpret. But once we've made an interpretation, it becomes a fact of life."

"In the national code, in one of those sections Wisconsin has deleted, the code talks about its scope, including the last sentence. 'They,' meaning the writers, 'believe code covers similar systems under the control of qualified persons, such as those experienced with industrial complexes or utility interactive systems.' You're familiar with the language?"

"That's the controversy," Cook had said.

"Again," Scott had said. "That's the national code, but has not been adopted in the Wisconsin code. Correct?"

"Right. I was involved in the deletion of that section."

He'd been involved all right. Involved in "interpreting," as he freely admitted. Like the ancient Greeks and Romans, interpreting animal entrails to divine their already-foregone conclusions. Cook simply interpreted the codes the way he pleased. Then he identified his interpretation as "Wisconsin law." In this way he dutifully satisfied his WPS masters.

Trevor Will returned to the jury box, still looking at the chart. "Now," he said, "what are our measurements showing here? First one was in October of 1988, in the parlor. The highest reading was .3 volts, .6 milliamps. Look how that compares. April 1997, highest level .025 volts, .05 milliamps. Vern Peterson did this one in July 1997: .002 volts, .004 milliamps.

"What measurements did Mr. Allen offer of current flowing through the cow? Do you remember? Larry Neubauer's measurements! The first one of those was supposedly taken in April 1997 and registered .914 volts from a hood over a bunk feeder. What do you know about that measurement? Nobody witnessed it except Mr. Neubauer. This is the most remarkable thing about it to me. Mr. Neubauer admitted he never told anybody about it for three years. He didn't write it down. Remember that? Mr. Allen testified he never heard of that measurement. If it was really as important as I heard this morning, don't you think he would have told Mr. Allen about it?"

Now I understood: Will was a modern Einstein, exposing the faked numbers and phony readings of the con artist Larry Neubauer.

"When," Will continued, "was the first time that measurement ever popped up anywhere? It was May of 2000. When Mr. Neubauer was trying to get himself hired as an expert witness for the lawsuit Russell Allen was planning against Wisconsin Public Service. And then suddenly, Mr. Neubauer remembered he had taken a vitally important measurement three years ago."

Diabolical, we were, the pair of us.

"And the second measurement? Mr. Neubauer took 800 milliamps on the parlor steel. You heard about that this morning. Again, it's not written down. It wasn't shared with WPS when they came out 10 days later. They didn't say a word. Why would Mr. Neubauer not say a word to Wisconsin Public Service Corporation if there was a legitimate measurement and he was genuinely concerned about helping Russ Allen? That's a question we all ought to think about. Why would he not reveal that measurement if it was legit?"

He stuck his hands in his pockets and waltzed over to the other end of the jury box. "Oh," he said, "I left out a couple of measurements Mr. Neubauer made, step potential measurements. They were so low he didn't write them down. They didn't concern him. In June of 2002 he goes back, the *day before* Wisconsin Public Service experts are coming to test the farm to see whether the claims in this lawsuit are valid. It's called 'discovery.' They've got to ask us for information, WPS has got to ask them. What in the world is Mr. Neubauer doing out at the farm in the middle of the night slopping around in manure the day before WPS gets there? What is he doing?" Will's eyes widened.

"What was the last measurement he took? .657 volts from a stall in the barn. Did he tell WPS about that the next day? No. So I want to leave you with this: You need to assess Mr. Neubauer's credibility. And his capability. You heard him testify differently than he testified in his deposition. You know, he flunked the electrician's exam seven out of eight times. You heard him throw around electrical words, and I'm no electrical engineer, but he didn't know—he used the word 'jewel' instead of 'amp' and misused the term watts and didn't know the abbreviation for resistance."

And he didn't have a doctorate in linguistics and he didn't speak Latin and he didn't leap tall buildings with a single bound. But maybe I'd been confused for the last month or so. Maybe Larry Neubauer was the defendant in this case and not WPS, and I was only now finding it out.

"The Allen farm," Will said, "was the highest and best work he had, his biggest customer, $250,000 of work. How many electrical code violations were there? Twenty-two. Mr. Allen was paying better than $10,000 a violation. And when it came time to fix it, Mr. Neubauer charged him for that, too.

"I'll put this question to you—you saw Mr. Vern Peterson on the stand. If you were having electrical work done in your house, would you hire him or Larry Neubauer? That's a pretty good guide to what measurements you ought to rely on in this case."

I had to admit Will was doing right by his client: Casting every doubt possible on my case, which was all he needed to do since he did not have to prove anything. The burden of proof was mine.

Now he sauntered over to his counsel table, glanced at his yellow pad, and then came back to the jurors. "Let's get to the third question: Was there enough current flowing through the cows to cause a drop in production? What do we know about that? We know that the United States Department of Agriculture has said after extensive experiments, not just in Cornell University but all over the world: it takes 4 milliamps, 2 volts of current through a cow to cause a production change. You know the rest, I'm not going to repeat it. There isn't a measurement in this case that comes close.

"We didn't hear a single witness from the plaintiff say, "I've got credible, peer-reviewed literature that demonstrates that at levels below that you get a production change in cows.'"

He conveniently neglected to mention the stacking of that particular deck, since it was the utilities that funded the studies written up in peer review. And so independent, maverick viewpoints were censored.

Curiously, Will stood up on his toes again. Then he began to pace slowly. "Let's talk now about what Mr. Allen's witnesses offered of why they think there is an effect here. Before going through that, I want to talk about what they didn't do.

"They didn't take cow-contact or step-potential measurements. Do you remember hearing about the meeting to talk about Russ' management and the isolation effort? Everybody knew about cow contact and step potentials, and yet not one person said to Larry Neubauer, 'Before we isolate, you better' take a complete set of measurements and then after do the same thing. If you want to run an experiment, wouldn't you want to know how the electricity changed? Isolating doesn't tell you a thing about current through the cow, unless you measure it. What do they offer instead? Flawed logic: We made a change in isolation, production went up, therefore it must have been the isolation.

"So," he said, prancing along the length of the jury box, "we've got flawed logic and we've got something else. We know that in late 1992 or early 1993 Mike Moore is at Russ Allen's farm, and they have a conversation. What does Russ Allen say? 'I want to get me a chunk of money like Bud Garrets did.' He was looking to sue WPS from 1992 but he had a problem. He didn't have a basis. We had checked out the farm and the farm was clean." The lies don't get any bigger.

Will quit prancing and stood looking among the jurors, his slender, manicured hands neatly folded in front. "In 1997 he hooked up with Larry Neubauer, and this was the perfect situation for them both. Perfect. We know from letters he wrote that Larry Neubauer didn't like WPS. He was trying to poison Russ' mind."

I laughed. This was priceless stuff.

"Why was it the perfect opportunity for Larry Neubauer? I'll give you 250,000 reasons. He had a prime candidate. This was a marketing opportunity. Russ thought he had a stray voltage problem. Larry was going to help him find one. They're fast friends. They've been flying all over the country doing this."

Just a couple of suit-filin', jetsettin' fools.

"Now." Will smiled. "Larry Neubauer took notes and these are some of the most revealing pieces of evidence in this case. It's easy to come up on the witness stand and present a certain appearance. But in candid, unguarded moments, talking to your friends, you let your hair down." He nodded several times in agreement with himself.

"In 1997 they isolate. Mr. Allen knows after you isolate you've got to have a bump in production if you're going to have a lawsuit. In February 1998, less than two months after the isolation transformer is put in, Russ Allen is already calling Larry Neubauer to talk about suing WPS. In the note that talks, about Jack Kevorkian's attorney. It says, 'Leaving for West Indies for two weeks.' I couldn't begrudge Mr. Allen vacation. He's entitled. But do you remember that he told us he had thousands of bills he couldn't pay? He was on the verge of bankruptcy? The farm was failing? That was the story given you folks. And he's taking a two-week vacation in the Caribbean? I don't think that squares, folks."

He let that hang in the air. Then he said, "Dr. Foust got fired because he didn't believe in stray voltage. Dr. Homan comes on the job knowing his bosses have gone on record as believing in stray voltage. You can't fight your boss. What better way to make a splash and yet still help the cows then to go along? Dr. Homan played along. He told Russ Allen what Russ wanted to hear."

This was very interesting, fun material…for a novel. Will could've made a mint writing fantasy.

Terry Homan didn't need my business—he had more work than he could handle.

"Do you remember Mr. Allen's direct testimony, 'I didn't decide to sue Wisconsin Public Service until spring of 1999 when my neighbors told me that Jerry Held was trying to dig up dirt on me'? His notes to Larry Neubauer prove that wasn't true. One here is dated December 30, 1998. Every decision that was taken since then was done with litigation in mind."

Even if that were true, which it wasn't, so what? Here we were. All the jury had to decide was if my claim was legitimate. Time frames and motives had zero to do with it.

"Now," he said, beginning to prance again, "I want to talk about the rings briefly. The ring goes through the pasture where the cows go for exercise. Mr. Allen is supposedly concerned because current is flowing through his cows. Actions speak louder than words, folks. How he could possibly send cows out there if he thought there was a problem, I can't comprehend. I'll tell you why he did it. He liked the ring there. Why? He liked the ring for this reason: He liked to tweak WPS with this thing.

"We talked a little about how the story changed to fit the facts. How about the start date of damages? Dr. Johnson didn't have the story straight till he got with Behr. There isn't any basis for saying damages started in 1978. He just picks a point and away they go.

"Honesty in little things is important. Think about all the witnesses that came in to testify from WPS. They all had depositions taken. How many had anything read to them where they contradicted their prior testimony? What happened with witnesses called by Mr. Allen? There was only one witness who didn't contradict his prior testimony.

"This system only works when people tell the truth. Do I think these people are perjurers? I'm not going to tell you that, but here's what happened. Every one of them, starting with Mr. Allen, who has several million reasons he'd like you to believe the revised version of his testimony, they're people hoping to help Russ Allen. And they see how they can testify in a way that will help Russ. And they convince themselves that's the way it really was. But they forget that when they were in that deposition room and didn't see all the angles, they told the truth."

If Trevor Will believed these things, he was a dark and cynical human being. If not, then he was desperate. I have to hand it to WPS witnesses, they stuck to their lies; then again, they were polished, they did it for a living.

Will paused. He went over to his table again, looked at his pad, and then returned to the jury.

"You heard much contradictory testimony," he said, going up and down on the balls of his feet as though doing toe-raises. "Stray voltage is blamed for everything. Production would have gone up faster if it hadn't been for—stray voltage. Production goes down—stray voltage. Cows are kicking in the parlor—stray voltage. Cows aren't kicking in the parlor—stray voltage. No matter what happens it can be blamed on stray voltage. And yet there are no measurements.

"Patti Bowers says no behavior changes in the parlor before and after isolation. You look at the milker communication sheets. Showed kicking and jumping just as much after isolation as before.

"Was WPS negligent? What you heard this morning was a gross misrepresentation. Why does WPS use this system? It's safe, reliable, economically better than the alternatives. It's used all over the country. When Mr. Zipse was on the stand talking about reconfiguring the distribution line, he said nobody else on this line is complaining. There are no other problems on this line. If there really was a problem with split bolts, don't you think other farmers or customers would be bringing it up to WPS? If this happened, you would have heard about it from the plaintiff. Just for the record, others are complaining.

"I submit you ought to answer the first question 'no.' Negligence, it ought to be 'no.' You may think, well, they should have been out tightening split bolts, but the question is, did failure to do that cause electricity to go on the Allen farm causing hurt cows causing decreased production? The answer is 'no.'

"Nuisance—look on the instruction. There isn't anybody else on the line that's complained. There isn't any electricity that prevents Russ Allen or anybody else from walking out on the farm. That's a simple one—I don't think you're going to get there. I think you're going to answer 'no' to the first question."

That was a large and dangerous assumption on his part.

"Now, damages. I would say there are words that describe Dr. Behr, and those are lottery tickets. Remember the line in his letter, that he's a 'maximizer'? His entire livelihood depends on spewing big damage numbers. If he stops, he's out of work. There may be a psychological reason to put that kind of ridiculous number up, $13 million: The thought that if the jury thinks that's too much, they'll still give us 2 or 3 million. Which is more than they're entitled to in the first place. That's at work too.

"When, if, you get to the damage question—I don't think you're going to get there—look at mitigation. I think you know Russ Allen did a lot of things that depressed production. If he had taken $250,000 that he wasted on Larry Neubauer and used it for a down payment on a barn, his production would be a lot higher than it is today. He ignored professional advice.

"What else changed over time? This notion that production went up. Bear in mind that both in Wisconsin and in Brown County 1998 and 1999 were some of the best years for farming ever, way above average. He rode the wave, plus he made other changes. If he had made them 10 years ago, he would have had a lot higher production. It is *not* a stray voltage problem."

He paused and licked his lips. "There's a lot more I could say. I hope I have given you what we told you we would, which was that the evidence would show the truth. You folks have heard all the evidence.

"I want to leave you with one thought about cause and effect. There's a story about a policeman who pulled up into the parking lot by Lambeau Field one day, and he sees a fellow jumping up and down. He watches, and this is going on for five minutes or so, and the cop finally comes up to him and says, 'What are you doing?' And he says, 'Well, it keeps the tigers away.' The cop says, 'There aren't any tigers around here.' And the guy says, 'See how well it works?'

"Ladies and gentlemen, when they tell you that stray voltage caused the problems on Mr. Allen's farm, think of that guy dancing in the parking lot at Lambeau.

"Thank you very much."

Will smiled, turned, tiptoed back to his counsel table, sat down and looked over his papers.

No sooner was he seated than Scott Lawrence rose to rebut him. And it was a good thing. I loved the idea of having the last word and letting the jurors go deliberate with my position fresh in their minds. It was plaintiff's right and Scott was smart to exercise it.

Scott Lawrence could not have been more different than Trevor Will, in looks as well as demeanor. Where Will was thin, elegant, furtive, intense and nervous, Scott was big, brawny, good-humored, gregarious and approachable. He was the kind of guy you wanted to have a beer with and talk Packers football, and I could sense that the jury liked him personally—everyone did.

He strode right up to them and undertook a demolition of Will. "Mr. Will would have you believe that Russell Allen and Larry Neubauer are litigious fellows that like to file lawsuits," he said, his Midwestern accent as sharp as a slice of Wisconsin cheddar.

He chuckled. "Dairy farmers are some of the least litigious people in society. They're only here because they need to be. We talked about other farms with the same problem—equipotential planes, low cow-contact voltages, electrical problems. That's the agenda behind this litigation. WPS doesn't want to hear that. It didn't want to hear that its old, dilapidated system puts huge amounts of current through dairy farms. As a result we've had so-called research, primarily at Cornell University, that underlies these 'standards' with flawed techniques—not in free-stall barns but by results due to damaged cows being removed as the research goes on, due to changing conditions. Contrary to all those controls you hear about from the scientists.

"WPS does not want to hear any of this. Unfortunately, there are damages in the dairy industry in Wisconsin, including Russell, as a result of that stance. Russell Allen never tried a lawsuit before in his life. If he and Mr. Neubauer had this insidious plan, do you think people at Larry Neubauer's office would write down these phone calls? Sure, you have thoughts of suing to get rid of current—cow behavior changed and production followed. But if they were planning something heinous, would they write it down? It doesn't make sense."

Scott leaned forward, legs back, and pressed both palms against the railing. "You've heard a concerted effort to say that Larry Neubauer and Russell Allen are bad guys. You folks will judge that. I don't think so. They're in a difficult problem that arises from an inadequate system that ought to be replaced."

He straightened up. "I want to remind you that the electrical industry wants everybody to consider cow contact measurements the be-all and end-all, but they are only a small piece of the picture. Much more significant are measurements of current coming into the farm, where you can make them on the neutral interconnection, which is what Mr. Neubauer did. He got very good results.

"Looking for cow contact voltages is like dumping a needle in a haystack and then trying to find the needle. The basic premises of their cow contact model are flawed: It doesn't explain what cows do. The model is just wrong. The cows—they tell you what's affecting them, not somebody's electrical model of how cows should respond. The best research that's been done on this problem is the kind done at Russ Allen's: get as much current removed as possible and see what happens to the herd.

"To accept Mr. Will's premise that isolation was done and then a lawsuit cooked up, you have to assume Mr. Allen was trying to be a below-average farmer for the last 20 years. It doesn't make any sense. The record, the testimony of people that work for Mr. Allen, doesn't bear that out at all."

Scott thrust his hands in his pockets and began walking back and forth in front of the jury. "Mr. Will was very critical of Mr. Neubauer's testing, but look what happened last year when WPS came to the farm and attempted to recreate the situation before isolation. They used bad testing for the first hour and a half. Who told them it was a bad test?

Mr. Neubauer, when he put his clamp around it. He told Mr. Hammarback and me and we told the WPS lawyers. The resistance of their test lead went down exactly 20 ohms when the new one was put in.

"This is important. Why on earth weren't they making sure they had a good test lead? Frankly, it offends me that they would criticize Mr. Neubauer for *his* test procedures in these circumstances. But what's important is the distribution system out there. It hasn't been replaced for many years and still needs to be.

"They simply don't want to get that 1935 wire that's been cobbled together with split bolts and connectors for the last 70 years out there. They have reasons: More cash flow, until the thing won't conduct electricity at all. But why on earth not replace it and see if that helps somebody? WPS would criticize the farmer for his own bad wiring, but doesn't want to apply the same principle to their own system."

Scott stopped and let that linger. "Russ Allen," he said, "has tried to make the best of a bad situation. He's lucky he's still in business. Many who've had this problem aren't. And, yes, he's knee-deep in debt."

When he said this a part of an old Tennessee Ernie Ford song popped into my head: *You load sixteen tons, what do you get? / Another day older and deeper in debt.* At this point I felt like I'd been toting around sixteen tons of WPS' lies all my life.

"He's thankful," Scott told the jurors, "that he's still got the family farm. He wouldn't if he didn't have problem-solving skills. Mr. Will's attacks on Mr. Allen and Mr. Neubauer are unfounded. It's up to you folks to weigh the credibility of the witnesses and let your verdict speak the truth.

"I don't think that Wisconsin Public Service will honestly address the problem at the Allen farm until you folks tell them to. We hope you will.

"I'm not going to go on about all the evidentiary points Mr. Will raised, but I would like to talk about one document in evidence, by Mr. Bodman, from 10 or 11 years ago. Dr. Johnson talked about it in his testimony. It's about some of the Cornell research underlying the voltage standards that the utility industry relies on. It talks about two studies in particular."

He displayed an item on the monitor. "This is Exhibit 1122. Mr. Bodman says each of the two studies reported in the USDA began with two additional cows. In each study at least two animals got so violent during the initial phase of the project that they were removed from the study. He goes on to point out that they were not included in the statistical base of these eight-cow studies—a *quarter* of the cows are out. This is a large portion of the research that the utility industry wants to rely on. It just isn't credible."

He walked back over to the middle of the jury box. "The effects of electricity flowing on grounded objects in cows' environment is complicated, but we know that when you take the electricity away, the animals get better. The important thing is to get rid of it and see if the cows improve. Which they do."

Scott stood there silently for several moments, making eye contact with each person of the jury. Then he nodded once, smiled and said simply, "Thank you."

And so we were at the point of decision, poised on the edge of a cliff. I could see the sharp, jagged rocks glistening down below. If I fell, it would be the end of me. There were times this trial had seemed to drag on forever but now it seemed to have started and ended in one blinding instant. Sitting there waiting for Judge McKay, I couldn't tell which impression was true. Maybe they both were. Right now, though, it felt as if we'd begun only this morning and were ending in the evening, in time for supper. I did remember how nervous I was that first morning I entered the Brown County Courthouse, May 19. Today was June 12. In those 24 days I'd aged 100 years. It had been a long day's journey indeed.

McKay sat up straight in his big, leather swivel chair. "Thank you, Mr. Lawrence," he said. "Thank you, Mr. Will and Mr. Lawrence."

Because my legal team felt that McKay had been biased against us throughout the trial, both Barry and Scott had asked him to add an advisory question to the jury verdict form: Did the power line at my farm require repair? Though mild and reasonable, the judge dismissed their request out of hand. To me, it was just further proof of what we already knew: McKay was in WPS' corner. My hope lay with the jury.

McKay swung around to face the jurors. All eyes in the courtroom focused on him. “Now, members of the jury,” he said. “This case is ready to be submitted to you for your serious deliberation. You will consider the case fairly, honestly, impartially and in the light of reason and common sense. Give each question in the verdict your careful and conscientious consideration. In answering each question, free your minds of all feelings of sympathy, bias or prejudice. Let your verdict speak the truth, whatever the truth may be.

“When you retire to the jury room, your first duty will be to select a juror to preside and write in the answers you have agreed upon. The vote of the presiding juror, however, is not entitled to any greater weight than the vote of any other juror. When deliberations are concluded and answers inserted in the verdict, the presiding juror will sign and date the verdict and all of you will return with the verdict to the courtroom.”

Bailiffs were sworn in and the jury went out to make its decision.

It was all over but the shouting.

Epilogue
RENDERING JUSTICE

Did I say that I was as nervous on the last day of trial as I'd been on the first? It's a lie. I'd never known tension the way I did this day, not even after experiencing my first heart attack and lying sweat-soaked and trembling on a hospital gurney.

My nerves were so strung out that I now knew what heroin addicts suffered when they went cold turkey: I felt a thousand leeches on my brain and twice as many piranhas devouring my spinal cord.

At noon, the jurors rose, stretched their legs, filed out of their box and headed for the jury room to begin to deliberate. As I watched them go, my arms and legs shook. I wanted to cry; it would've been a relief. But no tears would come. My eyes were bone dry. I felt dehydrated and very weak. Fortunately, Barry and Scott stuck around to keep me company.

I had no intention of leaving until the verdict came in, no matter how long it took. I was prepared to spend nights there, weeks. Friends urged me to go home and get some rest, but I couldn't leave. I couldn't eat, either. Neither could I doze or nap. All I could do was wait. And sweat it out, cold turkey.

There wasn't much to do around the Brown County Courthouse; it's not exactly an amusement park. On and off, I strolled around outside. The grounds were green and beautifully manicured, thick grass and lush shrubs in full bloom, and the weather was perfect—sunny and 80 degrees. Coming back inside and wandering around in the corridor, I ran across Scott Lawrence and tried to make small talk, started discussing the Brewers and the Packers. But my mind wasn't on it. I badgered poor Barry Hammarback, must've asked him ten times what he thought the jury would do.

"I'm hopeful," he kept saying, with more patience than I deserved. "Cautiously optimistic."

About an hour or so before the jury reached a decision, I was sitting in the courtroom with a friend of mine trying to find something to talk about to pass the time. Trevor Will and the deputy court clerk were chatting together not 10 feet away when suddenly Will turned to me and said from nowhere, "Mr. Allen, you are really a very nice man. I mean that sincerely."

He meant it sincerely? After the way he'd trashed and bashed me in court for the last four weeks? This really grated. Will was hypocrisy on the hoof and a human oil slick. I was in shock—how do you answer to that? I was so shocked I was speechless.

My feeling was that we'd put forward a persuasive overall case. Where I thought we'd fallen down, and this gnawed at me, was in our damage claim. I felt that the way we'd presented it was confusing for the jury, not nearly as compelling as it could and should have been. For that we were all to blame, especially me. I should've insisted on a much heavier and more precise emphasis on dollars, from my experts, my lawyers, and in my own testimony. But it was too late now.

Suppertime came, and still no word from the jury. I had no appetite. I just kept pacing, my lawyers right alongside me. I milled around the courthouse for a while, went into the empty courtroom and stretched out on a spectator bench, looking up at the high, domed ceiling and the gold chandeliers. I pictured my parents and thought about my sisters, Josie and Geri Lynn, and my brother, Les, Geri Lynn's Jessie, during happy times. Suddenly I ached from missing them. I closed my eyes and said a prayer for their souls.

Finally at 7:45 p.m. word came: There was a verdict.

My heart pounded like a jackhammer as we rushed into the courtroom—lawyers, bailiffs, judicial assistants, reporters, stenographer—and waited for Judge McKay to come in.

Then he did. We stood. As he entered, imperious in the black robe that accentuated his white mane, he sat down heavily in his leather chair, put on his glasses and said, "We have a verdict. We'll bring the jury in. Go ahead, please. Sit down, ladies and gentlemen."

We sat. The bailiffs escorted the jury to their box and the 13 seated themselves in their usual spots.

"I understand," McKay said to them, "that you have reached a verdict. Would the foreperson hand it to the bailiff, Ms. Wolf? Ms. Wolf, why don't you take the verdict, bring it over here, please. Thank you."

I sat with my head down almost in my lap. All the blood in my body had rushed there. I couldn't look up, couldn't look at anyone. I could barely hear what McKay was saying for the pounding inside my skull.

McKay started reading aloud to the assembly. "Russell Allen, plaintiff, versus Wisconsin Public Service Corporation, defendant. 'We, the jury, for our verdict answer the questions submitted as follows: Question 1, did stray electricity from Wisconsin Public Service Corporation's distribution system cause harm to the Allen dairy operation?' Answer, 'Yes.'"

My heart fluttered, skipped about six beats. I still didn't look up. Looking down had brought good luck. I didn't want to jinx it.

"Question 2, 'Was Wisconsin Public Service Corporation through its agents and/or employees negligent in its distribution of electricity?' Answer, 'Yes.'"

My fists tightened. I continued staring at the floor.

"Question 3, 'Was such negligence a cause of damages to the Allen dairy operation?' Answer, 'Yes.'"

I swallowed, hard.

"Question 4, 'Was Russell Allen, his agents or employees negligent in the use of electricity on his farm?' Answer, 'No.'"

Somewhere, Larry Neubauer was smiling.

"Question 6, 'Did Wisconsin Public Service Corporation distribute electricity to the Allen dairy operation in a manner that constituted a nuisance?' Answer, 'Yes.'"

"Question 7, 'Did such nuisance unreasonably interfere with Russell Allen's interest in the use and enjoyment of his farm?' Answer, 'Yes.'"

McKay looked up. "Question 5 which I skipped, was unanswered. Question 8 is unanswered," he said. "Question 9, 'What sum of money will fairly and reasonably compensate Russell Allen for damages sustained from stray electricity?' A, economic damages, $750,000. B, damages for annoyance, inconvenience, and loss of use and enjoyment of his property, $1 million dollars."

When I heard these figures, the breawent out of me, like I'd been kicked in the gut by one of my Holsteins. Now, I looked up. *Only a mil and three-quarters out of thirteen? What the hell were they thinking?* I asked myself, then answered: *That our damages presentation was shit. Damn, I knew it!*

McKay went on, "Dated at Green Bay, Wisconsin this 12 day of June, signed by the foreperson.

"Dissenting juror Brenda Klopp dissented on questions 1, 2, 3, 4, 6, 7, 9A and 9B. Jim Nitka dissented on questions 9A and 9B."

McKay removed his glasses, set them down and looked at the jury. "Ladies and gentlemen, have I accurately read your verdict?"

They answered with one voice: "Yes."

McKay nodded. "Members of the jury," he said formally, "your service in this case is completed. You do not have to answer questions asked about the case by anyone other than me. This includes the parties, the lawyers, the media, anyone. While I'm not prohibiting you from disclosing what happened in the jury room, you don't *have* to discuss the case with anyone or answer any questions about it. You may confer with me before answering any questions asked of you by anyone."

For certain the media vultures would soon descend upon them all. On me too, for that matter. But that was the least of my concerns. I'd be only too happy to tell them what stray electricity was doing to our state and our country.

"I want to thank you for your service," McKay said. "And you're excused."

All 13 left the jury box quickly and quietly and departed into the night.

Barry and Scott came over and we shook hands. Then they embraced me.

"See?" Scott said, smiling. "I told you."

Trevor Will made it clear he intended to file an appeal. Then he went around boasting that he'd really "won" the case because the jury award was less than the $13 million, only $1.75 million. The $1.75 million was a "victory" for a multi-billion dollar corporation, and, in a sense, it was.

The day after the verdict I went outside with my cutters and cut the down grounds again, hoping the third time was the charm.

Sometime within the next 60 days we had to return to court so McKay could rule on the jury's verdict. *Rule on a verdict?* That was truly bizarre. But no more bizarre than the case itself. For who could've imagined that in the United States of America in the 21st century, with our awareness of environmental issues at an all-time high, with a cabinet office like the EPA in place and the proliferation of groups like the Sierra Club, Greenpeace, PETA and a hundred others, that cows in the dairy states would become an endangered species? But, bizarre or not, that was the crisis we were facing.

At least now, after this, and even though the monetary award was paltry in comparison to the losses I'd suffered, I could see a way forward. The Allen farm was saved; we could go on. When, I wondered, would I see my money? And when would WPS fix its damned broken down system? What would happen to all the other farms poisoned by utilities?

Well, we went back into court for McKay's ruling, and it was a doozy. Instead of simply ratifying the verdict as he should have, McKay, in a breathtaking act of judicial illogic and lack of backbone—to say nothing of bias in favor of the utility—decided that although WPS had to pay me, it wasn't required to fix the electrical problem it had created. His ruling was a classic instance of judicial arrogance.

"Allen," it read, "has made an election of remedies here and has proceeded to obtain a jury verdict setting forth damages, if you will, for the nuisance that Allen perceives exists. And for this Court to consider an injunction subsequent to the awarding of damages for the nuisance seems not only inconsistent with Wisconsin law, but inappropriate.

"Now, if Allen were to say, 'Well, we don't want the money, we want the injunction,' then I think you have a different situation. But that's not what I'm hearing. That's not what I heard. For some reason I believe Allen thinks that they can have both, and I don't think that's the case."

He didn't think? That's not what he heard? He heard only what WPS whispered in his ear. All he needed to do was *ratify the jury's decision mandating a fix*. Who was McKay to nullify a jury decision? Yet that is what he did. Hell, after talking to one of the jurors a couple of weeks after the verdict, he wanted to write on the jury's verdict form that WPS had 30 days to fix it or it would cost them so much per day. How could the jury get it so right and not the judge?

The jury had ruled that WPS caused a nuisance. McKay was supposed to order WPS to stop harming my cows.

He didn't.

There was a huge irony here. Prior to the trial, which had been postponed for eight months and during which time my cows were dying in droves, I'd begged the court for an interim relief measure: To have WPS reconfigure their distribution system back to how it was in 1976, which would virtually eliminate all utility earth current from returning through the soil on my farm. By reconfiguring their system, the current would return to the substation at Denmark and not run through my property. All it would have taken was 300 feet of wire.

By asking for this, I was putting $13 million on the line if the reconfiguration didn't work; a mere eight months before trial, my case would've gone up in smoke. But I was willing to gamble on this throw of the dice because I was confident it would work. WPS must've thought that too, because they fought tooth and nail to prevent it from happening. And they succeeded. But it certainly showed their consciousness of guilt.

After the verdict we pleaded with WPS to fix the system. We made clear that an Appeals Court decision was very likely to rule in our favor and that it was in no one's interest to go back into court. But, after several attempts to get the fix failed, and falling behind $50,000 per month, the writing was on the wall. I had no choice. In November 2003, I auctioned off all my cattle.

It hurt badly. And yet I recognized that it was the best thing I could do for the animals. I felt better knowing that when I awoke in the morning, I would no longer have to look at dead cows.

I remember the first time someone asked me if I missed them. I thought about it and realized that yes, I did miss them. But I definitely didn't miss the problems that had afflicted them and us. What I do miss is the time when electricity was drastically reduced on the farm, like after I'd cut the down grounds, and we were sending 420 cows per day through the parlor. Like when Patti Bowers asked me if she could cull a cow with an udder so swollen with milk that she could not attach the milking unit to it. Like when our milk production reached 85 pounds per cow per day and everyone was happy, including the cows. That particular year was the only one since the 1976 reconfiguration that I miss.

WPS had said they were going to appeal the decision and did. The appeals court had remanded the case to the trial court, stating that McKay had blundered in his ruling and stipulating that I was clearly entitled to a fix, as per the jury, in addition to the monetary damages. They also ruled that it was "moot," that a fix wasn't necessary because I no longer had cattle on the farm. But why was that?

In a bizarre move, WPS appealed back to the same court in which they'd lost. Then lost again. Then decided to take a run at the Wisconsin Supreme Court. But they refused to hear the case. So, in August of 2005, WPS cut that check of $2,269,000 that included interest. Not much compensation for twenty years' worth of losses and sorrow. But it was what it was. I'd beaten WPS. A jury of my peers said so.

Nonetheless, WPS continued dragging its feet. I had to spend an additional $50,000 in legal expenses—not counting attorney fees—to force WPS to give me the fix. And I still have to go back into court. Knowing McKay was biased in the utility's favor as well as spineless, we asked to substitute the judge. Our request was granted and we got Atkinson instead. So WPS lost its protector. But the legal rigmarole goes on.

WPS again deposed me. I stated I wanted to return cattle to my farm. They asked to see my tax return and got wind of me writing a book. I gave them part of the tax return relating to the farm and denied their request to see this book that you now have in your hands.

When you are denied a request during discovery, you take it up with the judge. So, we went before Atkinson, where WPS argued that they needed to see my book and to find out if I have stocks and bonds.

We argued that I wanted to bring cattle back on the farm and needed to have soil unpolluted by electric current in order to do it.

Atkinson said this: Suppose Mr. Allen owned one billion dollars' worth of Google stock and someone was waiting in his driveway to purchase his farm. Wouldn't that farm be worth more if it was fixed?

WPS fought back by claiming that I didn't need cattle (who the hell were they to decide what any American citizen needed or didn't need?) and could build homes on my land. Then Scott Lawrence spoke a telling sentence: "There is a widespread perception that utility earth currents are harming people, and if WPS wants to go there with this, we will."

Judge Atkinson ruled that WPS was not entitled to my entire tax return, only the farm portion. And that they had no right whatsoever to see the contents of my book. The decision was just, fair.

After much foot-dragging and more legal shenanigans, WPS decided to take a stab at a fix. I wanted a distribution system that didn't use the earth to return the current back to the substation. They wanted something else. So they kept hiding behind the Public Service Commission of Wisconsin, which has been carrying their water for years. Finally, we told them to pick one of Don Zipse's suggested fixes.

They decided to eliminate all the 1930ish copper weld, dig up all bare current-carrying conductors within two miles of the farm, and reconstitute the system to mirror the way it was in 1976. *Me oh my,* I thought in wonder when I heard this. *Are they truly going to redirect the juice that killed my cows?* But then, they'd already stated in a letter to Judge McKay (prior to trial) that reconfiguring the system back to the Denmark substation would harm all the cows south and east of my place, mainly the Ossman farm cows.

I didn't think they would have the balls to do it and I feel sorry for what was about to happen to those animals and will make it clear to those farmers as to what is in store. But not until April of 2007. To obtain the fix, I agreed not to publish this book until then.

In October of 2006 WPS made the change. The buried ring in the ground of my farm finally showed zero. They redirected the current to other innocent farmers, and for that I am truly sorry.

And so I am forced back into court, suing WPS for losses and damages sustained from June of 2003, when the trial ended, through January of 2007. I'm asking for $7 million, plus punitive and treble damages—costs for a new herd, new parlor and new equipment, everything necessary to reconstitute the Allen Dairy Farm.

In addition to my legal team of Barry Hammarback and Scott Lawrence and Dan Murray, I plan to also retain Habush, Habush and Rottier, punitive damages experts. Interestingly, their offices are located in the same building in Milwaukee as those of Foley and Lardner, where my friends Trevor Will and Kurt Meckstroth hang their hats. I imagine the elevator between those offices is going to be busy.

They say that justice delayed is justice denied. That may or may not be true; I don't know. But there's another saying that applies here: Better late than never. I still hope for and expect justice in this next round, to get compensation for everything the utility has stolen from me and my family. Which, of course, includes my cows.

I'll get justice, too. Won't I?

Appendix A
GRAHAM/ACKERMANN SYSTEM

Central milking system with reduced stray current problems: United States Patent 4355595

Abstract: A central milking system with electrical units energized by an a.c. voltage source having an earth-grounded neutral conductor. Stray current inherently produced in the neutral conductor by neutral-to-earth voltage is prevented from being conducted to the teats of the cow being milked via the flow of milk from the cow by creating a highly electric-insulating path between all bodies or films of milk which extend to the cow and all conductive components which are connected to the neutral conductor.

Inventors: Ackerman, Roger A.; Ackerman, Kurt A.

Application Number: 244103

Filing Date: 1981-03-16

Publication Date: 1982-10-26

Assignee: Ackerman; Roger A. (Freeport, IL)

Current Classes: 119/14.03

International Classes: A01K 001/12

Field of Search: 119/14.01,14.03,14.05,14.06,14.08,14.43 310/89,91 307/327 417/423 R

US Patent References:

3469596; Sep., 1969 Branton; 119/14.
3810442; May., 1974; Jacobs et al.; 119/14.
4000718; Jan., 1977; Brown; 119/14.
4188910; Feb., 1980; Hocker; 119/14.

Other References:

Wm. Fairbank et al., Milking Parlor Metal Structure-to-Earth Voltages, 1978, pp. 1 to 11.
L. H. Soderholm, Stray-Voltage Problems In Dairy Milking Parlors, Dec., 1979, pp. 1 to 11.
Wm. Fairbank et al., Detecting Stray Currents in Milking Parlors, 1978, pp. 11 to 14.
R. D. Appleman et al., Neutral-to-Ground Voltage Problems With Dairy Cows, 1978, pp. 1 and 2.
Klenzade Products, Stray Voltage In Dairy Barns: A Shocking Problem, date unknown, pp. 1 and 2.
Mid-West Dairymen's Co., The Mid-West

Dairyman, Jun. 15, 1980, pp. 1 and 2.
Mid-West Dairymen's Co., The Mid-West Dairyman, Jul. 15, 1980, pp. 1 and 5.
University of Minnesota Agricultural Extension Services, Guidelines for the Installation, Maintenance, and Analysis of a Pipeline Milking System, 1979, pp. 1-24.

Primary Examiner: Chamblee; Hugh R.
Attorney, Agent or Firm: Leydig, Voit, Osann, Mayer & Holt, Ltd.
Claims:
We claim:

1. In a piped, central collection milking system which includes electrical units associated with an electrical power source having an earth grounded neutral conductor, said system further including electrically conductive components which are directly electrically connected to said neutral conductor, the improvement in said system comprising, means creating a highly electric-insulating path between (a) all electrically conductive components of the system which are directly connected to said neutral conductor and (b) all bodies or films of liquid milk which extend to the teat of an animal being milked.
2. In a piped, central collection milking system which includes electrical units associated with an electrical power source having an energizing conductor and having an earth grounded neutral conductor, said units having electrical elements connected electrically to said energizing conductor and having non-electrical elements directly electrically connected to said neutral conductor and not electrically connected to said energizing conductor, the improvement in said system comprising, means creating a highly electric-insulating path between all of said elements and all bodies or films of liquid milk which extend to the teat of an animal being milked.
3. In a piped, central collection milking system which includes electrical units associated with an electrical power source having an energizing conductor and having an earth grounded neutral conductor, said units having electrical elements connected electrically to said energizing conductor and having non-electrical elements directly electrically connected to said neutral conductor and not electrically connected to said energizing conductor, said system also including electrically conductive components which are conductively connected to the teat of an animal being milked by a body or film of liquid milk, the

improvement in said system comprising, means creating a highly electric-insulating path between each of said components and all of said non-electrical elements of all electrical units to which said component is physically connected.

4. A milking system as defined in claim 3 in which one of said electrical units comprises an electric motor, one of said conductive components comprising a pump having a pumping element conductively connected at least part time to the teat of an animal by a body of liquid milk, said pumping element being physically connected to a non-electrical element of said motor and being adapted to be driven by said motor when the latter is energized, said means comprising means for electrically insulating said pumping element from said non-electrical element of said motor.
5. A milking system as defined in claim 4 in which said pump further includes a conductive housing enclosing said pumping element and physically connected to said non-electrical element of said motor, said last-mentioned means comprising means for electrically insulating said housing from said non-electrical element of said motor.
6. A milking system as defined in claim 4 in which said motor and said pump include shafts adapted to be rotated to drive said pumping element when said motor is energized, said motor shaft constituting a non-electrical element of said motor, said last-mentioned means comprising means for physically interconnecting said shafts while electrically insulating said shafts from one another.
7. A milking system as defined in claim 4 in which said motor includes a housing, said pump including a housing enclosing said pumping element and physically connected to the housing of said motor, said motor and said pump including shafts adapted to be rotated to drive said pumping element when said motor is energized, said motor housing and said motor shaft constituting non-electrical elements of said motor, said last-mentioned means including means for electrically insulating said pump housing from said motor housing and further including means for electrically insulating said shafts from one another while physically interconnecting said shafts.
8. A milking system as defined in claim 4 further including a receiver for collecting milk from the teat of an animal being milked, said pump being operable when driven to pump milk out of said receiver, another one of said components comprising a conductive member disposed within the milk in said receiver and responsive to the level of

milk therein, another one of said electrical units comprising an electrical control operable to energize and de-energize said motor, said member being physically connected to said control and being operable to actuate said control to energize said motor when the milk in said receiver rises to a predetermined level and further being operable to actuate said control to de-energize said motor when the milk in said receiver falls to a predetermined level, said first-mentioned means further comprising means for electrically insulating said member from all non-electrical elements of said control while physically connecting said member and said control to permit said member to actuate said control.

9. In a piped central collection milking system associated with an electrical power source having an energizing conductor and having an earth grounded neutral conductor, said system including a receiver adapted to collect a flow of liquid milk from the teat of an animal being milked, a pump having a pumping element and having a surrounding housing each conductively connected at least part time to the teat of the animal by the milk in said flow, said pumping element being operable when driven to deliver milk from said receiver, an electric motor adapted to be energized by said power source and adapted when energized to drive said pumping element, said motor having elements directly electrically connected to said neutral conductor and not electrically connected to said energizing conductor, the improvement in said system comprising, means for electrically insulating said pumping element and said housing from said elements of said motor.
10. A system as defined in claim 9 further including a conductive member disposed within the milk in said receiver and responsive to the level of milk therein, an electrical control connected to said conductors and operable to energize and de-energize said motor, said member being physically connected to said control and being operable to actuate said control to energize said motor when the milk in said receiver rises to a predetermined level and further being operable to actuate said control to de-energize said motor when the milk in said receiver falls to a predetermined level, and means for electrically insulating said member from all parts of said control while physically connecting said member and said control to permit said member to actuate said control.
11. A method of avoiding electrically caused physical or emotional stresses in animals being milked by a piped central collection milking system having electrical units associated with an electric power source having

an earth grounded neutral conductor, said system further including electrically conductive components which are directly electrically connected to said neutral conductor, said method comprising the step of constructing or modifying the system to interpose a highly electric-insulative path between (a) all conductive components which are directly electrically connected to said neutral conductor and (b) all bodies or films of liquid milk which extend via reservoirs or conduits to the teat of an animal being milked.

Description:

BACKGROUND OF THE INVENTION

This invention relates to a piped, central collection system for milking animals such as cows.

In a typical system, a milking unit (e.g., a teat cup assembly and a claw) is suspended from the udder of the cow and includes air and milk tubes. An electrically controlled pulsator causes vacuum and atmospheric pressure to be introduced alternately into the air tube so as to effect massaging of the teats of the teat cup. The milk is drawn through the milk tube by vacuum and is delivered through a milk pipe to a receiver which separates the vacuum system from the milk flow system.

Associated with the receiver is a pump adapted to be driven by an electric motor and adapted to deliver the milk from the receiver to a bulk cooling tank. To prevent the receiver from overflowing while at the same time keeping the pump primed and avoiding excessive aeration of the milk, an electrical control automatically energizes the motor to start the pump when the milk rises to a predetermined level in the receiver and then automatically de-energizes the motor to stop the pump when the milk falls to a predetermined level short of the bottom of the receiver.

The motor, the control and other electrical units of the system are associated with an electrical power source (e.g., an a.c. voltage source) having an energizing conductor and having a neutral conductor which is grounded to earth at the electrical service entrance of the milking barn or parlor. For purposes of safety, the housings of the motor and the other electrical units are connected to the neutral conductor. In addition, other electrically conductive components of the milking parlor such as feeders, drinking tanks and water pipes normally are connected to the neutral conductor so that all such components will reside at substantially the same voltage level relative to one another.

In electrically operated milking systems of the foregoing type, voltage gradients between the earth and different conductive components which are connected to the neutral conductor often result in stray electrical currents passing through a portion of the cow's body. While such currents usually are small in magnitude (e.g., a few microamperes), they can cause emotional and/or physical stresses in the cow. As a result, the cow demonstrates reluctance to being milked and, even more importantly, holds her milk back and thus produces less milk and is more susceptible to mastitis.

The problems resulting from stray electrical voltage and currents in milking systems have been widely recognized. Many possible solutions for detecting and reducing such currents have been proposed as indicated, for example, by an article authored in 1978 by William Fairbank et al and entitled *Milking Parlor Metal Structure-To-Earth Voltages*, by another article authored in 1978 by R. D. Appleman and entitled *Neutral-to-Ground Voltage Problems With Dairy Cows* and by an article authored in 1979 by L. H. Soderholm and entitled *Stray-Voltage Problems In Dairy Milking Parlors.* One solution which has been widely attempted is to electrically bond all electrical units (e.g., motors and controls) and all conductive components (e.g., pipes, feeders and stalls) of the system directly back to the earth ground at the electrical service entrance. This makes all electrical units and conductive components reside at virtually the same voltage as the neutral conductor but does not eliminate different voltages from the various units and components to different earth points. Thus, the voltage to earth from the neutral conductor and a connected component at one earth location can be different from the voltage to earth from the neutral conductor and a connected component at another earth location because of the resistance of the earth between grounding points.

To eliminate the latter voltage differentials, it has been proposed to create an equipotential plane throughout the milk parlor, that is, to electrically shield the floor of the parlor from earth and to make the floor and all conductive equipment in the parlor reside at the same potential as the neutral conductor at the point at which the neutral conductor is earth grounded at the service entrance. If the floor and all conductive structures which the cow might contact with her hooves and with different parts of her body are at the same potential, stray current will not flow through the cow. It is, however, difficult and expensive to construct an equipotential plane throughout the floor of the parlor. Moreover, an equipotential plane requires that all outlet circuits in the parlor be equipped with ground fault interrupters for purposes of safety.

SUMMARY OF THE INVENTION

The primary aim of the present invention is to provide an effective, comparatively inexpensive and heretofore unrecognized solution to the problem of reducing stray electrical currents caused by neutral-to-earth voltage in an electrically operated milking system.

Specifically, the present invention is based on our discovery that stray current flowing through the neutral conductor of the system is conducted to conductive components which are (a) connected to the neutral conductor and (b) in contact with the liquid milk and then is conducted by the body of milk itself to the teats of the cow, and thence through the cow's body to earth via some earthed body portion such as a hoof. We have further found that such flow of stray current through the milk can be eliminated without constructing an equipotential plane in the parlor by creating a highly electric-insulating path between all conductive components which are connected to the neutral conductor and all bodies or films of milk which extend to the cow. As a result of the insulating path, stray neutral to earth voltage on the neutral conductor is prevented creating current to the cow's body by way of the milk and from the cow's body to earth.

Thus, an object of the invention is to provide a milking system in which the milk itself is highly electrically insulated from conductive components which are directly electrically connected to the neutral conductor.

A more detailed object is to provide a milking system in which the pump of the receiver is highly electrically insulated from the pump motor so as to prevent stray neutral-to-earth voltage on the housing of the motor from creating current flow to the pump and thence from the pump to the cow by way of the milk in the receiver and in the milk line.

Still another object is to provide a milking system in which the electrical control for automatically energizing and de-energizing the pump motor is highly electrically insulated from liquid level-responsive components which are in contact with milk in the receiver so as to prevent stray neutral-to-earth voltage at the housing of the control from sending current to the milk in the receiver and thence via milk in the milk line to the cow.

These and other objects and advantages of the invention will become more apparent from the following detailed description when taken in conjunction with the accompanying drawings.

BRIEF DESCRIPTION OF THE DRAWINGS

FIG. 1 is a perspective view which schematically shows a central milking system constructed in accordance with the present invention.

FIG. 2 is a circuit diagram and diagrammatic view which schematically illustrates the problem which we have discovered and solved by the present invention.

FIG. 3 is an enlarged fragmentary cross-sectional view taken axially through the milk receiver of the system shown in FIG. 1.

FIG. 4 is an enlarged fragmentary cross-sectional view taken axially through the milk pump and motor of the system shown in FIG. 1.

DETAILED DESCRIPTION OF THE PREFERRED EMBODIMENT

As shown in the drawings for purposes of illustration, the invention is incorporated in a piped, central collection system for milking cows or other animals. For the most part, the system has been shown schematically and only so much of the system as is necessary to an understanding of the invention has been illustrated and will be described. It will be understood that the system includes other conventional components which do not relate specifically to the present invention and which thus have not been disclosed.

In the typical system which has been illustrated, a milking unit 10 comprising a teat cup assembly 11 and a claw 12 is attached to and is suspended from the udder of each cow to be milked. Flexible air and milk tubes 13 and 14 made of synthetic rubber lead out of the milking unit and communicate with air and milk pipes 15 and 16, respectively, which extend along the milking parlor. It will be understood that several milking units communicate with the pipes 15 and 16 and thus those pipes serve several cows. The air pipe 15 may be plastic but the milk pipe 16 usually is made from stainless steel. The milk pipe is supported by metal posts (not shown) extending upwardly from the concrete floor 17 of the milking parlor and is isolated from the posts by electric insulators (not shown) in order to prevent electrolysis between the posts and the milk pipe.

A vacuum system 20 comprising a tank 21, a vacuum pump 22 and a motor 23 communicates with the air pipe 15 via a pipe 24. Connected into the air pipe 15 adjacent each air tube 13 is a pulsator 25 which basically comprises a valve 26 adapted to be shifted intermittently by an electrically controlled solenoid 27. When the valve of the pulsator is shifted first in one direction and then in the reverse direction, vacuum and atmospheric pressure are alternately introduced into the air tube 13 and the cup assembly 11. As a result, the cup assembly pulsates and massages the teats of the cow in a well-known manner to cause milk to flow from the teats.

Milk is drawn from the milking unit 10 and is delivered to a receiver 30 by vacuum. For this purpose, the milk pipe 16 communicates with the top of the receiver via a line 31. In addition, a vacuum line 32 extends from the vacuum system 20 and leads into the top of the receiver. The vacuum created in the receiver by means of the line 32 acts through the line 31, the milk pipe 16 and the milk tube 14 to suck milk from the milking unit 10 and into the receiver. The receiver serves to separate the milk and vacuum systems and usually is formed by a closed glass jar or by a closed stainless steel container. The receiver is mounted on an upright panel 33 which herein is made of stainless steel.

The milk in the receiver 30 is delivered to a bulk cooling tank 35 by a pump 36. In the present instance, the pump includes a stainless steel housing 37 (FIGS. 2 and 4) which communicates with the bottom of the receiver by way of a pipe 38. A stainless steel pumping element or impeller 39 is mounted rotatably within the housing and is connected to a metal shaft 40. When the impeller is rotated, the milk in the receiver is pumped through a line 41 (FIG. 1) and is delivered to the tank 35. To rotate the impeller 39, an electric motor 43 with a winding 44 (diagrammatically shown in FIG. 2) is mounted on the panel 33 and includes a rotary output shaft 45 drivingly connected to the impeller shaft 40, the housing 37 of the pump 36 being fastened to and being supported by the metal housing 46 of the motor. When milk flows into the receiver and rises to a predetermined level therein, the motor 43 is energized automatically to drive the pump 36 and cause the milk to be delivered from the receiver to the tank. When the milk falls to a predetermined level in the receiver, the motor is automatically de-energized to stop the pump. In this way, the pump is operated to draw milk from the receiver and yet the receiver is not completely emptied so as to avoid losing the prime on the pump and to avoid excessive aeration of the milk.

The various electrical units of the milking system are adapted to be energized by an a.c. power supply. A typical power supply system for a milking parlor is shown schematically in FIGS. 1 and 2 and includes a pole-mounted transformer 50 having a primary winding 51 connected to the electrical company substation and having a 230 volt center-tapped secondary winding 52 with two energizing conductors 53 and with a neutral conductor 54 coming from the center tap thereby forming two 115 volt circuits. In accordance with conventional practice, the neutral conductor 54 of the secondary is grounded to earth by a copper ground rod 55 connected to the neutral conductor and driven into the earth.

As shown in FIGS. 1 and 2, the conductors 53 and 54 lead to the electrical service entrance 56 of the milking parlor where the neutral conductor 54 again is grounded to earth by a copper rod 57. For simplicity, it has been assumed that the electrical units of the present system are operated from 115 volts a.c. and thus only one energizing conductor 53 and the neutral conductor 54 have been shown as extending into the milking parlor from the service entrance. Connected across the conductors is a control 60 which, at timed intervals, produces square wave d.c. voltage pulses across its output terminals, the pulses having a magnitude of about 24 volts. The pulses are routed to the solenoid 27 of the pulsator 25 and serve to repeatedly energize and de-energize the solenoid so as to shift the valve 26 and effect the intermittent application of vacuum and atmosphere pressure to the milking unit 10.

The winding 44 (FIG. 2) of the motor 43 also is connected across the conductors 53 and 54. In the present instance, the motor is energized and de-energized automatically by a control in the form of a switch 63 which is closed and opened automatically when the milk in the receiver 30 rises to and falls from predetermined levels. The switch 63 is connected to the conductors 53 and 54 in series with the motor winding 44 and includes a contactor 64 located within a metal housing or box 66. In accordance with conventional practice and for purposes of safety, the box 66 of the switch 63 and the housing 46 of the motor 43 are connected directly to the neutral conductor 54. Herein, the switch box has been shown schematically as being connected to the neutral conductor by a wire 68 (FIG. 2) while the motor housing has been shown schematically as being connected to the neutral conductor by a wire 69.

The motor 43 and the switch 63, of course, constitute electrical units. Each of these units includes electrical elements (e.g., the motor winding 44 and the switch contactor 64) which, when active, are connected electrically to the energizing conductor 53. In addition, each unit includes conductive non-electrical elements (e.g., the motor housing 46 and the switch box 66) which are connected electrically and directly to the neutral conductor 54 but which are not connected electrically to the energizing conductor 53. Indeed, the housing 46 and the box 66 are insulated or isolated from the energizing conductor 53 and are insulated or isolated from the winding 44 and the contactor 64, respectively, and thus no energizing current flows through the housing and the box.

While energizing current is not conducted through the non-electrical elements 46 and 66 of the electrical units 43 and 63, current of small magnitude (sometimes called stray current) can flow between the non-electrical elements and the neutral conductor 54 to which the elements are connected. Such current results from so-called neutral-to-earth voltage differentials which often exist at different locations throughout the milking parlor.

The cause of neutral-to-earth voltage differentials may best be explained by reference to FIG. 2. In that view, the combined value of the resistance of (a) the neutral conductor of the transformer primary 51 and of (b) the metal-to-soil interface resistance of individual ground rods between the primary and the electrical substation, has been schematically represented as being an equivalent neutral and ground return-path resistor R.sub.p at the transformer ground rod 55. As current returns to the substation via the primary neutral and earth, a small voltage drop is created across the resistor R.sub.p at the ground rod 55.

The resistor R.sub.f in FIG. 2 schematically represents an equivalent of the metal-to-soil interface resistance between the ground rod 57 at the service entrance 56 to the immediately adjacent earth point. The voltage drop across the resistance R.sub.p causes current to flow between and through the resistors R.sub.p and R.sub.f via the neutral conductor 54 and the earth path labeled Pe. This results in a small voltage drop across the resistance R.sub.f. The equivalent resistors R.sub.p and R.sub.f are those adopted in the above-identified article by Soderholm for purposes of explaining why a neutral-to-earth voltage may exist at various locations in a milk barn.

Because there is a voltage drop across the equivalent resistor R.sub.f, current of small magnitude may flow between the neutral conductor 54 and earth via any conductive structure which might be connected both to the neutral conductor and to the earth beneath the floor 17 of the milking parlor at any point remote from the ground rod 57. Since the resistance of the earth varies from location-to-location beneath the milking parlor, the voltage between the neutral conductor and earth at one point in the parlor may be different from the voltage between the neutral conductor and earth at another point in the parlor.

It is well recognized that the neutral-to-earth voltage which exists in a milking parlor can cause stray currents to be conducted through the cows which are being milked. If, for example, a cow is contacting a metal feeder which, for purposes of safety, is connected to the neutral conductor 54, stray current may be conducted from the neutral conductor to the feeder, through the relatively low resistance of the cow's lips and body and then to earth by way of the cow's hooves and the concrete floor 17. The consensus is that stray currents resulting from neutral-to-earth voltages of less than

0.5 volts in magnitude probably have no adverse affect on the cows. If a higher neutral-to-earth voltage exists, however, the tingle shock of a few microamperes applied to the cow may be sufficient to disturb the cow either emotionally or physically. The cow then demonstrates nervousness in the parlor and tends to hold back her milk. This results in decreased production and increased occurrences of mastitis.

Those skilled in the art have known of the general problem of milk cows being disturbed by stray voltages at low levels, as confirmed by the volume of literature on the subject. The art has proposed solutions to alleviate the difficulty by such approaches as reducing stray voltages on drinking tanks and stall posts or hardware. Yet, the art has been puzzled by continued cow disturbance and loss of milk production, and this leads to the conclusion that skilled technicians have been unable to see a specific aspect of the general problem. Our invention is predicated on our discovery and recognition that an electrical path through liquid milk will seriously disturb cows, and the invention solves that specific and serious aspect of the problem. For if all other paths for stray current through a cow's body are eliminated, or if source voltages for such paths are reduced drastically, the cow will still be upset and disturbed greatly if stray current passes via a liquid milk path through her tender teats.

Specifically, the present invention is based on our discovery that stray current resulting from neutral-to-earth voltage can be and is conducted to a cow being milked through the flow of milk itself. Moreover, we have found that the conduction of stray current through the milk to the cow can be eliminated by establishing a highly electric-insulating path between all films and bodies of milk which extend to the cow and all conductive components of the system which are directly connected to the neutral conductor 54.

FIG. 2 illustrates one manner in which stray current resulting from neutral-to-earth voltage can be conducted between the neutral conductor 54 and the cow by way of the milk. As mentioned previously, the metal housing 46 of the motor 43 is connected directly to the neutral conductor 54 by the wire 69. As the electricians say, the motor housing is "grounded", meaning it is tied to the neutral conductor 54 which is "grounded" to earth by the rod 57.

The motor housing 46 is usually conductively connected to the pump housing 37. In addition, the motor shaft 45 normally is conductively connected to the pump shaft 40 and to the pump impeller 39. As a result, the pump housing and impeller usually are conductively connected to the neutral conductor 54 and, for practical purposes, normally form an extension of that conductor. Accordingly, any neutral-to-earth voltage existing on the neutral conductor 54 also exists on the pump housing and impeller.

As shown in FIG. 2, the pump housing 37 and impeller 39 also are in contact with the milk being drawn from the receiver 30. Milk, containing calcium and other salts which produce ions is an excellent liquid conductor of electrical current. The milk in the pump housing is connected electrically by the flow of milk in the line 38 to the body of milk in the receiver 30. The body of milk in the receiver, in turn, is connected electrically to the milk in the milk pipe 16 by the stream of milk which flows from the pipe into the receiver through the line 31. Milk in the pipe 16, of course, forms a continuation of the flow of milk from the flexible milk tube 14 and the milking unit 10. The milk in the milking unit is in direct contact with the cow's teats and udder. Accordingly, a conductive liquid path usually is established between the neutral conductor 54 and the cow's teats via the wire 69, the motor housing 46 and the pump 36 and then by way of the body of milk which extends from the pump housing 37 to the cow' s teats. The cow's hooves are in contact with the concrete platform 17 which may be at a different earth potential than the ground rod 57 but which resides at some potential different from that of the neutral conductor 54 and the milk at the cow's teats. Thus, current created by neutral-to-earth voltage normally can flow from the neutral conductor to the cow's teats by way of the milk and then through the cow's body and hooves to the platform 17 and earth. If the cow happens to be drinking from a conductive water bowl on the platform or contacting a metal stall, still another conductive path through the cow to earth may be established.

We have discovered that current flowing to the cow via the conductive path established by the milk can be of sufficient magnitude to disturb the cow and to cause her to hold back her milk. Even if liquid milk is not physically flowing at a given time, a conductive path to the cow is nevertheless established by the milk film or residue on the walls of the pump housing 37, the receiver 30 and the milk pipe 16.

In carrying out the invention, the pump 36 is highly electrically insulated from the motor 43 so as to prevent current in the neutral conductor 54 and the motor housing 46 from being conducted to the pump and then to the milk. For this purpose, the pump housing 37 is separated from the motor housing 46 by four angularly spaced blocks 75 (FIG. 4) made of plastic or other highly-electrically insulative material. The blocks are sandwiched between adjacent ends of the pump and motor housings and are secured thereto by screws 76.

Means also are provided for insulating the pump shaft 40 electrically from the motor shaft 45 while rotatably connecting the two shafts. Herein, these means comprise a plastic coupler 78 (FIG. 4) having a hole 79 in one end for receiving the motor shaft 45. The latter is secured to the coupler by a key 80 and a set screw 81. Formed in the other end of the coupler is a hole 82 which receives a stub shaft 83, one end portion of the stub shaft being secured rigidly to the coupler by a pin 84. Thc other end portion of the stub shaft 83 is telescoped into the impeller shaft 40 and is secured thereto by a key 85 and a set screw 86.

By virtue of the plastic blocks 75 and plastic coupler 78, the conductive components forms by the pump housing 37 and the pump impeller 39 are insulated electrically from the conductive non-electrical elements (i.e., the motor housing 46 and motor shaft 45) of the motor. Accordingly, stray current from the neutral conductor 54 due to neutral-to-earth voltage is prevented from being conducted from the motor to the pump housing and impeller. Milk in the pump housing thus does not receive stray current via the neutral conductor and the motor and hence a source of stray current to the cow is eliminated. To prevent the panel 33 from establishing a conductive path between the pump and the motor, the latter is mounted on the panel 33 by a plastic board 87 (FIG. 1).

We have further found that stray current from prior level-responsive controls for receivers can be conducted to the cow by way of the milk. Accordingly, we have insulated all conductive components of the present control or switch 63 from the milk in the receiver 30.

Specifically, the switch 63 is actuated in response to the rising and falling of a stainless steel float 90 (FIG. 3) which is disposed within the receiver 30. The float is telescoped slidably over a stainless steel rod 91 which extends slidably through a cap 92 on the top of the receiver. The upper end of the rod 91 is connected to the lower end of another rod 93 whose upper end portion is connected to an actuator 94. The actuator arm is connected to and is electrically insulated from the switch contactor 64 and is disposed in conductive contact with the switch box 63.

When the level of milk in the receiver 30 rises, the float 90 slides upwardly along the rod 91 until the upper end of the float engages a pin 95 (FIG. 3) which is secured to the rod. As the milk continues to rise, the float acts against the pin and pushes the rods 91 and 92 and the switch actuator 94 upwardly. When the milk rises to a predetermined level in the receiver, the actuator 94 causes the switch contact 64 to close to energize the motor 43 and start the pump 36.

When milk is being pumped out of the receiver 30 at a rate faster than the inflow to the receiver, the float 90 slides downwardly along the rod 91. When the milk falls to a predetermined level, the float engages a disc 96 (FIG. 3) on the lower end of the rod. Such engagement pulls the rods 91 and 93 and the switch actuator 94 downwardly to open the switch contactor 64, de-energize the motor 43 and stop the pump 36.

In keeping with the invention, means are provided for preventing neutral-to-earth voltage on the neutral conductor 54 from causing current conduction to the milk in the receiver 30 via the switch box 66, the switch actuator 94, the rods 93 and 91 and the float 90. In the present instance, these means comprise a coupling 97 (FIG. 3) made of plastic or other highly-electric insulating material. The coupling is secured to the rods 91 and 93 by screws 98 and serves to connect the rods while insulating the rods from one another. Accordingly, stray current from the switch box 66 cannot be conducted through the actuator 94 and the rod 93 to conductive components such as the rod 91 and the float 90 which are disposed in contact with the milk in the receiver 30. Thus, there is no conductive path from the switch box to the cow via the milk.

It should be understood that, pursuant to the principles of the present invention, a highly electric insulating path also should be established between all other conductive components of the system which are directly connected to the neutral conductor 54 and all bodies of milk which extend to a cow being milked. For example, if a motor is associated with the bulk tank 35, all conductive components of the motor should be insulated from the tank and the milk therein. Also, any other conductive components which are connected to the neutral conductor and which are associated with the tank should be insulated from the milk.

From the foregoing, it will be apparent that the present invention brings to the art a new and improved milking system wherein all conductive components that are in contact with the milk extending to the cow are highly electrically insulated from other elements which are connected physically to the conductive components and which are connected directly to the neutral conductor 54. As a result, stray current produced by neutral-to-earth voltage on the neutral conductor cannot pass to the cow via the milk. By eliminating the possibility of the cows being shocked through the milk, we have found that the cows are more content during the milking operation, show no reluctance to drinking water, and produce greater quantities of milk. We have further found that our invention reduces the shocks to the cows to such an extent that there is no need of constructing a relatively expensive equipotential plane at the floor of the milking parlor. Thus, we have discovered a comparatively economical solution to the problems created in a milking parlor by neutral-to-earth voltage.

Appendix B

Lest anyone believe that the problem of stray current is confined merely to farms in the dairy states, the following are articles that appeared in a local New Jersey paper and document the human health hazards of uncontrolled electricity running rampant.

'Stray voltage' zaps Shore homeowners

Published in the Asbury Park Press 8/21/02

By JOSEPH PICARD
TOMS RIVER BUREAU

BRICK—The Smiths' new hot tub is hotter than they'd like—6.5 volts hotter, that is.

Eileen and Gary Smith of Driscol Drive also have a hot sprinkler system—it gives off 3 volts of electricity. So does their kids' swing set. Their front lawn emits 4 volts.

"Our electrician made the readings first," Eileen Smith said. "Then JCP&L made readings and got the same results."

The Smiths are suffering from a condition called stray voltage. The condition and the term both gained notice in the Midwest during the 1980s and 1990s when dairy farmers sued power companies, at times successfully, for damaging their herds. Cows were getting jolted with electricity when they touched barn walls or fence posts, and some juries found that the stray voltage was caused by deteriorated or insufficient power company equipment, namely, neutral lines to carry the charge back to the power station.

According to Peter Johner, a spokesman for Jersey Central Power & Light Co., "stray voltage" means electrical ground current that a neutral wire can't carry. The current that can't be accommodated then strays into the ground and connects with whatever material will conduct it, he said. The metal frame of a hot tub, for instance.

The Smiths became aware of the problem in mid July. Gary Smith went to tend to the tub. He was standing in bare feet outside the tub, wiping down the inside walls, when he felt his arm go weak. When he removed his arm from the water, he felt a tingling sensation.

"He thought he might be getting a heart attack," Eileen Smith explained. "So he didn't tell me right away."

Two days later, Gary Smith tried the hot tub again. When he felt the current, he called to his wife. She also received a minor shock. Immediately afterward, she called an electrician. Off limits When it was determined that the problem was "stray voltage," the Smiths pressed the power company for a solution. Since mid July, the Smiths have not used the hot tub or the swimming pool.

"It's not that you'd get a shock in the pool," Smith explained. "But coming out of the pool dripping wet and barefoot, you can get a jolt from the ground."

The Smiths' three young sons—Zachary, 8, Nicholas, 6 and Tyler, 3—are no longer allowed to walk outdoors barefoot or wet.

"It's a bummer," Nicholas said. "I got shocked by the pool. Now we can't go swimming."

The swing set is also off limits since the Smiths were advised that electrical power can surge without warning and greatly increase in voltage. The Smiths also fear long term health effects from the current.

When the Smiths thought the utility was dragging its feet, they circulated a letter to their neighbors, outlining the situation. They learned that other residents also had problems.

"It's been a problem here for over 10 years," said Lee Cole. "I used to complain to the electric company, but they never did much. Years ago, I stopped complaining."

Cole said that she no longer uses her pool but that guests who do occasionally complain of feeling an electrical current. She did not open her Jacuzzi this year because of recurrent shock problems, she said.

"Last summer, my husband noticed a current in the pool," said Lorraine Kulik, also of Driscol Drive. "He tested with a volt meter, and there was a slight charge." Kulik said she and her husband surmised it must be some excess of electricity brought on by so many appliances in the neighborhood. But the Smiths have alerted them to stray voltage.

"It's just a slight charge, but there should not be any charge," Kulik said. Carolyn Sommers has begun a petition drive to press JCP&L for a solution. "People are now aware of the problem in the neighborhood, and we will be watching JCP&L, expecting them to correct this danger in a timely manner," Sommers said. Repair under way That's exactly what JCP&L is doing, according to Johner.

"We learned of the situation, and an engineering team was sent to study the problem," Johner said. "We determined that the cause of the stray voltage, in this case, is a bad neutral wire."

Johner said JCP&L decided to replace 7,000 feet of neutral wire in the vicinity of the Smiths' house rather than search for the specific fault. He said the neutral wire being replaced is elevated, above ground wire but controls action through the entire circuit, which also runs underground.

"They started work on Monday," Johner said, noting that the company expected the job to be completed by tomorrow.

Johner added that JCP&L was ad dressing a recent problem several customers were experiencing and has no record of a chronic stray voltage problem in the area.

The Smiths yesterday remained skeptical that the company's solution will solve the problem. They said they are consulting with attorneys on possible legal options.

More 'stray voltage,' this time in Lacey

Published in the Asbury Park Press 8/24/02

By JOSEPH PICARD
TOMS RIVER BUREAU

LACEY—Another apparent case of stray voltage has some parents in the Bamber Lake area worried for the health of their 22 month old son.

Donato and Michelle Sierchio had their house at 2207A Lacey Road built last year. This July, they had an in ground pool installed. About two weeks ago, the Sierchios' 22 month old son, Donato, was in the pool when he started crying.

"We determined that something was hurting him when he touched the cement," the elder Sierchio said. "Then I touched the cement while wet and got a shock."

The Sierchios called their electrician. He shut off all power to the house and discovered there were still 4 to 7 volts of electricity being emitted from the pool surfaces and elsewhere on the property.

"He assured us that the voltage was not coming from our house," Sierchio said. "I contacted our power supplier."

Jersey Central Power & Light Co. twice sent someone to the Sierchios' house. Both times, JCP&L told the Sierchios that their problem was not being caused through any fault with its equipment.

The Sierchios believe their problem is a condition called "stray voltage." Stray voltage gained notice in the Midwest in the 1980s and 1990s when dairy farmers sued power companies, sometimes successfully, for damaging their herds.

Cows were becoming ill after getting jolted with electricity when they touched barn walls or fence posts, and some juries found that the stray voltage was because of deteriorated or insufficient power company equipment—namely, neutral lines to carry the charge back to the power station.

According to Peter Johner, a spokesman for JCP&L, "stray voltage" is electrical ground current that a neutral wire can't carry. The current that can't be accommodated then strays in the ground and connects with whatever material will conduct it, he said.

Stray voltage is not necessarily caused by power company equipment and can come from some other source, as court documents from Michigan and Wisconsin make clear.

The Sierchios read an article in Wednesday's Asbury Park Press about a stray voltage situation in Brick. JCP&L officials have said that the 4 to 6.5 volts being emitted from the hot tub and swing set of the Smith family at 501 Driscol Drive are being caused by a bad neutral line. The company is attempting to correct the situation.

After reading the Press article, the Sierchios again contacted JCP&L. "I'm thinking of the health risks," Donato Sierchio said. "If these currents are making cows ill, then what are they doing to our child?"

He said JCP&L consented to return to the house to do further tests. Frank and Maritza Pennisi of 2 Nancy St. in Howell also are experiencing what appears to be stray voltage.

Jim Markey, an area manager with JCP&L, said yesterday that company policy prohibits talking publicly about specific customer problems, so he could not answer questions about the specific situations at the houses of the Sierchios and the Pennisis. As for Driscol Drive, Markey said the installation of the new neutral line was completed yesterday.

"We are now doing a follow up investigation to determine if the condition has been remediated," Markey said, adding that JCP&L responds to all complaints regarding stray voltage.

Stray voltage solution elusive

Published in the Asbury Park Press 8/27/02

By JOSEPH PICARD
TOMS RIVER BUREAU

BRICK—Things are still too hot on Driscol Drive.

Jersey Central Power & Light Co. worked throughout last week to correct a condition known as stray voltage or stray ground current affecting several homes on Driscol, off Burnt Tavern Road. The problem apparently is causing pools, hot tubs, fence posts, sprinkler systems, swing sets, even houses and lawns, to emit up to 7 volts of electricity and give residents both shocks and worries.

But JCP&L spokesman Peter Johner admitted yesterday that the fix did not work. "Replacing 7,000 feet of neutral wire with new neutral has not remediated the problem," Johner said. "We are still investigating, still trying to find out what's causing this."

"Not only have they not fixed the problem, it's gotten worse," said Eileen Smith of 501 Driscol Drive, who called public attention to the situation two weeks ago. "Before they tried to fix it, our irrigation system pipe was giving off 3 volts. Now, it's giving off 6 volts."

"Our pool is now giving off an extra volt or two," said Mike Kulik at 504 Driscol. Johner said he'd been told by customers that the voltage has increased, but he did not yet have confirmation from JCP&L's own technicians.

Stray voltage gained notice in the Midwest in the 1980s and 1990s when dairy farmers sued power companies, at times successfully, for damaging their herds. Cows were becoming ill after getting jolted with electricity when they touched barn walls or fence posts, and some juries found that the stray voltage was released by deteriorated or insufficient power company equipment—namely, neutral lines that are supposed to carry the unused energy back to the power station.

According to JCP&L's Johner, "stray voltage" means electrical ground current that a neutral wire can't carry. The current that can't be accommodated strays in the ground and connects with whatever material will conduct it, he said. Expert: Utilities at fault Duane Dahlberg, a retired physics professor from Minnesota, has been studying stray ground current since 1983. He also has been an expert witness against power companies in stray voltage lawsuits. He said the problem is present everywhere in the country.

"Historically, utilities have used the earth to carry current, without much thought of what the consequences might be," Dahlberg said yesterday.

Dahlberg explained that most power companies use what is called an open neutral line.

When this line carries current back to the power station, it gets substantial assistance from the earth, which will carry current adjacent to the neutral.

"Power companies know the earth will carry the current, so they expanded their systems, bringing more current in, but did not add requisite amounts of neutral cable," Dahlberg said. "When the charge becomes more than the open neutral can hold, the ground current strays. "

Johner said that JCP&L uses mostly open neutral cable.

Dahlberg added that a five wire system, like that used in Denmark, would eliminate ground current. In addition to the three hot wires, there's an insulated neutral wire and a separate ground wire. The insulated neutral carries the charge back to the power station without employing the earth, and the separate ground is there only for emergencies.

"The power companies know this, but they have expanded their open neutral systems so much that it will be extremely expensive to tear up the ground and replace those neutrals with insulated cable," he said.

Johner pointed out that not all cases of stray ground current are caused by the power company.

"There can be other sources," he said. "In the case of Driscol Drive, we may find out that it is not caused by our equipment."

"I think JCP&L is looking for a reason to walk away from this," Eileen Smith said. "Last week, they said it was their fault. Now, it appears they are backpedaling. I think they don't know what to do."

Not an isolated case

In addition to Driscol Drive homes, at least three other residences in Ocean and Monmouth counties, all customers of JCP&L, have complained of stray voltage on their property in recent weeks—the Pennisi family of Nancy Street, Howell; the Sierchios of Lacey Road, Lacey; and the O'Learys of Penn Place, Lacey.

Each of the affected families thinks the power company is not properly responding to their distress calls. JCP&L said that it is investigating.

"The JCP&L engineers told us they didn't know where it was coming from," Michelle O'Leary said, regarding the 4 volts of electricity emitted by her concrete swimming pool. "But if they think I'm going away, they are wrong. I have young children, an 8 year old boy and a 6 year old girl, who may be exposed. No one can use the swimming pool. I am not going to let this rest."

Some of those people affected are consulting their physicians regarding possible health risks. Some, too, are consulting attorneys.

Stray voltage problems in Brick not new, some say

Published in the Asbury Park Press 8/28/02

By JOSEPH PICARD
TOMS RIVER BUREAU BRICK

The stray voltage problem that is plaguing a township neighborhood is nothing new, according to a Driscol Drive resident.

Lee Cole of 515 Driscol Drive said she and her guests have been experiencing electric shocks in her pool, hot tub and indoor shower for 10 years.

"I contacted the power company when it first happened," Cole said. "They came out, told me not to worry, that it was only a little bit of electricity. But people were getting shocks. They still are."

Cole said she is not the type to make a big fuss, so after seeking remediation for a short period; she gave up trying and accepted the situation.

"But now I'm glad my neighbor is such a fighter and has brought attention to this problem," Cole said.

Cole's next-door neighbors are Eileen and Gary Smith who, about two weeks ago, called public attention to the issue of stray voltage, or stray ground current, which has charged their hot tub with 6.5 volts and their swimming pool with 4 volts. Others in the area have experienced jolts as well.

Jersey Central Power & Light Co. has said that the problem may involve its equipment and has been trying to remedy the condition for the past 10 days. But last week, when told of Cole's 10-year-old complaint, a power company spokesman said the company had no record of a chronic ground current problem on Driscol Drive.

Cole's friend, Eugene Scalzi of Union, who is also an electrician, said he tried to remediate her voltage problem five or six years ago. The utility, at that time called GPU Energy, was contacted and came out to the house, Scalzi said yesterday.

Harold Savage, a Bloomfield-based attorney, said he recalled writing a letter for Cole to the power company about 10 years ago, complaining of the voltage problem.

"Yes, I remember writing the letter," Savage said yesterday. "I don't know if I still have a copy."

When told that several people remembered the power company being informed of the stray voltage problem at Cole's house years ago, JCP&L spokesman Peter Johner said that he had been misunderstood.

"I meant that I myself was unaware of any stray voltage problem on Driscol Drive before this July," Johner said. "The company may have a record of complaints."

Johner said he would try to learn if Cole or anyone else from that area had filed stray voltage complaints in years past.

Fixing blame not easy

According to Johner, "stray voltage" means electrical ground current that a neutral wire can't carry. The current that can't be accommodated then strays in the ground and connects with whatever material will conduct it, he said.

Johner also said that the power company is not necessarily the cause of stray current and that a business with its own electrical transformer may cause it.

Duane Dahlberg, a retired physics professor from Minnesota who has been studying stray ground current since 1983, said the source is almost always the power company.

Dahlberg explained that most power companies use what is called an open neutral line. When this line carries current back to the power station, it gets substantial assistance from the earth, which will carry current adjacent to the neutral. Power companies know the earth will conduct current for them, so they have expanded their systems without adding requisite amounts of neutral cable.

"When the charge becomes more than the open neutral can hold, the ground current strays," Dahlberg said.

Johner said that JCP&L uses mostly open neutral cable.

Dahlberg said that insulating the neutral cable would correct the problem. But to rip up and re-work entire systems would be so great an expense that power companies are trying to avoid the necessary task, he said.

JCP&L replaced 7,000 feet of above-ground neutral cable last week to try to remedy the Driscol Drive situation. The attempt failed. Houses, hot tubs, pools and lawns are still emitting volts. JCP&L said it is continuing to look for the source of the voltage.

State investigates

People in Lacey and Howell also have complained of stray voltage. JCP&L says it is investigating, although affected residents, worried about possible health effects, are getting impatient.

"They're acting like it's not their problem," said Michelle O'Leary of Penn Place in the Forked River section of Lacey. "I'm contacting the mayor and writing my legislators. I am not going to let this rest."

The state Board of Public Utilities, the oversight agency for New Jersey's utilities, also has begun an investigation.

"We are in contact with JCP&L and will stay on this until the situation is resolved," said Tracy Munford, spokeswoman for the BPU.

Stray voltage problem: Lakewood farm jolted since 1998

Published in the Asbury Park Press 8/29/02

By JOSEPH PICARD
TOMS RIVER BUREAU

LAKEWOOD—A 5 1/2 acre farm on Cross Street has been emitting as much as 24 volts of electricity from structures and items on the property for at least four years, said property owner Chris Earle.

"It can come from any place on the property at any time," Earle said. "Outdoor faucets, the shower. On moist or rainy days, I've gotten zapped standing at the kitchen sink. We've gotten readings as high as 24 volts. It averages around 13 volts." Earle said Jersey Central Power & Light Co. has been aware of the problem from the start.

"They've been out here several times," Earle said. "The engineers they contracted also got measurements of 24 volts. They said it was stray voltage coming up my well. They kept telling me they're going to do something about it but nothing ever got done. Once, they sank three 40 foot copper rods in the ground to ground the current. It didn't work. Then I just got pushed around from official to official."

JCP&L is trying to remediate a stray voltage problem affecting about a half dozen houses on Driscol Drive in Brick, where ground current is charging pools, hot tubs, railings and lawns with up to 7 volts of electricity. Since reports started appearing in the Asbury Park Press a week ago, a residence in Howell and two residences in Lacey, in addition to the Earles' farm, have reported similar problems.

Earle and his wife, Maria, have lived at 830 Cross St. for 18 years and have two daughters, ages 19 and 7. They never had the problem in the house that originally was on the property; that home burned down six years ago and was rebuilt. Earle said JCP&L initially blamed the new house's electrical wiring for the voltage problem, but his electrician shut everything down and still found live current on the property.

"My daughters are afraid to use the shower," Earle said. He added that he has consulted with the family physician about the possible health effects of exposure to such levels of electricity.

"My doctor basically did not know the answers," he said. Duane Dahlberg, a retired physics professor from Minnesota who has been studying stray ground current since 1983, said the Earles' doctor is not alone among physicians and scientists.

"There has been almost no research on the effects of exposure to electrical current on animals or humans," Dahlberg said. "We have only anecdotal evidence."

JCP&L spokesman Peter Johner said JCP&L has been getting asked about possible health effects of low level voltage. He said he also has no answers.

Johner said that JCP&L has no record of the Earles' complaining of stray voltage since at least September 2000. Earle said he got frustrated and stopped calling. "It ended with them not returning my last three phone calls," he said.

Several of those experiencing stray voltage are expressing doubts about the power company's ability to correct the problem. These people are consulting physicians and attorneys, as well as contacting legislators, in search of assistance. The state Board of Public Utilities, the oversight agency for New Jersey utilities, has opened an investigation into the matter.

Asked whether JCP&L has a historic problem with stray voltage, Johner replied, "I've worked for this company for 3 0 years, and I just heard of stray voltage in July." Johner said stray voltage means electrical ground current that a neutral wire can't carry.

The current that can't be accommodated then strays in the ground and connects with whatever material will conduct it, he said.

Johner also said that the power company is not necessarily the cause of stray current and that a business with its own electrical transformer may cause it.

But Dahlberg said that the source is almost always the power company and that the solution would be to insulate the neutral cables that carry unused current back to the utility. He said power companies are reluctant to do this because of the great expense of replacing miles of cable, much of it underground.

JCP&L: 'We'll solve stray voltage'

Published in the Asbury Park Press 8/31/02

By JOSEPH PICARD
TOMS RIVER BUREAU BRICK

Residents of Driscol Drive experiencing a stray ground current problem are worried that Jersey Central Power & Light may forget their plight come winter.

But JCP&L says its engineers will stay on the job until the source of the trouble has been identified.

"The weather will be getting cooler and people will be shutting off their air conditioners," said Gary Smith, 501 Driscol Drive. "The load on the power system will lessen and stray voltage readings will go down. I can imagine JCP&L saying that the voltage problem has been solved and packing up and going away. But next summer, when the weather gets hot and the air conditioners go on, we will have this same nightmare all over again."

JCP&L spokesman Peter Johner says, however, the company is in the fight for the duration.

"We know there's a problem on Driscol Drive, and we are determined to find out what's causing it," Johner said. "We will not be leaving until we have located the source of the stray voltage. Then, if it's our fault, we will fix it. If the source is from elsewhere, we will notify the proper party so they can fix it."

Mayor Joseph C. Scarpelli said the township will make sure the power company keeps its word and locates the problem.

"This is a quality of life issue for these residents," Scarpelli said. "It's a problem that has to be solved. I've urged JCP&L to continue to do everything they can to solve it."

Stray voltage gained notice in the Midwest in the 1980s and 1990s when dairy farmers sued power companies, at times successfully, for damaging their herds. Cows were getting ill from jolts of electricity when they touched barn walls or fence posts, and some juries found that the stray voltage was due to deteriorated or insufficient power company equipment—namely, neutral lines to carry the charge back to the power station.

According to Johner, 'stray voltage' is ground current that a neutral wire can't carry. The current that can't be accommodated then strays in the ground and connects with whatever material will conduct it.

He added that the power company is not necessarily the cause of stray current, and that a business with its own electrical transformer may cause it.

About six families on Driscol Drive are experiencing electric shocks from items in their yards and, at times, in their houses. At least six other residences—houses in Howell, Toms River and Lakewood, two houses in Lacey and a farm in Lakewood—have reported conditions that appear to also be stray voltage, although JCP&L, while it has responded to these other complaints, only considers the Driscol Drive dilemma stray voltage.

In the latest case, Bill and Davenie Wallace, 1663 Hidden Lane, Lakewood, say they have been experiencing shocks and other electrical problems for 10 years. They say JCP&L has done remedial work there from time to time—the latest this month—but that problems persist.

"We are considering seeking legal advice," Davenie Wallace said yesterday.

Gary and Eileen Smith of Driscol Drive, who first drew public attention to the situation there two weeks ago, are frustrated with the power company and, they said, close to retaining a law firm.

The Smiths have also received a call from Gov. McGreevey's office, saying he or a representative would like to meet with them to discuss their problem.

"They said they'd call back to make arrangements," Eileen Smith said.

Stray voltage draws more complaints; JCP&L investigating

Published in the Asbury Park Press 9/05/02

By JOSEPH PICARD
TOMS RIVER BUREAU

DOVER TOWNSHIP—Another Ocean County household is complaining of stray voltage, saying Jersey Central Power & Light has not been able to correct the problem since 1999. Bob and Mary Marcantonio of Grand River Boulevard in Dover Township have been experiencing electric shocks from their in ground pool since they first used it on Memorial Day weekend three years ago.

"We've had the power company out here four times," Bob Marcantonio said. "They've investigated, tried different things, but the shocks haven't gone away."

Reports began surfacing two weeks ago on a neighborhood in Brick, where people complained about electric volts emitting from items in the yards and houses of about six residences. Since then, seven other families in Ocean and Monmouth counties, including the Marcantonios, have reported similar situations.

Some of the people complaining, including the Wallaces of Hidden Lane in Lakewood and Lee Cole of Driscol Drive in Brick, say their voltage problem dates back at least 10 years.

JCP&L has had engineers and crews working to correct the problem in Brick for the past two weeks—so far, without success.

"We are doing everything in our power to find the source of the stray voltage," said JCP&L spokesman Peter Johner, regarding the Driscol Drive neighborhood in Brick. "If it is due to our operation, we will fix it. If it's from some other source, we will notify whoever it is. But we are not giving up until we find out what's causing this."

Johner took the same line with the other individuals cases—that JCP&L will continue to search until the source of the trouble is found.

But the Marcantonios say that JCP&L appears to want to forget about them.

"They tried to blame the problem on the wiring for our pool," Marcantonio said. "But we've had our electrician out here, the same guy who wired the pool, and he went over everything. He shut off the power to the house and still there were volts coming from the pool area. He said the stray current is not due to his wiring."

According to Johner, "stray voltage" is electrical ground current that a neutral wire can't carry. The current that can't be accommodated then strays in the ground and connects with whatever material will conduct it, he said.

Critics of power company practices say that companies use open neutral wires which use the earth, as well as the wire, to carry current back to the power station. They like stray voltage experts Duane Dahlberg of Minnesota and Donald Zipse of Delaware say using insulated neutral cable and a separate ground wire would eliminate stray voltage, but that utilities are reluctant to do this because of the expense.

"They'd rather continue to endanger the public," said Zipse, a life fellow of the Institute of Electrical and Electronic Engineers.

But Doug Dorr, a researcher with the Power Electronics Applications Center in Knoxville, Tenn., called the critics' view of the source of stray voltage "pessimistic." The center is an adjunct of the Electric Power Research Institute, which is used as a think tank for the power industry.

Dorr said loose or broken equipment, or poor wiring, is more likely to be the source of stray voltage.

"The problems in New Jersey are not isolated," Dorr said. "Utilities run into stray voltage problems all over the country, and usually correct them."

Critics agreed that stray voltage problems are widespread, but disagreed that most of them have been corrected.

Zipse, who has studied the problem in Brick, has offered to locate the source of the ground current for $15,000. So far, he has gotten no takers from residents, the township or JCP&L.

The state Board of Public Utilities is working with JCP&L to help figure out the cause of the stray voltage, BPU President Jeanne M. Fox said yesterday.

Fox said the state doesn't have much experience with the issue. Except for recent complaints from Brick residents, Fox said regulators have not received any complaints in the last five years.

"It hasn't had much of a history in the state of New Jersey," Fox said. "It is clear that this is an annoyance to people, and it could be more than that."

Lawyer hired in stray voltage case

Published in the Asbury Park Press 9/07/02

By JOSEPH PICARD
TOMS RIVER BUREAU BRICK

The couple who first drew public attention to the problem of stray voltage hired an attorney yesterday, saying they need help dealing with their power company.

Gary and Eileen Smith of Driscol Drive have retained the Lakewood based law firm, Bathgate, Wegener & Wolf.

"We felt we were unable to resolve this problem by ourselves," Gary Smith said yesterday.

William Wolf confirmed that his firm has been hired by the Smiths.

"It is our intention to notify Jersey Central Power & Light to rectify the situation regarding the Smiths, and to provide them any other appropriate relief," Wolf said. In the middle of August, Eileen Smith contacted the Asbury Park Press to say that she had between 4 and 7 volts of electricity coming from her hot tub, sprinkler pipes, lawn and swimming pool, and that JCP&L did not seem to know what to do about it. Smith contacted neighbors and learned some had similar problems. After the Smiths' story was published, seven other households in Ocean and Momnouth counties reported stray voltage problems.

According JCP&L spokesman Peter Johner, "stray voltage" is return electrical current that a neutral wire can't carry, The current that can't be accommodated then strays in the ground and connects with whatever material will conduct it, he said.

But, whereas Johner and JCP&L contend that stray voltage most likely results from equipment failure, power company critics believe utilities use the earth to less expensively conduct current, which strays when the system is overloaded.

JCP&L has been attempting to correct the situation in the Smiths' neighborhood for the past three weeks. So far, company engineers have not located the source of the problem, although the utility has promised not to give up until the source is known.

But the Smiths are skeptical of the power company's sincerity.

"If they know the problem is caused by overloading their neutrals, then all this searching for a source is a sham," Gary Smith said.

The Smiths have three young sons. Eileen Smith said that, by taking legal action now, it will make it easier for her children to seek legal relief when they are no longer minors, if any health problems result from exposure to low levels of electricity.

Appendix C

Forty-nine of the fifty states in the U.S.A. employ an electrical distribution system similar to that of Wisconsin. The lone exception is the state of California, in which the Public Utility Commission, wisely, says the following. Is it too much to hope that other states will follow suit?

California's Public Utilities Commission - General Order No. 95

On January 19, 1994 the California Public Utilities Commission in their General Order 95 generated a resolution SU-25 resulting in Rule 33.2

33.2 Ground or Earth as a Conductor
Ground or earth shall not be used as a normal return or circuit conductor. In direct current supply systems or in single phase or polyphase supply systems, a neutral or any other conductor shall be used under normal use as a return or circuit conductor; however, the grounding of the neutral or any other conductor is not permitted as a normal return or circuit conductor. The neutral or any other conductor is permitted to be grounded only for the purposes of stabilization and protection.

Appendix D

Here is the best description of what a multigrounded neutral distribution system is, the way it works, and why it's dangerous. Written by world-class authority Donald Zipse, who testified at length during the trial, it tells everything you need to know about how the system causes sickness and slowly electrocutes all creatures great and small.

My first case was a very attractive middle aged lady, probably a size 4 or smaller dress, who had a new bathroom added to her house. In the area where she lived the plumbing had to have copper piping. About 40 feet of copper pipe was installed in contact with the dirt under the flooring. This acted the same as an Equipotential Plane or a ground electrode.
When she took a shower she occasionally received an electric shock. The available electric current was sufficient to have killed her.

- Paper #1- the shocking Swimming Pool details the problem with swimming pools. Just like dairy cows the swimming poll has two sources of stray current from the utility. Directly connected through the primary to secondary transformer connection, to the service entrance panel in the home, now connected to the equipment grounding conductor (The green Wire) to all electrical connected metallic boxes and swimming pool.
The second way is the multiple connection of the neutral along the street of road of the current carrying neutral of the multigrounded neutral distribution system, to earth 4 times per mile. The current flows uncontrolled over the earth
- Hot Tubs-especially if located outside on the earth or on concrete base.
- Bat tubs
- Lakes-North Carolina case. Utility know that when you stepped into the lake you received an electric shock. Knew about it for years. Person stepped off jet ski in water up to waist and received an massive electrical shock. Result was case where stray current has disabled a person, while in the water from his waist down. The person has medical problems with his stomach being upset all the time and medicine does not settle it. His bowels feel like he has to go all the

time. He had for three years Erectile Dysfunction along with other medical problems

• Dogs and cats laying on the earth subjected to stray current-Dog when I petted him his hair fell out. When he went over to the metal water dish on the concrete porch he jerked his head out of the water and went over to a plastic flower pot and drank.

The utilities will spare no expense to disqualify a person who will tell the truth. The utilities will "tamper with the witness." Example: Utility lawyer goes to the poor hurt by electricity person's lawyer and tells the plaintiffs lawyer that the witness is a "maverick" and the he is disruptive. "You do not want to use him he will cause you problems."

Plaintiff's lawyer goes on the web and finds out that the witness is a normal person.

next the utility lawyer goes to the Professional Board of Registration and get the board to file an affidavit with the court labeling the witness a criminal because he is not registered in the state and he is "working" in the state.

Medical doctors do not have to get a license. The witness has not design anything that the public or persons in the private sector would use that could hurt them. The Board's duty is to protect the private and public areas from construction errors and protect the public and private sectors for harm.

Appendix E

Following are actual incidents of electrocution documented by Donald Zipse. Also, he points out steps utilities take—like character assassination—to muzzle people who dare to tell the truth about the destruction electrical distribution systems are causing.

The Hazardous Multigrounded Neutral Distribution System And Dangerous Stray Currents

With **Epilog**

Paper No. PCIC-2003-03

Donald W. Zipse, P.E.
Life Fellow, IEEE

Zipse Electrical Engineering, Inc.
PO Box 7052
Wilmington, DE 19803-0052

671 Kadar Drive
West Chester, PA 19382-8123

Abstract – The multigrounded neutral system is the predominant electrical distribution system used in the United States. It permits an uncontrolled amount of electric current to flow over the earth unrestrained, posing the potential of harm to the public and to animals causing electric shocks and is presumed responsible for undetected electrocutions. To compound the problem of uncontrolled current over the earth, utility practices places harmful current from lightning strokes directly into customers homes.

As today's electrical loads continue to escalate, more current is forced into and over the earth. The question electrical engineers and the public must answer is what is the acceptable safe level, if any, of the passage of uncontrolled flow of electrical current over the earth and into our homes, office buildings and industrial facilities.

The decisions to "save money" by the adoption of the multigrounded neutral electrical distribution system outweighed the public's safety are disclosed in this paper and the electrical distributions systems that are safe are presented.

The history of the development of the multigrounded neutral distribution system is covered. Examples of the dangers associated with the multigrounded neutral distribution system and the ungrounded distribution system are disclosed. The methods of converting these dangerous and hazardous distribution systems into safer systems are presented.

Index Terms – Circuit, ground, grounding, earth, multigrounded neutral distribution system, neutral

Definition of Terms:

The definitions are predominately those used in the United States unless otherwise noted.

Circuit: Webster's Dictionary definition, "A path or route, the complete traversal of, which without local change of direction, requires returning to the starting point. **b.** The act of following such a path or route. [1] **3.** *Electronics* (IEEE Definition): **a.** A closed path, followed or capable of being followed by an electric current." The beginning can be arbitrarily selected.

A conductor or system of conductors through which an electric current is intended to flow [1]

Earth: A conducting body of varying resistances, often used in place of a conductor. Earth is commonly used as a reference point for building and structure wiring systems. The term is used interchangeable with the term "ground".

Earthing: A connection to earth.

Electric Shock: Stimulation of the nerves and possible convulsive contraction of the muscle caused by the passage of an electric current through the human or the animal body. May or may not result in electrocution.

Electrocution: To kill with electricity.

Fault current: (general) A current that flows from one conductor to ground or to another conductor owing to an abnormal connection (including an arc) between the two. A fault current flowing to ground may be called a ground fault current [2].

Ground: A conducting connection, whether intentional or unintentional or accidental, by which an electric circuit or equipment is connected to the earth or to some conducting body of relative large extent that serves in place of the earth [3]. (Also, see Grounding)

Ground current: Current that flows through the ground, earth, equipment ground conductors, etc.

Ground electrode: A conductor buried in the earth and used for collecting ground current from or dissipating ground current into the earth.

Ground fault: See *Ground fault current* below.

Ground fault current: The ground current resulting from any phase-conductor-to-earth fault.

Ground fault current (Normal): The flow of ground fault current should be brief, until the protective device opens. This brief flow of current is considered normal.

Ground fault current (Abnormal): The ground fault current, which is continuous, resulting from any phase conductor coming into contact with a grounded conductor or grounded equipment or to earth, or as the result of a neutral-to-ground fault, is objectionable and the fault should be removed, corrected or repaired as soon as possible.

Ground return circuit: "A circuit in which the earth or an equivalent conducting body is utilized to complete the circuit and allow the current circulation from or to its current source." [3] Connected to earth or to some extended conducting body that serves instead of the earth, whether the connection is intentional or unintentional or accidental [4].

Grounded Conductor: "A conductor that is intentionally grounded, either solidly or through a noninterrupting current-limiting device. [1]

A (current carrying) conductor that is intentionally grounded. This can be the neutral or one of the phase conductors in a Corner-of-the-Delta Grounding system [5].

Grounded: Connected to or in contact with earth or connected to some extended conductive body that serves instead of the earth. [1]

The NEC definition is: "Intentionally connected to earth through a ground connection or connections of sufficiently low impedance and having sufficient current-carrying capacity to prevent the buildup of voltages that may result in undue hazards to connected equipment or to persons."

Grounded, solidly: Electrical Systems. Connected directly through an adequate ground connection in which no impedance has been intentionally inserted [4].

Grounded system: A system of conductors in which at least one conductor or point is intentionally grounded, either solidly or through a noninterrupting current-limiting device. [1]

Grounding: "A permanent and continuous conductive path to the earth with sufficient ampacity to carry any fault current liable to be imposed on it, and of a sufficiently low impedance to limit the voltage rise above ground and to facilitate the operation of the protective devices in the circuit." (Note: Can be thought of as being associated with the "green" wire when associated with the NEC.)

Grounding conductor: A conductor that is used to connect the equipment or the wiring system with a grounding electrode or electrodes. [1]

Normally a non-current conductor used to connect electrical equipment or the grounded circuit of a wiring system to a grounding electrode or electrodes. Part of the equipment grounding system.

Grounding electrode: A buried metal water-piping system, or metal object or device buried in, or driven into the ground so as to make intimate contact. The grounding conductor is connected to the grounding electrode. This is usually the reference connection point for a grounded wiring system [5].

Multigrounded/multiple grounded system: A system of conductors in which a neutral conductor is intentionally grounded solidly at specified intervals. A multigrounded or multiple grounded system may or may not be effectively grounded. *See:* effectively grounded. [1]

Neutral: (IEEE Definitions) (1) (rotating machinery) The point along an insulated winding where the voltage is the instantaneous average of the line terminal voltages during normal operation.

Neutral conductor: A system conductor other than a phase conductor that provides a return path for current to the source. Not all systems have a neutral conductor. An example is an ungrounded delta system containing only three energized phase conductors.

Neutral Point: (IEEE Definitions) (2) (A) (power and distribution transformers) The common point of a Y connection in a polyphase system. (B) (power and distribution transformers) The point of a symmetrical system which is normally at zero voltage.

Stray current: The uncontrolled flow of continuous electric current over and through the earth.

Stray voltage: An incorrect term associated with the measurement of the voltage developed from the flow of stray current.

System, electrical: The portion of the electrical conductors, constituting a voltage level, which exists between transformers, and if the last transformer, generating the lowest system voltage, the conductors extending from the transformer to the load.

Zipse' Law: "In order to have and maintain a safe electrical installation: All continuous flowing current shall be contained within an insulated conductor or if a bare conductor, the conductor shall be installed on insulators, insulated from earth, except at one place within the system and only one place can the neutral be connected to earth."

I. INTRODUCTION

There is no difference between the fundamentals in low voltage and high voltage as Ohms Law, Kirchhoff's Laws, etc. apply equally to each.

Examples of the protective grounding used in low voltage, 600-volt and below, applications will be described and used to explain the hazards involved with the present day multigrounded neutral distribution system, used in the United States. This will allow the reader to see the parallels between the safe low voltage distribution system and the dangerous medium voltage multigrounded neutral distribution system.

There will be references to animals in this paper. Cows and pigs are much more sensitive to electricity than humans. If animals are dying because of stray current and they are as in one case at the rate of five cows per day for over a year until it was discovered that the reason was stray current, how much increase in electrical load will it take until the human animal is affected? This could happen in the work place or in the home. Reports are constantly being received by the author about showers, swimming pools [5] [6], and hot tubs shocking persons.

The term ground will be interchanged with earth in the hope that in the future the United States will adopt the European terminology as the term earth is much more descriptive and it will be one more step in harmonizing and unification of terms.

II. HISTORY

When Thomas Alva Edison started his electric illuminating company and began the electrical distribution system, he used only one insulated (from earth) conductor and the earth for the return conductor. He filed his patent application on February 5, 1880. This uncontrolled flow of electric current over the earth resulted in shocking horses and his employees as they dug along side of the underground distribution system. Horse pulled traction companies' employees working on the tracks received electric shocks, especially when separating the track joints.

This prompted Edison to devise the three-wire distribution system, similar to what we use today in our homes. However, Edison insulated all three conductors. This allowed Edison to know exactly where all the current was at all times. In his "SPECIFICATION forming part Of Letters Patent No. 389,280, dated August 30, 1887" he states on page 2, "the conductors of which are well insulated from each other and from the earth." Edison goes on to give several reasons for not using the earth for a return conductor.

Likewise, when the Underground Railway was being developed in London, they also elected to eliminate any stray current by using a four (4)-rail track system, two rails for supporting and guiding the coaches, one supply rail and the return conductor rail. The latter two rails were insulated from earth.

Major debate raged over whether an electrical system should be connected to earth – ground. Like Edison's electrical systems, which originally were not connected to earth, the ungrounded electrical system flourished until approximately 1913. In 1913, the National Fire Protection Association's (NFPA) National Electrical Code (NEC) made mandatory the connection to earth of any electrical system of 150 volts or less as measured to earth. However, when more than one connection to earth exists on the same electrical system, current can flow uncontrolled over the earth, metallic piping, equipment and through the earth causing problems with personnel safety, electrical equipment, etc., [6], [7].

Over time the utilities in the United States have failed to learn from Edison, as today the utilities use the earth for a partial ground return path permitting the uncontrolled flow of parallel current over not only the earth but over the adjacent underground metallic piping and other conducting materials resulting in damaging the health of dairy cows, reducing their milk output and harming humans and other animals. [6][7][8][9]

III. THE PROBLEM

In low voltage applications, 600-volts and under for safety purposes the NFPA's NEC requires in residential, commercial and industrial facilities, starting approximately in the early 1950s, a third wire, the equipment grounding conductor. In some cases, the metallic conduit is used for the "equipment grounding conductor" However the equipment grounding conductor should be installed in each and every circuit. This wire was added to electrical appliances such as electric drills, heaters, washing machines, etc. The equipment grounding conductor is connected to the neutral at just one point within the house, at the service entrance panel. The equipment grounding conductor is connected to earth. However, as will be discussed later, the neutral is also connected to earth a second time at the transformer, which results in a parallel circuit. The green color or bare conductor can be connected both intentionally and unintentionally to earth at many places, which is the object of the equipment grounding conductor.

The function of the equipment-grounding conductor, the green or bare conductor is **not** to carry any continuous flowing electric current. The key word here is "continuous." The only time the equipment grounding conductor and or earth carries any electric current is when an electrical fault occurs and then the object of the equipment grounding conductor is fulfilled by allowing the fault current to flow back to the source with very little resistance or impedance in the circuit. This allows the protective device, such as a fuse or circuit breaker to operate and turn off the electricity thus ending quickly the hazardous electrical fault condition.

For over 50 years, the 3-wire concept has been used to insure electrical safety in 600 volts and below electrical circuits, which are used in homes, commercial and industrial establishments:

- Phase conductor (usually black colored) carrying electric current to the load or appliance
- Neutral conductor (usually white colored) carrying all the return electric current
- In addition, the equipment grounding conductor (usually green in color) and/or earth only carrying electric current when an electrical phase-to-ground fault occurs which will insure adequate flow of fault current to operate the protective device such as to blow the fuse or open the circuit breaker.

This concept of restricting electrical current flow over the equipment grounding conductor and/or earth to only momentary fault current is valid for all electric circuits of any voltage when safety is a major concern. This is true whether low voltage or high voltage, whether residential, commercial or industrial wiring or electrical distribution wiring.

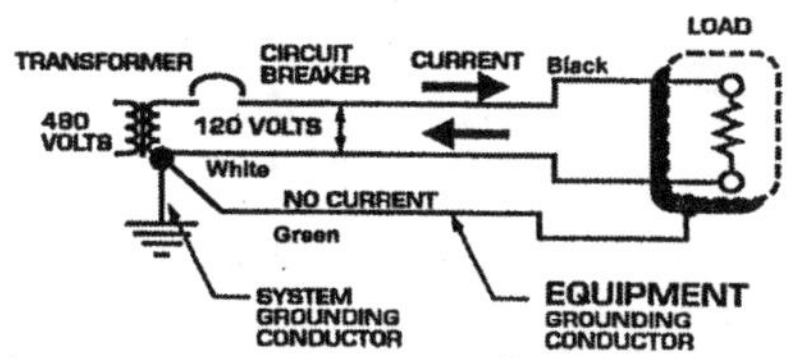

Figure 1 Circuit diagram showing the black phase conductor carrying current, the white neutral conductor carrying return current and the equipment grounding conductor carrying no continuous current.

The majority of the electrical utilities in the United States have elected to save costs of electrical distribution installations by combining two conductor functions into one conductor.

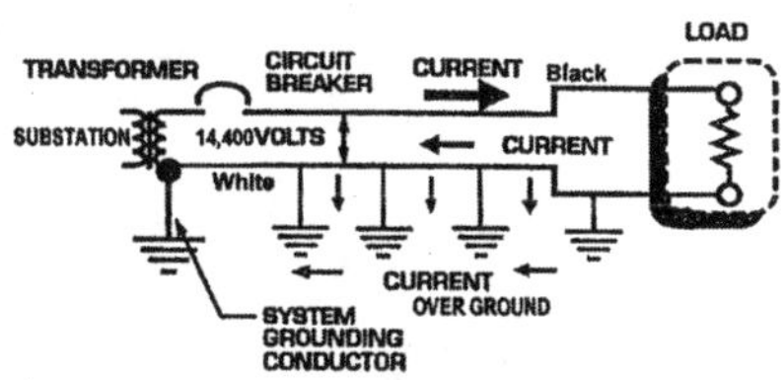

Figure 2. Diagram showing the electrical substation with a transformer, protective device, the circuit breaker, the two electrical distribution conductors, the multiple connections of the neutral to earth and the flow of uncontrolled electric current over the earth.

By combining the white wire, the neutral return current conductor carrying continuous electric current with the non-continuous current carrying equipment grounding conductor the utilities have saved billions and billions of dollars over 50 or more years as will be shown later.

The problem is that the utilities have used the earth as a partial return for the neutral current allowing the neutral current to continuously flow uncontrolled over the earth, as will be shown, which has caused irreparable and irreversible harm to dairy herds and pig farms and has caused electric shocks to humans and most likely may have resulted in electrocutions.

IV. IS IT "STRAY VOLTAGE" OR "STRAY CURRENT"?

Before one can fully comprehend the magnitude of the problem one has to understand the difference between "Stray Voltage" and "Stray Current" and determine which is the correct term. When voltage is measured, the measurement is between two points.

Remember Ohm's Law which states in order to have a voltage we must have a current flowing through a resistance; Voltage = Current times Resistance. Thus, it is very easy to measure voltage as compared to measuring current. In order to measure current one method is to place a sensing coil around the conductor or if possible the animal.

The question that begs an answer is which is the determining factor, voltage or current. In order to answer that question we need to look at the human animal sensitivity to voltage and current.

A. Human Sensitivity to Electricity

It has been reported by Edward Owen, a student of Professor Charles F. Dalziel, University of California that Professor Dalziel "required" his students to participate in experiments to measure the human animal's response to voltage and current by placing their feet into a bucket of salt water and holding onto a conductor. Professor Dalziel then applied varying amounts of current and measured their response. As recalled by the author from presentations by Professor William B. Kouwenhoven, Electrical Engineering, Johns Hopkins University, he used fresh cadavers to measure electric current necessary to revive the heart in order to develop the defibrillator. These experiments and others produced Table 1.

It has been shown that it takes approximately 35 volts across dry skin to force electricity into the human body. Less voltage or electrical pressure is required for a woman's dry skin.

As shown in the Table 1, for a 60 Hertz alternating current (ac) at 0.4 milliamps (mA) or 0.0004 Amps (A) a male will feel a slight sensation on the hand. For a woman it only takes 0.3 mA. A painful shock requires 9 mA for a man and for a woman 6 mA. One must note these low values.

Time plays a function in the equation of electrical current and injury. The longer a person is subjected to an electrical current flow through the body, the more likely an injury will occur.

"The most damaging path for electrical current is through the chest cavity. In short, any prolonged exposure to 60 Hz current of 20 mA or more may be fatal. Fatal ventricular fibrillation of the heart (stopping of rhythmic pumping action) can be initiated by a current flow of as little as several milliamperes. These injuries can cause fatalities resulting from either direct paralysis of the respiratory system, failure of the rhythmic heart pumping action, or immediate heart stoppage." [7]

It is clear that it is the current that causes a reaction in humans, not the voltage. The voltage is the driving force and there is a threshold below, which there is an inability to drive any current through the human body.

Comparison of the human animal with the cow animal shows many similar conditions. Both are mammals with a blood system and both have an electrical communication system, nerves.

Professor Dalziel in 1946 states, "Perhaps the most serious misconception concerns the effects of voltage versus the effects of current. Current and *not* voltage is the proper criterion of shock intensity." [10] It is a shame the U.S. Department of Agriculture, the American Society of Agriculture Engineers and others who coined and use the term "stray voltage" in the 1970s failed to do adequate research on the subject.

B. Measuring Electric Potential, Voltage

Electric current flowing into the earth from a ground rod at a pole will flow into the earth in all directions. The value of current, in amperes, can be different for various directions. The resistivity of the soil is not uniform. This is explained in the IEEE Technical Paper titled: "Analysis of Grounding Systems in Soils with Hemispherical Layering" [11]. The paper discusses grounding systems located inside or near hemispherical soil heterogeneities.

The amount of current flowing into the earth will depend upon the resistance of the earth. Voltage measurements for an individual ground electrode at an electric pole are not taken uniformly around the pole, but usually only one voltage measurement is taken for each pole in an arbitrary direction and at an arbitrary distance.

Supposedly, **the object of taking a voltage measurement is to determine if any adverse effects exists due to a stray voltage being present. As has been shown it is current that is responsible for adverse reactions of humans and animals, not voltage.**

It is usual for utility personnel to make stray voltage measurements with one reference lead connected to the ground rod or the down ground conductor while the other reference lead, a probe is forced into the soil some distance from the pole. As is shown in Figure 3, where the second reference probe is placed will determine the voltage drop across the earth. The voltage drop will be a function of the current flow between the ground rod and the reference probe times the resistance of the soil between the two points, the rod and the probe.

As is shown in the Figure 3, depending where the rod is placed could determine the voltage. What if there were additional current flows from an adjacent pole flowing in or across the same reference distance, rod to point A. The value of voltage would not be an indication of current flow from the rod at the pole under investigation into the earth.

With only a voltage measurement being made the value of such information is useless and usually a waste of time and energy. However, this is the standard used to determine if a problem exists.

	Direct Current		Alternating Current			
			60 Hz		10kHz	
	Men	Women	Men	Women	Men	Women
Slight sensation on hand	1.	0.6	0.4	0.3	7	5
Perception "let go" threshold median	6.2	3.5	1.1	0.7	12	8
Shock – not painful and NO loss of muscular	9.	6.	1.8	1.2	17	11
Painful shock – muscular control loss	62.	41.	9.	6.	55	37
Painful and severe shock breathing difficult	90.	60.	23.	15.	94	63

Table 1. Sensitivity of Humans to Electric Current in Milliamperes

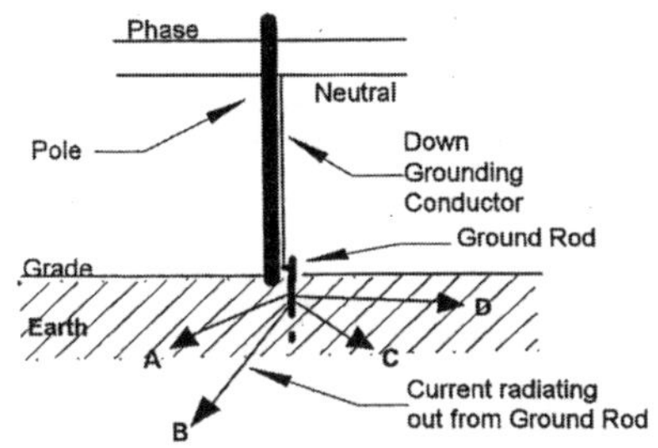

Figure 3. Current emanating from the ground rod - magnitude as a function of earth resistance and electrical load.

Ohm's Law has three unknowns, Voltage, Current and Resistance. Algebra shows that in order to solve any equation one must know or have values for one less unknown than the total number of unknowns. With Ohm's Law there are three unknowns, thus to solve Ohm's Law one must have values for any two items; voltage and current, current and resistance or voltage and resistance. Measuring just the voltage basically does not show anything except the value of voltage between those two points. That is all; nothing else can be stated about a voltage measurement between two points if only the voltage was measured.

Making a voltage measurement only gives the value of voltage across that one spot. The one voltage measurement does not tell you the value of current or resistance of the circuit under investigation. Voltage decreases as resistance is encountered. However, current remains constant throughout the circuit.

No matter where one measures the current in the circuit shown in Figure 1, the current will remain the same. **The current will not change depending on where one measures the current in the circuit. This concept is critical in dealing with stray current, not the magnitude of voltage once the voltage threshold is sufficient to breakdown the barrier to current flow is exceeded.**

V. HISTORY OF ELECTRICAL DISTRIBUTION AND EARTH RETURN

In order to fully comprehend the detrimental effects of allowing the earth to continuously conduct electric current it is imperative that one understands the history and how the utilities elected to use an unsafe electrical distribution method. It was shown that **earth return had been tried in the early days of electrical distribution and found to be too hazardous and harmful to not only the human population but also to animals.**

A. *Experience of Thomas A. Edison and Earth Return of Electrical Current*

Thomas Alvin Edison patented the first electrical distribution system. The patent was filed on February 5, 1880 in the United States and given the number 2,282. It was titled, "System of Electrical Distribution".

Edison tried several approaches to distributing electricity. Tom Shaughnessy, PowerCET Corporation writes, "Early on, Edison implemented a floating approach for his DC systems after several events demonstrated the adverse effects of stray DC currents flowing throughout buildings and neighborhoods. Once, a horse was shocked when it walked on "electrified soil" near Edison's Pearl Street generating station and laborers working on his underground distribution system believed there was a "devil in the wire."' As a result of some of these problems, Edison apparently adopted a three wire system which did not rely upon earth return." [12]

With seven years of experience, Edison in August 30, 1887 in his Specifications forming part of Letters Patent No. 389,280 on page 2, line 34, states:

"All of such conductors from the generators at the station to the lamps are made in pairs—one for the outgoing current and the other for the returning current of electricity, the circuits throughout the system being complete or round metallic circuits, **the conductors of which are well insulated from each other and from the earth.** (Emphasis by the Author) The use of the earth for one half of the circuit would largely increase the difficulties arising from the grounding of the conductors or the crossing of the conductors among themselves or with the conductors of other circuits to such an extent that a system so constructed would be impracticable."

Edison recognized the hazards involved in using the earth as a return path. "The Edison Three-wire System" is described in the Instruction book No. 8148, published by the General Electric Company in Dec. 19, 1900 and consists of a three wire grounded mid-point at one location only, which supplies phase to phase voltage of 220 and 110 volts for lighting. This is the basis of the electrical distribution in homes today. (See figure 5.)

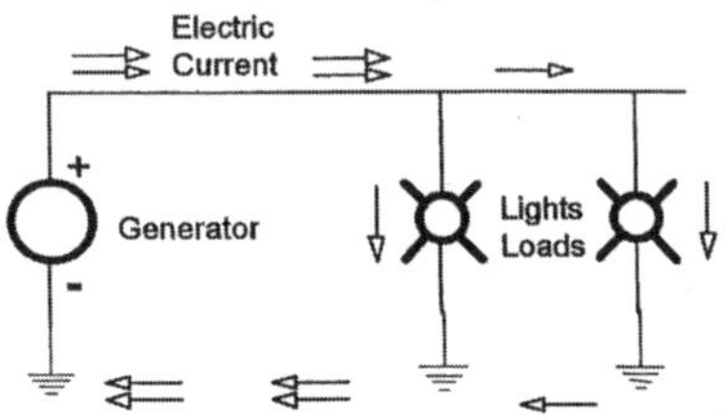

Figure 4. Direct Current Electrical System with the Earth Used as Current Return Path

With a balanced electrical load, that is loads "A" equal to loads "B" the current in the neutral will cancel and the neutral will not carry any current. The maximum amount of current the neutral will carry will be if the load is connected to one generator only, and only that load, is turned on.

However, an important concept can be learned from figure 5. Figure 5 shows the "A" and "B" loads opposite each other, which conveniently cancels the neutral current. Let us assume that the loads are not opposite each other as is shown in Figure 5 and that the "A" loads are extended beyond the end of the "B" loads as would occur in an extension of a single phase circuit.

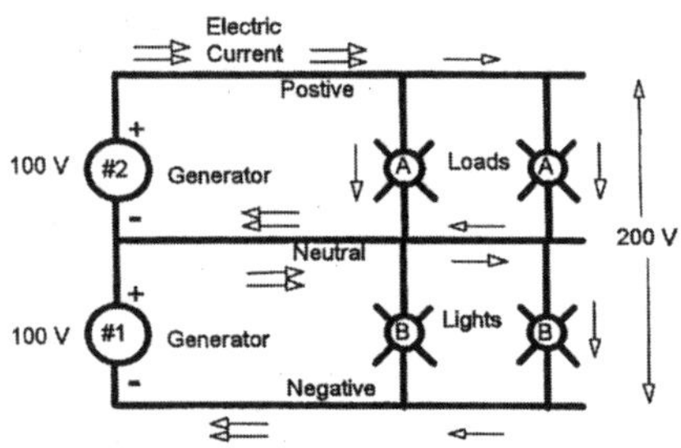

Figure 5. 3-Wire Direct Current Electrical System Conductor Used as Current Return Path

In the above case, the neutral would be carrying neutral return current over part of the circuit. If one were to measure the current on each conductor at the source, it would be a balanced load. However, out on the system the load would not be balanced. We will need this concept later. It is also important to recognize that so far we are dealing with all the current being contained within a conductor insulated from earth.

B. Alternating Current Grounding Practices

The Westinghouse-Tesla system of alternating currents was not always solidly grounded (earthed). The practices of grounding were first introduced for AC systems in 1885 by Elihu Thomson. One must note that today European and Latin American countries may ground (earth) only at the power source.

Mr. H. H. Dewey states that from the beginning of power transmission in 1888 or 1890 to 1910 or 1912 that there was no general practice as regards the question of operating electrical systems either with grounded neutral or with the neutral isolated.

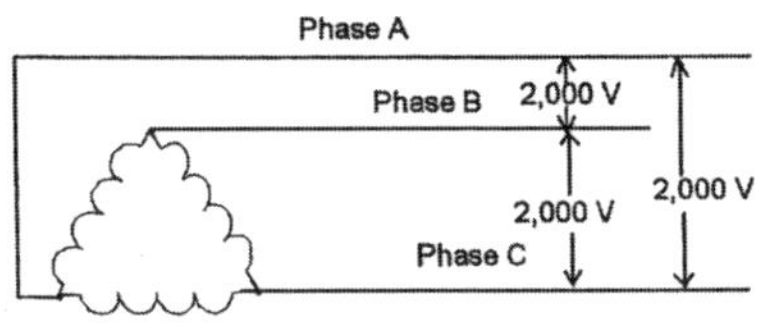

Figure 6. Electrical System Secondary Source Connected in Delta Ungrounded – Not Connected to Earth

> ". . . there was no decided difference in the results of operating either with the neutral grounded or ungrounded.. As they (the electrical systems) grew in extent, voltage surges began to make themselves felt over wide areas, resulting in breakdown of insulation at various points. In seeking a remedy, many engineers grounded the neutral of their generators or transformers." [13]

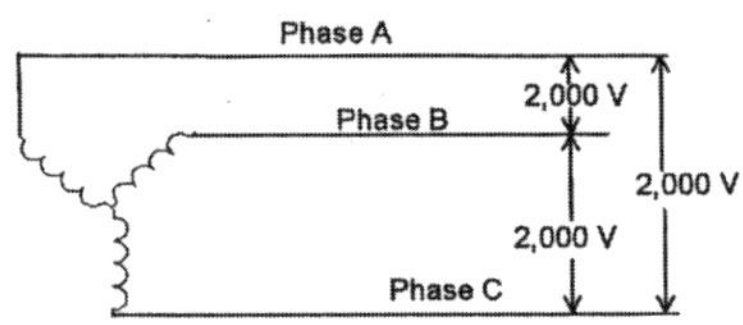

Figure 7. Electrical System Secondary Source Connected in Wye Ungrounded – Not Connected to Earth

The lack of sufficient insulation to withstand the voltage surges resulted in grounding the electrical system at the generator or at the transformers. If more than one transformer were connected to the system for transmission purposes, sometimes each transformer was grounded.

Any arcing fault on an ungrounded delta connected electrical system will result in the electrical system voltage rising to two, three or to as high as six times the normal voltage from conductor to earth, culminating in insulation failure [4]. Dewey goes on to write, "The evidence was not conclusive at that time and perhaps is not conclusive today (1923) but there are certain fundamental principles that seem

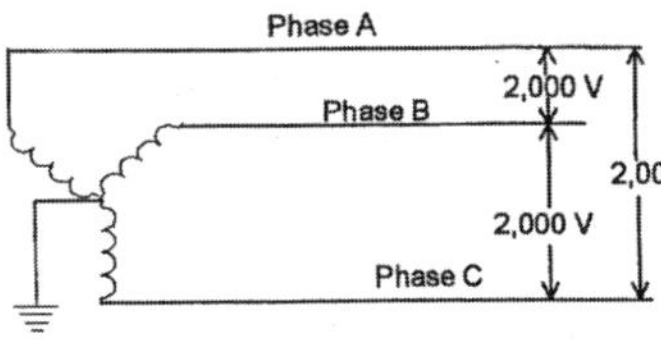

Figure 8. Electrical System Secondary Source Connected in Wye Grounded – Connected to Earth at One Location Only – Uni-Grounded

to be reasonable definite and most of them tend to show important advantages in favor of the grounded neutral system."

C. *Utilizing the Neutral*

In a wye transformer connection the middle point, the neutral point, can be connected to earth as is shown in Figure 8 without using the neutral for any electrical loads. Figure 9 shows the neutral also connected to earth, but the neutral conductor is extended along with the phase conductors. The configuration shown in figure 9 allows electrical loads, transformers to be placed between any of the three phase conductors, phase-to-phase and/or phase-to-neutral.

This connection, phase to neutral will force electric current to flow over the neutral back to the transformer. So far, this electrical connection is acceptable, as long as the neutral is insulated or treated as being potentially energized, but modifications will be made in the future that will negate safety for the public and animals.

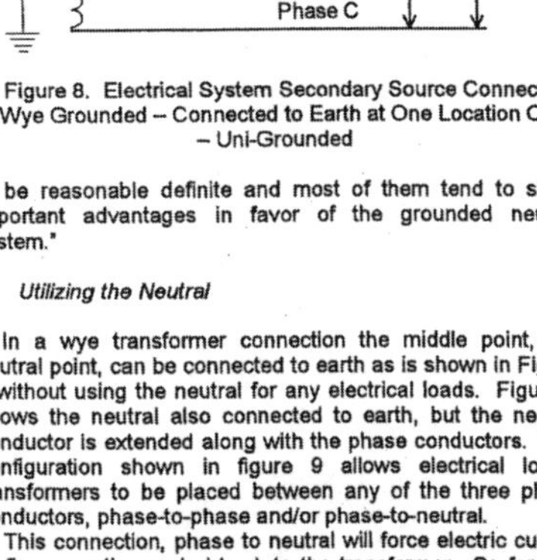

Figure 9. Electrical System Secondary Source Connected in Wye with Neutral Conductor Grounded and a Grounded Conductor Carried with Phase Conductors – Connected to Earth at One Location Only – Uni-grounded

In 1914 J. P. Jollyman, et al., writes about the Pacific Gas and Electric Company's practices which include the utilization of the neutral in distribution systems. Transformers are wound for only 57.7 percent of the insulation voltage requirements for a full voltage, phase-to-phase transformer. This amounts to a substantial savings. In addition, there is only one phase bushing on the transformers. This is additional savings over the costs of fully rated two bushing transformer and full transformer insulation rating. The savings in costs equate to additional profits.

The authors continue to cite additional savings in construction of the distribution line. The pin type insulator, the insulator screwed onto a wooden pin mounted in a cross arm, was the only type that was available in 1904. Savings were generated by applying only 57.7 percent of the voltage rating to the insulator. Instead of having an ungrounded electrical system with 25,000 volts between phases and earth, they now had, with a grounded neutral, only 14,400 volts to earth from any phase. There was still between phases 25,000 volts, but the pin insulator only was stressed to 14,400 volts.

This appears to be a sensible move, but this electrical configuration led later to other problems. Increasing the insulation rating would have been one of the correct methods to resolve the problem along with other solutions that will be discussed later.

D. *Wisconsin Rural Electrification*

One of the conditions for harmful stray continuous electric current flow is the electrical load. **As the electrical load increases the detrimental effects, also increase.** In order to show why dairy farmers are the first to encounter the stray current problem and are now coming to the forefront of the stray voltage (incorrect term) or stray current problem, we will need to go back and look at the growth of both the rural electrical industry and the dairy industry.

Today the problem manifests itself in the dairy and pig industry since they are very sensitive to stray current as compared to humans but as the load increases and maintenance decreases, tomorrow it will be humans. Unfortunately, it appears that tomorrow is now making itself known with the multitude of electric shocks in showers, hot tubs and swimming pools.

Mr. Post, an early AIEE member [14] writes about the eastern and southeastern 11 counties served by Milwaukee Electric Railway & Light Company where there were 21,000 farms. On January 1, 1926 there were only 2,740 farms receiving electric service, 13 percent of the total. The object of the utility was to discourage the use of motors and to encourage the use of lights and electric ranges. Operation of motors was prohibited from sundown to 11 p.m.

Note that in Figure 10, in 1926 there are two phase conductors supplying the two bushing transformer. The secondary neutral is connected to earth at a remote distance from the pole where the lightning arrester is connected to earth. This Wisconsin utility started out wiring transformer and secondary electrical systems correctly in 1926.

"Up to 1923 only about 120,000 of the 6,250,000 farms in the United States had been wired to use electricity." By 1930 according to a radio speech made by the General Electric Company's President over radio station WGY, ". . . shows that only about one-tenth of the farms are connected to public service electric lines and in the second place, that only a small part of the power used on any farm is electric power."

This low usage of electricity continued until late 1930s. There was an increase in the number of farms connected to

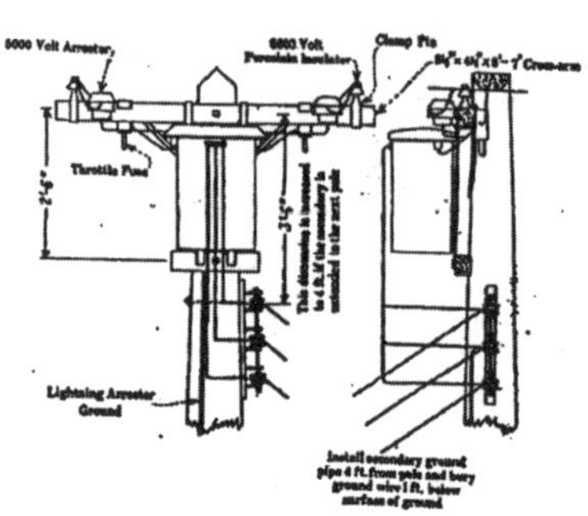

Figure 10. Pole Mounted 2-bushing Transformer with Lightning Arrester Separately Connected to Earth with No Primary Connection to Secondary Conductors, circa 1926.

the electrical distribution systems, however there were few electric motors used on the farms. The motors that were used were for water supplies and cream separators. [15]

VI. USING OTHERS PROPERTY WITHOUT COMPENSATION

A. Lightning Arrester Primary to Secondary Wiring Connection

In Chicago in 1932, the transformers were failing. By this time, the NEC was requiring grounding of the neutral in the electric service to homes. The earthing connection was made to the metallic water lines.

The Utilities Research Commission of Chicago and the Engineering Experimental Station at the University of Purdue conducted an investigation of surge protection of distribution circuits as to why transformers were failing. Test involved the interconnection of the primary lightning arrester ground and the grounded neutral of the secondary main circuit.

The conclusions reached were, "Measurements of the voltage between primary phase c lead and secondary neutral have shown that the ***interconnection of the secondary neutral with the lightning arrester ground is, in general, beneficial to the transformer.*** In particular, with a low resistance secondary neutral ground and a high resistance lightning arrester ground, the interconnection reduced the above voltage by 30 to 50 per cent." [16]

Instead of lowering the lightning arrester's resistance to earth by installing additional ground rods or other methods, at additional costs to the utilities, the utilities elected to save the additional cost by using the customers' connections to earth. This practice placed the industrial facility, the homeowners and the farmers in danger from excessive lightning current flow over the neutral conductor and the ground conductor in their facilities and the potential of high voltages during the lightning arresters' operations.

This connection permitted the hazardous electrical current from the operation of the lightning arrester to flow into the customers' homes, into the homeowner's ground rod, through the metallic water piping without their knowledge or consent which benefited the utility without proper approval or compensation of the owner of the secondary wiring system.

B. Primary Neutral to Secondary Neutral Connection

The extension of this invasion of the customers' wiring systems was when the utilities connected the primary neutral to the secondary neutral at the transformer and allowed primary neutral current to flow unimpeded into and over the customers' homes, into the homeowner's ground rod, through the metallic water piping into their showers, hot tubs and bathtubs and into the industrial facility without the owners' knowledge or consent. To comprehend this situation we need to start with what is a transformer.

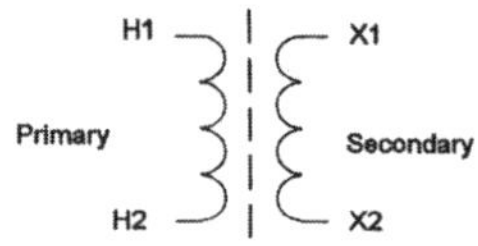

Figure 11. Transformer Electrically Isolated

Figure 11 shows what is considered by the majority of electrical engineers as a standard electrical transformer connection or wiring. Note that on the left side is the primary winding with two terminals, H1 and H2. For the purposes of this report, only a step down transformer will be discussed.

The lower voltage windings are referred to as the secondary. In the case at hand, a primary voltage of 14,400 volts is applied to the primary winding and the secondary voltage for sake of discussion will be 120 volts. Note that Figure 11 shows a dividing dashed line separating the primary windings from the secondary windings.

There are no solid electrical connections between the primary windings and the secondary windings. The vast

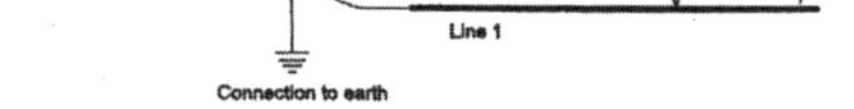

Figure 12. Phase to Neutral Primary Connection with Secondary winding Center Tapped – Primary and Secondary Electrically Isolated

majorities of electrical engineers are aware of and consider this the correct wiring for a transformer configuration.

A poll of electrical engineers NOT associated with utilities were asked if Figure 12 was the electrical wiring diagram for the transformer located on the pole outside their home, or the wiring diagram for the pad mounted transformer located on the ground near their home or if they had an underground electric transformer vault, the correct representation for the transformer?

Eighty-five percent of the electrical engineers working in industrial and commercial facilities, answered yes. The correct answer is Figure 13 with the solid electrical connection from the primary neutral to the secondary neutral. Such a connection is unthinkable and unheard of in the industrial, commercial and residential electrical engineering fields. Only electrical engineers working in the utility distribution area are aware of such a non-standard transformer connection and the disastrous results of allowing hazardous, stray electrical

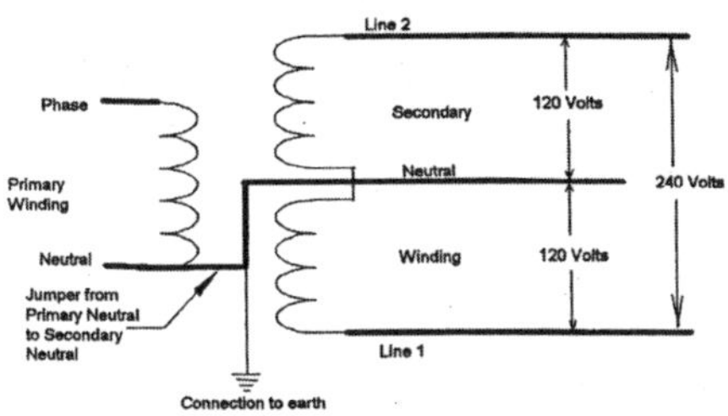

Figure 13. Phase to Neutral Primary Connection with Secondary winding Center Tapped and the Primary Neutral Solidly Connected to the Secondary Neutral.

current to flow continuously uncontrolled over the earth, metallic piping, building steel, into industrial, commercial facilities, homes and farms.

This is verified by the Institute of Electrical and Electronic Engineers' Standard 141, titled, "IEEE Recommended Practice for Electrical Power Distribution for Industrial Plants" and IEEE Standard 241, titled, "Recommended Practice for Electric Power Systems in Commercial Buildings" [17].

Why should one industry, the utility industry, be allowed to use transformers with such a non-standard transformer connection considered unsafe and/or unknown of by the vast majority of electrical engineers just because it saves money?

VII. UNDERSTANDING THE CONTINUOUS FLOW STRAY CURRENT

To completely understand the flow of stray current a person needs to 1) understand the definition of stray current and 2) realize how the current flows over the primary to secondary neutral connection and into industrial and commercial facilities, residences or farms and 3) the multigrounded neutral distribution system with uncontrolled flow of continuous electric current over the earth and into and through industrial and commercial facilities, residences or farms.

A. Stray Current

It has been shown previously that it is the current that is dangerous to the human animal and to the animal kingdom. Voltage is only the pressure that pushes the current. Voltage does not burn the body. It is the current that burns the body, sets the heart into fibrillation, halting the pumping action of the heart, resulting in death of both.

As long as we know that the mighty Mississippi River is contained within its banks, everyone is comfortable, satisfied and content. Likewise, the flow of all electric current must be contained within an insulated conductor just like the banks of the Mississippi. When the mighty Mississippi overflows its banks there is a problem and we become concerned. Just because we cannot see the flow of electric current over the earth is no reason that we should not be concerned when the electric current over flows its "banks".

Any continuous flow of electric current and the key word is continuous flow, must be contained within a conductor insulated from earth and not permitted to flow uncontrolled over and through the earth, metallic water pipes, building steel, etc.

When an electric fault occurs it is normal and acceptable for the fault current to flow over the ground path, whether that be a green or bare ground conductor or the earth, only for so long as it takes to open the circuit protective device (circuit breaker, fuse, recloser or other device). If the wiring system has been correctly installed, that period of time is very short; normally less than a second before the protective device, opens and halts the flow of fault current.

Thus we can now **define "stray current" as the continuous flow of any current, other than momentary fault current, over the earth, metallic piping, building steel, into houses and farms, etc. which is thus objectionable and undesirable to the continued good health of humans and animals.**

B. Parallel Current Paths

A lineman was asked to place an ammeter on the primary phase conductor of the conductor shown in Figure 13. He yelled down the reading, "42 amps". "Would you read the neutral conductor current please?" he was asked. "22 amps" he called down. "What happened to the other 20 amperes?" "Oh! That is normal" he replied. This exchange actually took place. [18]

Let us look closely at Figure 12, the primary winding of the transformer. The current enters the conductor on the conductor labeled "phase". Now the current flows just like water in a coil of hose, around and comes out the end marked "Neutral". All of the electric current that went into the transformer on the primary phase conductor, all of it came out on the primary neutral terminal of the transformer. See Figure 12.

The same is not true of the Figure 13. The current coming out of the transformer at the primary neutral terminal now has two paths that the "return current" can take. Some of the current will flow back on the neutral primary conductor and some will flow onto the neutral connection to earth on the secondary side of the transformer. In the example above with the lineman, 20 amperes flowed out the connections to earth and was uncontrolled and flowed continuously over the earth, through metallic pipelines, into industrial facilities' metallic piping, building s[illegible], commercial buildings' steel and metallic piping, residences, metallic piping and over the earth, etc resulting in hazardous conditions to humans and animals.

One of the methods used to "eliminate" "stray voltage" is to increase the size of the neutral conductor in utilities distribution circuits. It is assumed by some that increasing the neutral size will "prevent" current from flowing over the earth since the larger neutral has less resistance.

Confusion may also exist over harmonics and their associated problems. One low voltage distribution system solution for 3rd harmonics is to increase the neutral. Harmonics are not applicable in this discussion of stray current. The following will clarify the fallacy of increasing the neutral size.

Using Figure 14, let us assume that leaving the transformer at the neutral is 10 amperes of electrical current. There is a path using the neutral conductor, path A to B and another path A to C through the earth to another earth connection D to B. Since the resistance through the conductor path is 1 Ohm and equal to the resistance through the earth, 1 Ohm also, the current will divide equally. There will be 5 amperes through each path. This is based on Ohm's Law.

If we keep the resistance of the conductor constant, path A to B and change the resistance of the path A, C, D to B through the earth the current flow through the earth values will change. Using Figure 15, we will keep the resistance of the conductor from A to B constant at 1 Ohm. The resistance through the earth, "Z" will change. This change will affect the amount of current through both the paths A to B through the conductor and the path through the earth, A to C to D to B.

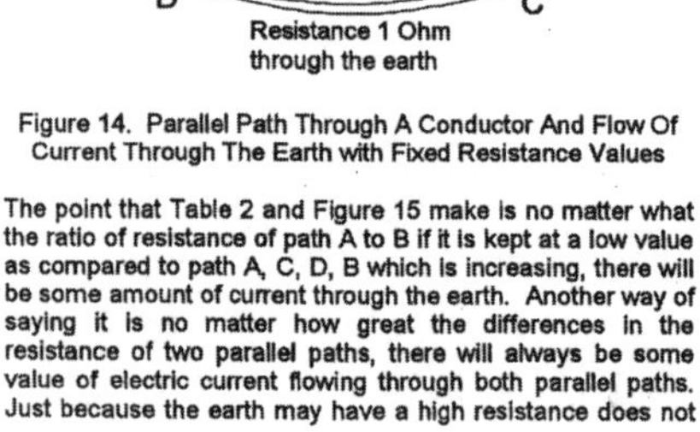

Figure 14. Parallel Path Through A Conductor And Flow Of Current Through The Earth with Fixed Resistance Values

The point that Table 2 and Figure 15 make is no matter what the ratio of resistance of path A to B if it is kept at a low value as compared to path A, C, D, B which is increasing, there will be some amount of current through the earth. Another way of saying it is no matter how great the differences in the resistance of two parallel paths, there will always be some value of electric current flowing through both parallel paths. Just because the earth may have a high resistance does not mean that there will be no current flow over and through the earth.

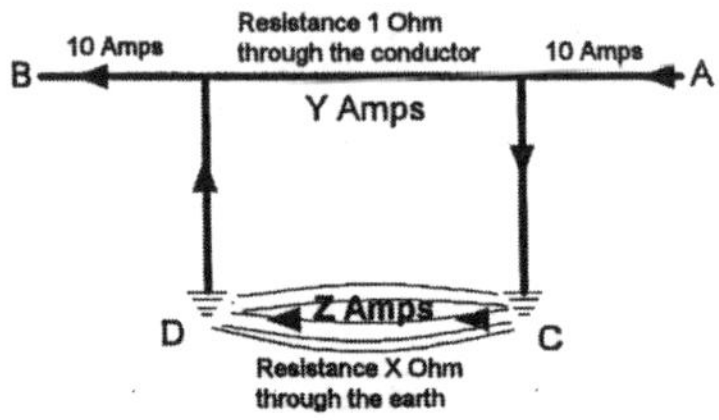

Figure 15. Parallel Path Through A Conductor And Flow Of Current Through The Earth with Variable Resistance Values

Resistance X Ohms	Current through Conductor Path A to B	Current through the Earth Path A, C, D, B
1 Ω	5.00 A	5.000 A
10 Ω	9.09 A	0.910 A
100 Ω	9.90 A	0.100 A
1000 Ω	9.99 A	0.010 A

Table 2. Current through a Parallel Path Using Figure 15.

Just to put into perspective for a male 1.1 mA is the level of perception "let go" threshold median.

When there exists more than one electrical connection to earth, no matter what the resistance of the earth is, there will always be some amount of hazardous electric current flowing continuously over and through the earth uncontrolled. Referring to Table 1, it can be seen that any current exceeding 9 milliamperes (0.009 A) can be hazardous to humans.

VIII. DANGEROUS PRACTICE OF COMBINING THE EQUIPMENT GROUNDING CONDUCTOR AND THE NEUTRAL CONDUCTOR INTO ONE CONDUCTOR SERVING TWO FUNCTIONS

The next step in the evolution of electrical distribution would have been Figure 16. Figure 16 shows a substation symbolized by the secondary wye transformer configuration with the neutral connected to earth. In addition to having the neutral connected to earth the neutral is carried on the poles along with the three phase conductors.

Somewhere along the distribution line, the three phases are dropped in favor of just one phase conductor. This is changing from three-phase distribution to single-phase distribution circuit. A typical load is shown in the form of a transformer connected from a phase conductor to the neutral conductor.

In figure 16, the current flow is shown originating at the substation transformer starting at the neutral connection in the center of the wye, labeled "N". The current flow continues to

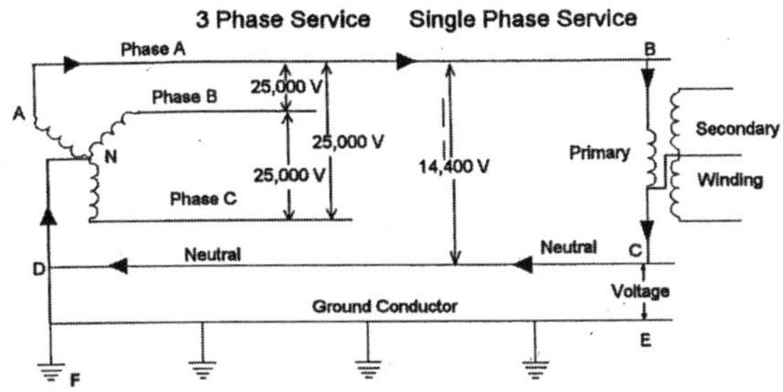

Figure 16. Wye Connected Electrical System Grounded at the source, Uni-Grounded With Transformer Connection To Earth At Only One Location With Both A Neutral And Ground Conductor Carried With The Phase Conductors With Phase-To-Neutral Loads (Which Generate A Hazardous Neutral Voltage –To–Earth, Point C to E) And A Ground Conductor

point A to point B through the transformer to point C onto the neutral and continuing on the neutral to point D and back to the beginning point, the center of the substation transformer N.

The neutral conductor has resistance. With current flowing through the neutral conductor, which has resistance, a voltage will be developed between point "C" and point "D". The voltage on the neutral at point "C" to earth point "E" can be high enough, depending on resistance and amount of current, to cause harm to a person, especially a lineman. Even though this wiring configuration contains all the continuous current flow within a conductor insulated from earth, there is a problem with a voltage build up from the neutral to earth along the length of the neutral conductor especially if it is on a pole and bare.

Since there is a voltage developed between the neutral conductor and the ground conductor, the neutral conductor should be treated as an energized conductor and installed on insulators, which add additional costs to the electrical distribution system and result in additional costs for the utilities. This type of electrical system allows the use of the less costly phase-to-neutral transformer.

Figure 16 is referred to as a 5-wire electrical distribution system consisting of one, two or three phase conductors, an insulated (from earth) neutral and an "equipment grounding conductor". This is exactly like what has been used in low voltage electrical distribution system in industrial and commercial facilities since approximately the 1950s.

Again, costs were the determining factor in resolving the problem with the difference in voltage between the neutral and the earth. Instead of installing, the neutral insulated from earth and treating the neutral as an energized conductor, the utilities took the least costly way and multigrounded the neutral thus reducing the build up of harmful potential to the linemen to less than approximately 25 volts at the mid-location between grounding points.

IX. MULTIGROUNDED NEUTRAL

The least costly solution to the problem with a voltage build up between the neutral and the ground conductor as shown in Figure 16 was to combine the function of the neutral and the ground conductors into one conductor. This combination of the neutral conductor and the ground conductor is called "multigrounded neutral".

When a fault of a primary conductor to earth occurred it was necessary to have low impedance paths back to the substation transformer in order that sufficient electrical current would flow to operate the protective device, the fuse, circuit breaker or recloser. In order to insure a low resistance, impedance path back to the source it was necessary to earth the multigrounded neutral conductor just like when there were two separate conductors, the neutral and the ground as shown in Figure 16.

Figure 17 shows the multigrounded neutral system. In Figure 16 all the current is contained within an insulated conductor; See path "N" to "A" to "B" through the service transformer to "C" to "D" and back to the starting point "N". With a multigrounded neutral distribution system it is necessary to have an electrical connection to earth at least 4 times per mile to keep the voltage on the multigrounded neutral from exceeding approximately 25 volts making it safe for the linemen should they come into contact with the neutral and the earth. See NESC Rule 096 C in the section on the National Electrical Safety Code.

With the multigrounded neutral conductor connected to earth at least 4 times per mile and at each transformer and lightning arrester there are now multiple paths over and through the earth that the hazardous electric current can flow over continuously, uncontrolled. The solid arrows in Figure 17 indicate this hazardous current flow. This flow of hazardous electric current, which is shown with the solid arrows, is created as primary current only. Additional primary current may be produced by unbalanced electrical loads.

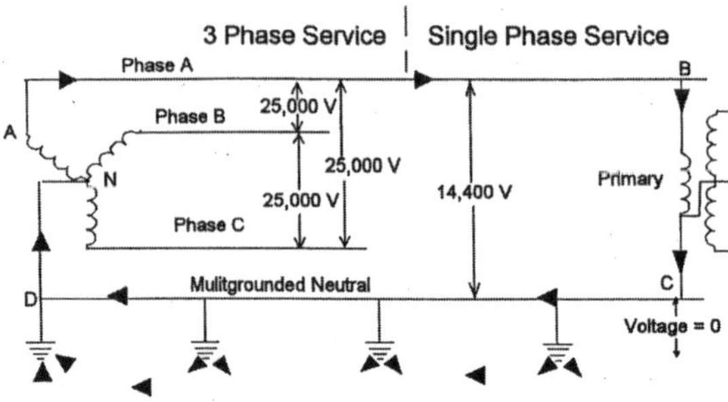

Figure 17. Wye connected at the source primary electrical system multigrounded neutral showing the continuous current flow through the earth when utilizing phase-to-neutral loads

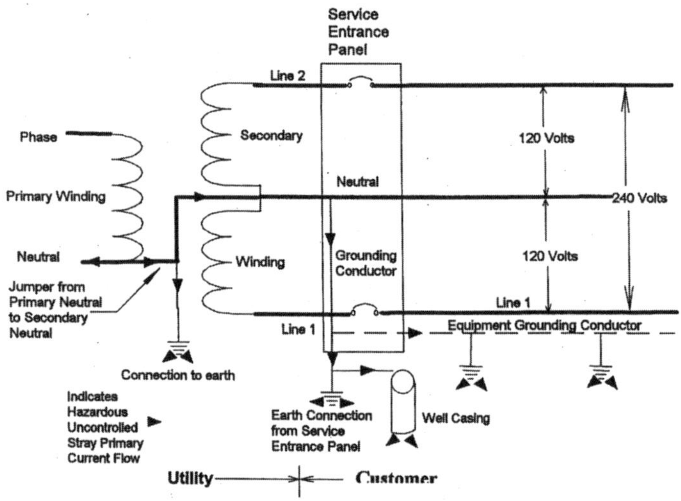

Figure 18. Wiring diagram showing the hazardous uncontrolled stray primary current flow over the customer's property.

A. Hazardous Electrical Current Flow

Figure 17 shows just part of the primary current flow into the earth. The path that this current flow takes through the earth cannot be determined. We cannot put an isotope on each electron and trace its path as it flows uncontrolled through the earth.

Public Service Commissions and Utilities are obligated to provide adequate electrical service without any adverse effects. It is irresponsible to permit stray uncontrolled electric current to flow into and over private property. This act is hidden from view, as the public is very unaware of what is being done by the utilities. It is not advertised on TV nor is information included in the bill sent to the utility customers.

B. Primary Neutral Current Entering through the Service Transformer (S)

As was explained previously in the section titled, "Lightning Arrester Primary to Secondary Wiring Connection" in 1932, The Utilities Research Commission of Chicago and the Engineering Experimental Station at the University of Purdue conducted an investigation of surge protection of distribution circuits. The conclusions reached were, "Measurements of the voltage between primary phase c lead and secondary neutral have shown that the ***interconnection of the secondary neutral with the lightning arrester ground is, in general, beneficial to the transformer.***"

This connection and the adverse effects are shown in Figure 18. The primary current leaves the transformer winding and has a fork in the path. One path is back to the substation over the multigrounded neutral conductor. In many cases, the neutral conductor is of a smaller size than the phase conductor supplying the current and the multigrounded neutral conductor has many high resistance twist connections which increases the impedance of the line forcing more and more current into the earth uncontrolled.

In addition, the neutral current can now flow over the jumper between the primary neutral conductor and the secondary neutral conductor as is shown in Figure 18. The NESC requires the transformer's neutral to be connected to earth at the transformer. This is the first connection to earth in Figure 18. Just because the NESC has a Rule that requires something to be accomplished does not make it safe or correct as will be covered later. In fact, a judge made such a ruling in a case.

The distribution line and the product, electrical energy, are inseparable and thus, become one. A Court has concluded that electricity is a product by holding as follows: "It is no answer for a company engaged in the public sale of such a highly destructive agency as electricity to say that precautionary measures which may have been adequate to meet the test of due care 50 years ago when the lines were installed, must be regarded as sufficient in modern industrial society."[1]

The National Electrical Code (NEC) requires the neutral in the service disconnect and overcurrent panelboard to be connected to the earth also. Now the secondary neutral is connected to earth a second time. A parallel connection of the neutral to earth now exists permitting hazardous electric current to flow continuously uncontrolled over the earth.

The equipment-grounding conductor has to be connected to everything that is metal and contains an electric conductor as per the NEC. The well casing offers another path to earth, the third parallel path contributing to the flow of hazardous uncontrolled electric current over the earth.

The discussion thus far has been about the hazardous stray continuous flowing primary neutral current from the electrical system owned and operated by the utility. This hazardous stray primary neutral current flowing continuously over the earth is the major contributor to the flow of uncontrolled stray current.

In addition to the primary electrical stray current, there are additional stray current sources flowing from the secondary side of the service transformer.

C. Fee for Using the Utilities Poles by Other Utilities

The communication and cable companies when mounting their conductors on utility companies' electric power poles compensate the utility company for their use.

Assume that Mr. Utility Customer wanted to communicate with his brother's house located on an adjacent street by utilizing an electronic communication device and the existing power conductors of surreptitiously. There would be no interference with the normal electric power distribution system. Such action would no doubt be considered by the utility company as utilization of their electric lines without adequately compensating. Alternatively, consider a person hanging his own conductors on the utility company's poles, the utility company would want compensation for the use of their poles.

The original wiring installed in and on Mr. Customer's property after the electric service meter was paid for and owned by Mr. Customer. That wiring consists of the neutral conductor being connected per the National Electrical Code to a driven ground rod supplied and owned by Mr. Customer. The internal electrical wiring on Mr. Customer's property has the neutral connected to the metallic water piping system and is owned by Mr. Customer.

The utility has usurped Mr. Customer's internal wiring system for the enhancement of their electrical system in order to reduce the utility's costs for the benefit of the transformer. The National Electrical Code (NEC) requires the neutral in the service disconnect and overcurrent panelboard to be connected to the earth through the grounding conductor. Now there is a second connection to earth. A parallel connection of the neutral to earth now exists permitting current to flow continuously uncontrolled over the earth.

The equipment-grounding conductor has to be connected to everything that is metal and contains an electric conductor as per the NEC. The well casing offers another path to earth.

The discussion thus far has been about the stray continuous flowing primary neutral current from the electrical distribution system. This stray primary neutral current flowing continuously over the earth is the major contributor. There are additional stray current flows from the secondary side of the service transformer.

D. Total Primary and Secondary Hazardous Uncontrolled Continuous Stray Current Flows

Figure 19 shows the magnitude of the points where the hazardous uncontrolled continuous flow of primary and secondary current can originate. The single phase-to-neutral transformer electrical circuit originates at the "N" in the center of the transformer winding. The current flows to point "A" to point "B" and through the primary winding of the service transformer.

As the current exits the transformer winding there is a fork in the circuit and one branch continues to "C to "D" and back to the beginning. However, the other path is the primary neutral to secondary neutral, point "K". From point, "K" the

[1] Black v. Public Service Electric & Gas Co.

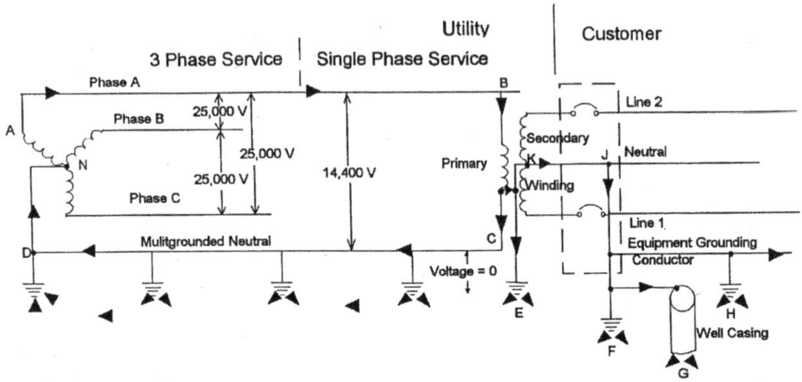

Figure 19. Wye Connected Electrical System Multigrounded Neutral with Primary Neutral to Secondary Neutral Electrical Service Connection Showing the Hazardous Stray Uncontrolled Primary Current Flow.

current can flow to the ground rod at the base of the pole, point "E" and into the earth, uncontrolled.

The other path is to point "J" down to another junction point where the stray current can flow into the earth at point "F" or into the earth at the well casing, point "G". There is another branch allowing the stray current to flow onto the "equipment grounding conductor and into the earth at every point where the equipment grounding conductor touches earth.

Figure 19 also shows the secondary electrical system where the primary and secondary continuous flowing stray uncontrolled current entered the earth at points E, F, G and H, but not limited to these points.

X. A RECOMMENDED SAFE METHOD FOR ELECTRICAL DISTRIBUTION

Figure 20 shows the source transformer at the substation as a wye connected transformer, uni-grounded. The phase conductors are insulated from earth just as the present day phase conductors are. The ground conductor is earthed at multiple places. This multiple earthing of the ground conductor allows any fault current that enters the earth, should a phase conductor fall to earth, to flow over the lower impedance path of the ground conductor and over the earth back to the neutral point of the transformer. This improves the safety of the electrical distribution system.

The load transformers are electrically connected phase-to-phase not in the present method of phase-to-neutral. This requires a two bushing transformer and fully rated insulation in the transformer.

The lightning arresters are connected to the grounding conductor, which has a low impedance connection to earth because of the four connections to earth plus the connection at each transformer can and lightning arrestor.

To accomplish this improved electrical distribution system requires the installation of another insulated phase conductor, which should have been installed in the beginning. The present "multigrounded neutral" conductor becomes the ground conductor.

Since the lightning arrester now has a low impedance earth connection due to the multiple earthing connections of the ground conductor, it is no longer necessary to use the earthing systems of the customers to obtain a low impedance connection to earth for the lightning arrestor successful operation. With the now required two bushing, full phase-to-phase voltage across the service transformer there will be no primary neutral current or primary neutral terminal on the service transformer.

Since there is no primary neutral the non-standard, transformer electrical connection of the primary neutral to secondary neutral cannot be made. Thus, the flow of primary current into the secondary side and into the customer's property will not exist. With this electrical distribution system, that does not have any primary neutral connections to earth, the customer will no longer have hazardous stray continuous electrical current from the primary distribution system flowing over his property or nearby earth.

XI. TRANSFORMER CONNECTED 3 PHASE, 5-WIRE GROUNDED WYE WITH LOADS PHASE-TO-PHASE OR PHASE-TO-NEUTRAL CONNECTIONS

Figure 16 shows an alternate method that is also safe for the public and animals. It is known as the 5-wire system. This system has been shown to reduce stray currents. [19]

This second type of system that is safer than a multigrounded neutral distribution system is a 3 phase, 5-wire grounded wye with all loads either phase-to-phase or phase-

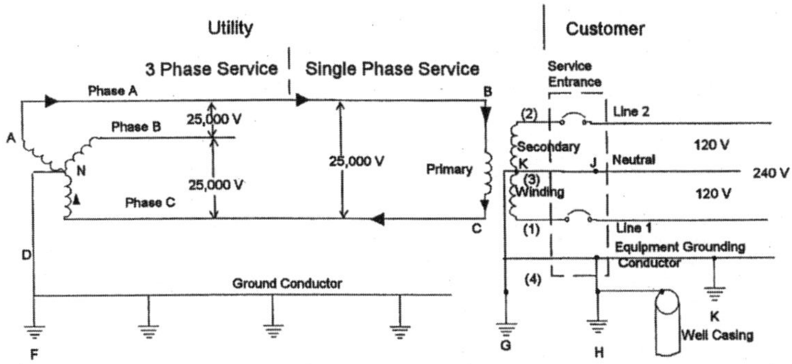

Figure 20. 4-Wire, Wye Connected Electrical System Grounded at the Substation Only and at the Service Transformer Secondary Side Only with Complete Electrical Isolation from Earth and All Continuous Electric Current Contained in Conductors Insulated from Earth.

to-neutral connections. The fourth conductor is a neutral insulated from earth. This type of electrical distribution has been in existence within industrial and commercial facilities for over 50 years. In the case of utilities, the neutral is treated as an energized conductor

This electrical configuration is considered one of the most desirable and safest type of electrical distribution systems for the simple reasons that whenever a phase comes into contact with the earth or the ground conductor, the fifth wire, the ground conductor, which is connected to earth at frequent intervals and is directly connected back to the transformer allows the placement of protective devices that can detect low levels of fault current to trip immediately.

Depending on the configuration of the circuit the fault current detection could be as low as approximately 25 to 50 amperes or lower and can be made to open the circuit removing the voltage from the circuit as compared to the 100, 150, 200 amperes or more current required to operate the protective device.

The utilities saved billion and billion of dollars over the years by utilizing the multigrounded neutral distribution system. The cost to fix or solve the problem is much less since the transformers can be re-used. The cost to add a single bare neutral is all that is required. The existing multigrounded neutral conductor would be converted to the ground conductor and the required NESC four ground rods per mile already exist. The costs to correctly wire the circuit or circuits system wide would be minimal, considering only one additional conductor would have been required. It has been estimated that the costs would be approximately $ 10 000.00 per mile and could be accomplished in the same fashion that the three wire receptacle was phased in – over a period of years.

It is reported by Mr. Lawrence C. Neubauer, Master Electrician who has tested many dairy farms in California that the utilities in the dairy areas of California do not use multigrounded neutral distribution system. There are no reports of stray current from the primary system.

There are no prohibitions in the NESC against such an electrical installation. Electric Power Research Institute (EPRI), an organization of the electric utilities, conducted extensive testing of the 5-wire electrical distribution system. The experimental facilities of the utilities have installed and tested such a safe electrical distribution system in New York State and have made the report available to their EPRI members.

B. Make the Secondary Service Conductors Safe

Initially, if utility companies had used four (4) service drop conductors, Figure 20, conductors (1), (2), (3) and (4), the secondary neutral earthed at the transformer as it should be, and the secondary neutral an insulated conductor, this would have resulted in a very safe electrical service drop without any continuously flowing stray electric current over the earth.

By bringing the insulated phase and insulated neutral conductors plus the equipment-grounding conductor to the customer there will be no need to re-earth the neutral in the service disconnect and overcurrent panelboard. The re-earthing of the neutral a second time in the service panel results, as has been shown, in parallel continuous stray electrical current paths.

XII. ARGUMENTS TO CONFUSE

There will be presented arguments that will try to show the system as shown in Figure 20 cannot work because of various excuses. One such argument will state lightning strikes will destroy the transformer unless the transformer is connected to the secondary circuit. Or that the primary neutral must be connected to the secondary neutral in order to "stabilize the system". This is pure poppycock, a new technical term.

In the intervening years since 1932, the utility industry has forgotten the reason for the resultant connecting of the primary neutral to the secondary neutral. The intermediate step was the adoption of the multigrounded neutral distribution system, which connected the lightning down connection to earth to and through the multigrounded neutral conductor, which was connected to earth multiple times, and the multigrounded neutral conductor was connected to the secondary neutral since it was *"beneficial to the transformer"*. In the industrial world of electrical distribution, Figure 20 is the method used to install electrical distribution systems every day for over 50 years or more.

It will be granted that transformers were not as well constructed in the 1920s as they are today. Failures were common due in some part to poor lightning arrester design and insulation methods as cited above. Transformers today rarely fail and when they do the fail, the mode of failure is not one where the primary voltage is applied to the secondary winding causing the users to have high voltage on their electrical circuits.

In fact, in a recent Virginia Division of Energy Regulation, State Corporate Commission action concerning VEPCO where VEPCO dropped a 25,000-volt line onto a 12,470-volt distribution line the voltage at the service entrance was calculated to have risen only to 334 volts maximum. The point is even with the primary connected to the secondary there was still a voltage rise at the customer services. Not everyone suffered electrical damage even after this happened four times. There were no transformer failures reported.

Like many things in the past, the utilities were aware of the cost savings and may not initially been aware of the safety concerns. It can be argued that once a practice is begun, it is often difficult to undo what has been done. However, it can and has been undone.

Proof that this non-multigrounded neutral electrical distribution system functioning safely can be found in California's Public Utilities Commission's General Orders – 95.

A. California's Public Utilities Commission's General Orders – 95

On January 19, 1994 the California's Public Utilities Commission in their General Orders – 95 generated a resolution SU-25 resulting in Rule 33.2.

33.2 Ground or Earth as a Conductor
Ground or earth shall not be used as a normal return or circuit conductor. In direct current supply systems or in single phase or polyphase supply systems, a neutral or any other conductor shall be used under normal use as a return or circuit conductor; however, the grounding of the neutral or any other conductor is not permitted as a normal return or circuit conductor. The neutral or any other conductor is permitted to be grounded only for the purposes of stabilization and protection.

In California, there is another term "messenger" that is used that has a different meaning and application elsewhere in the rest of the utility companies. California applies the term messenger correctly.

21.11 Messenger
means stranded wires in a group and which generally is not a part of the conducting system, its primary function being to support wires or cables of the conducting system; sometimes called "suspension strand".

The messenger has been corrupted into serving as not only a messenger, but as a combined neutral and ground conductor as will be discussed in detail later.

B. Transformers – Two Bushing Type – Two Phase Conductors

Two bushing transformers are required in California. Two bushing phase-to-phase transformers have full voltage

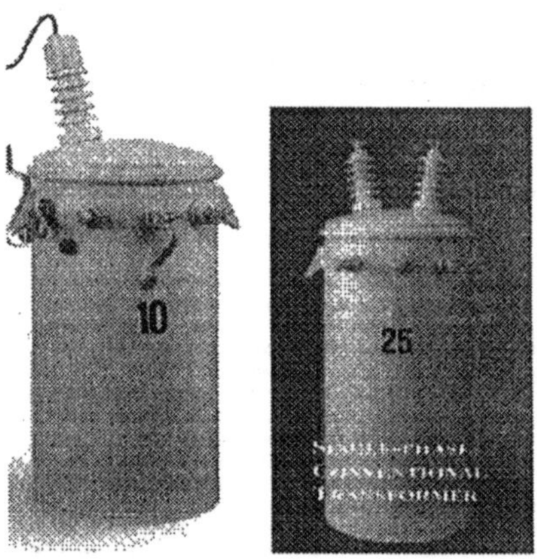

Figure 21. Left – Single Bushing Transformer With Ground Strap Connecting Secondary Neutral To Metal Can Of Transformer (Shown just above the "10"), Which Is the Primary Connection. Right – Two Bushing Transformer.

insulation rating where as a phase-to-neutral transformer has only 58 percent insulating rating, resulting in a savings in material cost in insulation and bushings.

A two bushing transformer eliminates any neutral current, thus there are no hazardous uncontrolled stray continuous flowing current over the multigrounded neutral conductor to earth since there is no need for any multigrounded neutral conductors

XIII. CONFLICT BETWEEN NATIONAL ELECTRICAL SAFETY CODE AND THE NATIONAL ELECTRICAL CODE

The National Electrical Safety Code Rules govern the electrical installation up to the service point, which include the service transformer. The service point is usually the electric meter or the line of demarcation between the utility and the customer.

"**NESC Rule 011.Scope**

B. The NESC covers utility facilities and functions up to the service point.

NOTE: The National Electrical Code. (NEC.), NFPA 70-1999 covers utilization wiring requirements beyond the service point."

There is lack of trust between the two code making bodies, the NESC and the NEC. Both bodies want the neutral to be connected to earth, ground. The most logical place to make this electrical connection to earth is at the transformer. The IEEE industrial and commercial electrical engineers state this in the IEEE Standard 141 [17].

The NEC fearing the utility companies will not ground the neutral at the transformer and the utilities supplying only a 3-wire service using the messenger for triple duty, requires the neutral to be grounded a second time at the service disconnect and overcurrent panelboard. This second grounding of the neutral on the secondary side of the transformer results in a parallel path condition and the problem of hazardous stray secondary neutral currents.

In addition, the utilities insist saving money by combining the neutral conductor with the ground conductor and the messenger support wire; three functions into one conductor.

It would be acceptable to combine the ground conductor and the messenger, but not include the neutral conductor. The neutral conductor must be insulated like the phase conductors, from earth except at the transformer.. In the service disconnect and overcurrent panelboard the neutral is connected again to the earth using the grounding conductor. See Figure 22.

A. Joint NESC and NEC Meeting Rejected

To resolve the conflict between the NESC and the NEC over the neutral being grounded by both code-making bodies at different locations a letter was sent to the Chair of both organizations requesting a joint meeting. This suggestion was based on the joint meeting of the NEC with the Canadian Electrical Making Code body to have uniform adoption by both codes of the European method of hazardous location installation, the IEC's Zones method. Both bodies rejected the suggestion for the NEC and the NESC meeting.

The members to the National Electrical Safety Code making body have their way paid by their utility employers. This is a clear conflict of interest. When one considers that utility employees have lived with the errors that were made in the distant past it should be no surprise to a person that utility employees would hesitate to try and change something management would be opposed to if it would cost additional funds.

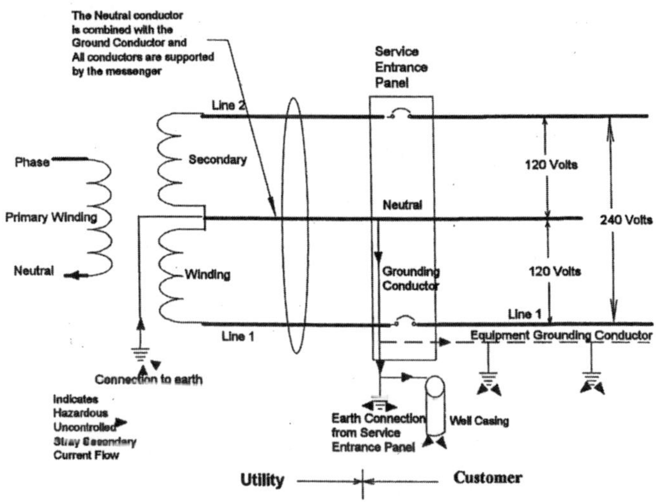

Figure 22. Conflict between the NESC and the NEC over the Grounding of the Neutral.

B. Proposals to the NESC to Fix the Service Drop Conductors Rejected

Contributing to the conflict is the fact that the conductors that run from the utilities' pole to the house consists of only two-phases, also known as hot or energized conductors and a messenger to support the two-phase conductors.

A neutral must be delivered to the house. The utilities have saved millions and millions of dollars by combining the neutral with the messenger when the neutral should be, for safety purposes, an insulated conductor.

In addition, the grounding of the transformer generates an equipment-grounding conductor that must also be carried into the house. In summary, the utilities should have supplied the following conductors in the service to any residence:

- Insulated either two or three phase conductors. (Line 1 and Line 2 in Figure 22.)
- Insulated neutral.
- A bare wire connected to the grounding conductor at the transformer and used as a messenger to support the above-insulated conductors.

Figure 23 shows the correct number of conductors installed between the transformer and the home. A proposal was submitted to both code making bodies, NEC and the NESC that would permit the correct number of service drop conductors if the homeowner paid the difference. In other words, there would be no additional cost to the utility, for the correct installation of service conductors that would prevent the flow of continuous electric current over the earth and/or home. This proposal was soundly rejected by the NESC, receiving only one vote, that of this writer.

No doubt, it was surmised that if such an admittance of incorrect electrical installation were acknowledged by the adoption of such a Rule change, the utilities would be inundated with legal action by homeowners. No such legal action to this writer's knowledge has occurred in California.

C. Proposals to the NEC to Fix the Service Drop Conductors Rejected

Proposals and comments have been made to the National Fire Protection Association's National Electrical Code Making Panel to change the service entrance conductors as described above. The proposals and comments were soundly rejected by the NEC with the excuse that it is the domain of the NESC.

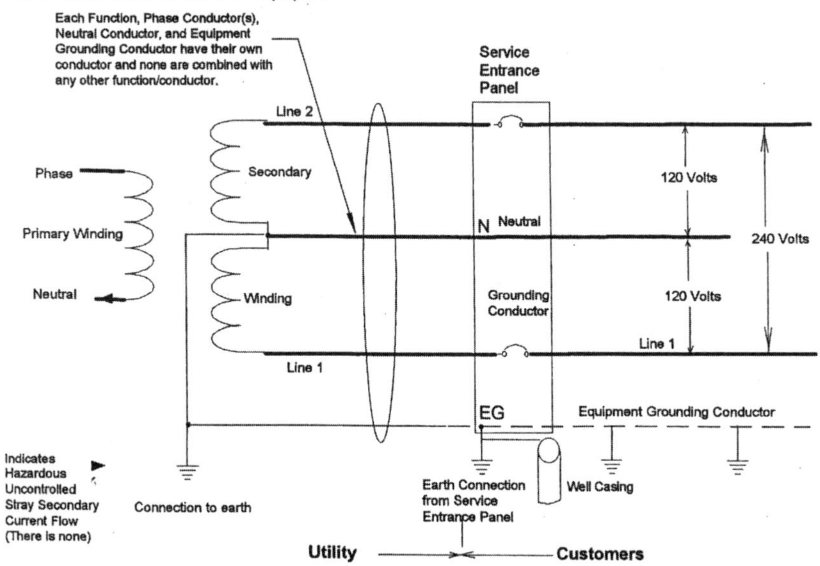

Figure 23. The Correct Number of Conductors for the Utilities Service Entrance

However, the NEC does require and has specifications for the conductors that connect with the utility's service drop conductors. Each, the NESC and the NEC say that they would consider making the change if the other code making body would do so first. Yet, both bodies refuse to meet in a joint meeting to iron out the conflicts.

D. Proposals to the NEC to Prevent adoption of Multigrounded Neutral Rejected

Before the NEC's 2002 code edition, Section 250.184 read as follows:

Section 250-184. Solidly Grounded Neutral Systems

(a) Neutral Conductor. The minimum insulation level for neutral conductors of solidly grounded systems shall be 600 volts.

Exception No. 1: Bare copper conductors shall be permitted to be used for the neutral of service entrances and the neutral of direct-buried portions of feeders

Exception No. 2: Bare conductors shall be permitted for the neutral of overhead portions installed outdoors.

FPN: See Section 225-4 for conductor covering where within 10 ft (3.05 m) of any building or other structure.

(b) Multiple Grounding. The neutral of a solidly grounded neutral system shall be permitted to be grounded at more than one point for the following:

1. Services
2. Direct-buried portions of feeders employing a bare copper neutral
3. Overhead portion installed outdoors

(c) Neutral Grounding Conductor. The neutral grounding conductor shall be permitted to be a bare conductor if isolated from phase conductors and protected from physical damage.

During the NEC 2002, cycle there was proposed expansion of NEC Section 250.184.

(D) Multigrounded Neutral Conductor. Where a multigrounded neutral system is used, the following shall apply:

(1) The multigrounded neutral conductor shall be of sufficient ampacity for the load imposed on the conductor but not less than 331/3 percent of the ampacity of the phase conductors.

Exception: In industrial and commercial premises under engineering supervision, it shall be permissible to size the ampacity of the neutral conductor to not less than 20 percent of the ampacity of the phase conductor.

(2) The multigrounded neutral conductor shall be grounded at each transformer and at other additional locations by connection to a made or existing electrode.

(3) At least one grounding electrode shall be installed and connected to the multigrounded neutral circuit conductor every 400 m (1300 ft).

(4) The maximum distance between any two adjacent electrodes shall not be more than 400 m (1300 ft).

(5) In a multigrounded shielded cable system, the shielding shall be grounded at each cable joint that is exposed to personnel contact.

This writer made a presentation to the NEC Making Panel # 5 at the beginning of the 2002 cycle proposal stage explaining the fallacies of such action. Since a subcommittee made the proposal and the proposal was taken from the NESC, it is believed the committee's proposal was received with validity where as an individual opposing the NESC was not.

In addition, one must realize that the NEC Making Panel # 5 is extremely slow in making correct decisions. This was evident in the adoption of the same four wire as described above for electric ranges and dryers. Beginning in 1944 because of the war effort electric ranges and dryers were allowed to combine the neutral and the equipment grounding conductors to save copper "for the duration". There was never a definition of the term, "for the duration" and the duration lasted 50 years.

In 1950s, the NEC adopted the safety policy of the three-wire grounding system prevalent in homes, commercial and industrial facilities. House trailers were wired using the incorrect method of combining the equipment-grounding conductor with the neutral. This action resulted in many electrical deaths and electrical shocks when someone touched the metal sides of the trailer and was standing on wet ground.

The NEC Making Panel handling Articles 550 and 551, trailers and mobile homes about 1968 made mandatory wiring that had insulated phase conductors (2) and a separate insulated neutral conductor and a bare or insulated equipment-grounding conductor. This separation of the neutral and the equipment-grounding conductor eliminated electrocutions and electrical shocks from the metal siding of trailers and mobile homes of persons standing on wet earth or in water and touching the metallic siding.

The very next NEC cycle, 3 years later, the panel responsible for marinas made the very same change. However, NEC Making Panel # 5 resisted making the same change for electric ranges and dryers for over 21 years. The author was told that NEC Panel # 5 had not heard of anyone being electrocuted thus there was no need to make any change in the wiring for ranges and dryers and continued to allow the combining of the neutral and the equipment grounding conductors. However, persons were receiving electric shocks from electric dryers. Finally, after over 21 years, in 1996, NEC Making Pane # 5 made the change and now the NEC requires that all new electric ranges and dryers be wired with insulated phase conductors (2), a separate insulated neutral conductor and an equipment-grounding conductor. Safety at last prevails with trailer homes, marinas, ranges and dryers. However, the service conductors still pose a problem.

There are several points to be made from the above discussion.

The logical progression from trailers to marinas to ranges and dryers has been accomplished. The next step is the service entrance conductors.

The majorities of the members of the National Electrical Safety Code Making Panels are employed by or sell services to the utilities, and thus are reluctant to make code changes that will cost their employers or clients' additional funds.

The majorities of the members of the National Electrical Safety Code Making Panels have probably been employed only by utilities and thus are very comfortable with the existing multigrounded neutral electrical distribution system since it is the only electrical distribution system with which they have worked and with which they are familiar.

XIV. LEGISLATION AND REGULATION

It is interesting to note the courts thinking on code and legislature enactments. Wisconsin legislation states:

"Sec. 288C. **Compliance With Legislation or Regulation.**

"Compliance with a legislative enactment or an administrative regulation does not prevent a finding of negligence where a reasonable man would take additional precautions."[2]

A New Jersey court has stated:

"An uninsulated high voltage power line carrying a deadly current must be considered one of the most dangerous contrivances known to man. In the pursuit of its business, by installing and maintaining such a line, a utility company must use care commensurate with the risk of harm, i.e. a high degree of care, to others who in the course of their ordinary and lawful activities might suffer injury, death or damage therefrom. In the utilization of such deadly wires along, adjacent to, across or over public highways, streets or ways, or private property, the (utility) company must be aware not only of the killing quality of its high-voltage lines, but also need to adjust the degree of care exercised in its control and maintenance of such wires."[3]

It was not until the period of 1942 to 1950 that the ability to determine the amount of current humans are sensitive to began. Excluding the EKG studies, which began about 1921, it was in the 1950 to 1974 period that the amount of electric current controlling the heart became known. These studies resulted in the defibrillator being developed.

Unfortunately, the use of the earth as a partial return path was permitted in the April 15, 1927 Edition of the National Electrical Safety Code although the wording is vague.

NESC Rule 012 C General Rules - 1927

C. For all particulars not specified in these rules, construction and maintenance should be done in accordance with accepted good practice for the given local conditions known at the time by those responsible for the construction or maintenance of the communication or supply lines and equipment.

"NEC Rule 215 C Use of Ground as Part of Circuit - 1927

In urban districts, supply circuits shall not be designed to use the ground normally as the sole conductor for any part of the circuit.

Recommendation. — It is recommended that such use be avoided in rural districts."

"NESC Rule 215. C. Use of Ground as Part of Circuit - 1927

"The use of the ground for the return portion of telegraph circuits and other communications and signaling circuits has long been customary and is for many purposes satisfactory, . . . For supply purposes a similar arrangement has been sometimes used, but it has objections from both the service and the accident standpoints, and its use is considered undesirable."

At that time, the NESC also lacked the understanding of hazardous levels of current as is evident from the following:

"(c) Current in grounding conductor. - Where multiple grounds are used there is a possibility of circulating currents between the different ground connections, arising from unbalanced loads, improper connection of grounding wires, and for other reasons. It is advisable to ascertain the amount of this current flow when the grounds are made in order to make certain that it is not great enough to be objectionable. A fraction of an ampere, or even several amperes on circuits of large capacity, may not be a serious matter, but cases can easily arise where the flow would be sufficient to be disturbing to the service."

As we see from above the amount of electric current to cause harm to humans and animals is extremely small, not "even several amperes on circuits of large capacity". We know today that that statement is false. Nevertheless, the warning was there.

A. *Failure of the National Electrical Safety Code to Protect the Public*

The National Electrical Safety Code states in the beginning of the Code the purpose.

NESC Rule 010.Purpose - 1997

"The purpose of these rules is the practical safeguarding of persons during the installation, operation or maintenance of electric supply and communication lines and associated equipment.

"These rules contain the basic provisions that are considered necessary for the safety of employees and the public under the specified conditions."

B. *Removal of or Disconnection of Grounding Down Connections*

In order to protect from the objectionable hazardous stray continuous flow of electrical current the NESC allows removal of or disconnection of grounds.

"NESC Rule 092 - 1997

"D. Current in Grounding Conductor

"Ground connection points shall be so arranged that under normal circumstances there will be no objectionable flow of current over the grounding conductor. If an objectionable flow of current occurs over a grounding conductor due to the use of multi-grounds, one or more of the following should be used:

1. Abandon one or more grounds.
2. Change location of grounds.
3. Interrupt the continuity of the conductor between ground connections.

"Subject to the approval of the administrative authority, take other effective means to limit the current. The system ground of the source transformer shall not be removed. The temporary currents set up under abnormal conditions while the grounding conductors are per-forming their intended protective functions are not considered objectionable. The conductor shall have the capability of conducting anticipated fault current without thermal overloading or excessive voltage buildup."

Thus, it is acceptable to remove the parallel connections once it has been shown that irreparable damage has been inflected.

---

[2] Kemp v. Wisconsin Electric Power Co., 44 Wis.2d 571, 579 (1969), quoting Restatement, 2 Torts 2d, p. 39.

[3] Black v. Public Service Electric & Gas Co.

Section 250-6(b) and (c), NEC – 1999, also states:

"**(b) Alterations to Stop Objectionable Current.** If the use of multiple grounding connections results in an objectionable flow of current, one or more of the following alterations shall be permitted to be made, provided that the requirements of Section 250-2(d) are met.

•Discontinue one or more but not all of such grounding connections.

•Change the locations of the grounding connections.

•Interrupt the continuity of the conductor or conductive path interconnecting the grounding connections.

•Take other suitable remedial action satisfactory to the authority having jurisdiction.

(c) Temporary Currents Not Classified as Objectionable Currents. Temporary currents resulting from accidental conditions, such as ground-fault currents, that occur only while the grounding conductors are performing their intended protective functions shall not be classified as objectionable current for the purposes specified in (a) and (b)."

The currents that are continuously flowing currents over the earth, are objectionable. On the other hand, non-objectionable currents are those that flow over the earth momentarily until a protective device such as a fuse, circuit breaker or recloser operates to stop the current flow.

XV. UNGROUNDED ELECTRICAL DISTRIBUTION SYSTEMS

Another very dangerous to the public and to utility linemen is the ungrounded electrical distribution system. With only the capacitance connection to earth there is less likelihood that the normal ungrounded electrical distribution system will operate the protective devices at a low value of current, when a person comes into contact with the energized lines.

An example is a case where a 12 year old boy riding his bicycle on the payment after a storm when an energized conductor snapped and wrapped around his neck. He laid on the ground with the energized conductor around his neck smoking while his parents watched helplessly for between 20 and 45 minutes. The utility linemen had to cut the line before it was safe to attend to his lifeless body. [20]

Another case involved a lineman who was accustomed to multigrounded neutral distribution system and during storm duty was sent to another utility where the distribution system was ungrounded. He fuzzed the line and getting no response somehow touched the line and was electrocuted.

XVI. GFCI – PROTECTION AGAINST ON SITE STRAY CURRENT

The GFCI contains two electronic circuits. One circuit measures the amount of current flowing on the phase conductor and the amount of current on the neutral conductor. If the amount differs by 0.005 to 0.006 amperes, the GFCI trips, stopping the flow of current. It is assumed the unaccounted for current could be flowing through a person and thus protection from electrical injury is provided.

The second circuit in a GFCI detects when a neutral to ground fault occurs. The designers of the GFCI with pressure from Europe where whole house ground fault detection is required installed in 1986 the second neutral fault detection circuit.

Since any current flowing over the earth would be objectionable, unwanted and harmful the GFCI trips and opens the electrical circuit stopping the flow of harmful neutral current over the earth. A proposal has been submitted to the 2005 NEC to have dairy and pig farms install GFCIs on all 120-volt circuits.

XVII. CONCLUSIONS

It appears that whenever the utilities were faced with the choice of public safety or saving costs they chose saving costs over the safety of the public with respect to the multigrounded neutral distribution system and the modifications that were made that were "***beneficial to the transformer***".

In order to have an electrical distribution system that is safe from the great and continuous risks of harm from stray current flowing uncontrolled over the earth in unrestrained amounts the following must be accomplished:

i. The multigrounded neutral distribution system must be eliminated and replaced by the 4-wire, see figure 20 with phase-to-phase loads or the 5-wire, see figure 16 with either phase-to-phase or phase-to-neutral loads type distribution system, which will eliminate using the earth for a return current path and will contain all the return current on an insulated conductor as specified by Zipse's Law.

ii. The hazardous connection of the primary neutral to the secondary neutral is no longer needed to allow previously poorly earthed lightning arrestors to function since the ground conductor is earthed at four times per mile providing adequate earth connection for the lightning arrestors.

iii. No longer will the messenger be allowed to functioning as a messenger, a neutral and a ground conductor. The service drop conductors will consist of one, two or three phase conductors, an insulated neutral, and the messenger that can also serve as the earthing conductor only.

iv. No longer will the neutral be connected to the earth at the service disconnect and overcurrent panelboard. With a separate insulated neutral, which will be connected to earth at the transformer and a ground conductor from the transformer coming into the service disconnect and overcurrent panelboard at the residence, it will not be required to connect the neutral to the earth again at the service disconnect and overcurrent panelboard.

v. Consideration should be given to the installation of GFCIs on all 120-volt circuits that could become or could develop a neutral-to-ground fault.

If the public wants to be able to enter a swimming pool or a hot tub or to take a shower without the fear of receiving an unwanted electric shock or to be a victim of an electrocution, then the public must rise up in letter writing to their Public Service Commissions and their legislators both local and federal and demand an electrical distribution system free from to flow of uncontrolled hazardous stray current.

EPILOG

Since the paper was written back in November 2002 two multigrounded neutral distribution systems have been tested and found to have insufficient stray current flow to harm dairy cows. The conclusion reached is if the multigrounded neutral distribution system is installed correctly, operated and maintained properly in accordance with the NESC's Rule 092D and Rule 214A they can be acceptable for Dairy areas.

XVIII. REFERENCES

[1] Institute of Electrical and Electronic Engineers' National Electrical Safety Code, ANSI C-2, Piscataway, NJ, Edition 2002.

[2] IEEE Std 100-1996, " The IEEE Standard Dictionary of Electrical and Electronics Terms", Piscataway, NJ

[3] ANSI/IEEE Std. 80, IEEE Guide for Safety in Substation Grounding, New York, NY

[4] ANSI/IEEE Standard 142, "Recommended Practice for Grounding of Industrial and Commercial Power Systems" (Green Book), Piscataway, NJ.

[5] NFPA Standard 70, "National Electrical Code" Quincy MA.

[6] D. W. Zipse, "Electrical Shock Hazard Due to Stray Current", (Subtitled, "The Shocking Shower") in IEEE I&CPS *Conference Record,* May 1999.

[7] D. W. Zipse, "Are The National Electrical Code and the National Electrical Safety Code Hazardous to Your Health?", (Subtitled, "The Shocking Swimming Pool") in IEEE I&CPS *Conference Record*, May 1999.

[8] State of Wisconsin, Circuit Court, Brown County, Case No. 00-CV-1441, Russell Allen –v– WPSC

[9] http://www.ag.state.mi.us/speclit_div/speclit_hm.htm, select "Dave Stetzer" exhibits for a short list of site investigations, Michigan Attorney General case against Consumers Power in the PSC.

[10] Charles F. Dalziel, "Dangerous Electric Currents", AIEE Transactions, August – September 1946, Vol. 65, pp. 579 – 585.

[11] J. Ma, F. P. Dawalibi and W. K. Daily, "Analysis of Grounding Systems in Soils with Hemispherical Layering", PES Transactions.

[12] Tom Shaughnessy, PowerCet Corporation, 3350 Scott Boulevard, Building 55, Unit 1, Santa Clara, CA 95054, Vol. No. 3, Summer 1997.

[13] H. H. Dewey, General *Considerations in Grounding the Neutral of Power Systems,* AIEE, April 1923, pp. 405-416.

[14] G. G. Post, Important *Features of a Successful Plan for Rural Electrification,* AIEE Transactions, May 1926, pp. 515-521.

[15] Electricity On The Farm, Feb. to Aug. 1930, G. E. History, Schenectady Museum, Schenectady, NY.

[16] C. Francis Harding and C. S. Sprague, Interconnection of Primary Lightning Arrester Ground and the Grounded Neutral of the Secondary Main", Transaction AIEE, 1932, pp.234-238

[17] IEEE Standard 141, "IEEE Recommended Practice for Electrical Power Distribution for Industrial Plants", Piscataway, NJ,

and

IEEE Standard , 241, titled, "Recommended Practice for Electric Power Systems in Commercial Buildings, Piscataway, NJ,

[18] Present-Day Grounding Practices on Power Systems – third AIEE Report on System Grounding, AIEE 1947, Vol. 66, pp. 15251551.

[19] Short, et al, *Five Wire Distribution System Demonstration Project*, IEEE Transactions on Power Delivery, Vol. 17, No. 2, April 2002.

[20] Edie Buchanan Administrator of the Estate of Michael C Yandek, Deceased v. FirstEnergy Corporation, et al., Case No.: 408609, Judge: TIMOTHY McCORMICK. Cleveland, OH

XII. BIBLIOGRAPHY

ANSI/IEEE Std. 100, IEEE Standard Dictionary of Electrical and Electronic Terms, New York, NY

AIEE Committee Report, "Application of Ground Fault Neutralizers, *Electrical Engineering*, vol. 72, July 1953, p. 606

IEEE Standard 81, "IEEE Guide for Measuring Earth Resistivity, Ground Impedance and Earth Surface Potentials of a Ground System in Part 1: Normal Measurements."

John P. Nelson. "High Resistance Grounding Of Low Voltage Systems" IEEE – IAS – PCIC 96-3 *Conference Record,* Sept. 1996.

John P. Nelson. "The Grounding of Power Systems Above 600 Volts: A Practical Viewpoint", IEEE – IAS – PCIC 03-X *Conference Record,* Sept. 2003

VITA

Donald W. Zipse (S'58-M'62-SM'89-F'94-LF'97) was graduated from the Williamson Free School of Mechanical

Trades with honors where he gained practical experience in electrical construction and in power plant operation. He received his electrical engineering degree from the University of Delaware and went to work for Cutler-Hammer as an area sales engineer. He spent 16 years with ICI America, Inc in their Central Engineering Department as a company wide electrical specialist.

For the next 14 years, he was with the FMC Corporation in their Engineering Service organization, functioning as an Electrical Engineering Consultant, responsible for providing electrical design of new facilities and consulting service to the total corporation, both chemical and mechanical groups.

He is a registered Professional Engineer. He represents the IEEE on the National Electrical Code Making Panel #14, Hazardous Locations as well as the Lightning Standard NFPA 780 and is a member of the International Association of Electrical Inspectors. He serves on the National Electrical Safety Code Grounding Subcommittee.

He has served on many IEEE committees, participated in the color books (IEEE Recommended Practice), and standards groups, including the Standards Board and the Standards Board's Review Committee. He is a member of the IEEE COMAR, Committee On Man And Radiation and Standards Correlating Committee #28, Non-Ionization Radiation. Mr. Zipse received the Standards Medallion for his work in and promoting standards.

He has published countless technical papers on such diverse and controversial subjects as Unity Plus Motors, Computers, Neutral to Ground Faults, NEC Wire Tables, Health Effects of Electrical and Magnetic Fields, Measuring Electrical and Magnetic Fields, Lightning Protection Systems: Advantages and Disadvantages, the NESC and the NEC: Are Dangerous to Your Health? Electrical Shock Hazard Due To Stray Current and has participated on National Electrical Code panels and in teaching the Code.

For the last ten years, he has been President of Zipse Electrical Engineering, Inc., an electrical forensic engineering consulting firm. For the past seven years, he has been primarily involved as a forensic engineer and expert witness in cases resulting from electrical accidents and electrocutions and for the last two years he has been involved in legal cases concerning stray current and dairy cows.

Appendix F

Here are descriptions of the horrific effects of stray current on animals, as culled from university research by renowned veterinarian Richard Schulte. Below these are details of actual occurrences at a Michigan kennel. The symptoms and effects of the current on dogs and cats are identical to those which sickened and killed those Kentucky thoroughbred racehorses.

SYMPTOMOTOLOGY OF STRAY VOLTAGE
By Dr. Richard Schulte, DVM

THE JOURNAL OF DAIRY SCIENCE: PUBLISHED A PAPER WRITTEN By R.D. APPLEMAN AND R.J. GUSTAFSON IN 1985 ENTITLED THE SOURCE OF STRAY VOLTAGE AND EFFECT ON COW HEALTH AND PERFORMANCE. IN THE ARTICLE IT STATES THE " COMMONLY CITED CATTLE RESPONSES INCLUDE: 1) INTERMITTENT PERIODS OF POOR PRODUCTION; 2) UNEXPLAINED POOR PRODUCTION: 3) INCREASED INCIDENCE OF MASTITIS; 4) ELEVATED SOMATIC CELL COUNTS; 5) INCREASED MILKING TIMES; 6) INCOMPLETE MILK LETDOWN; 7) EXTREME NERVOUSNESS WHILE IN THE PARLOR; 8) RELUCTANCE TO ENTER THE MILKING PARLOR; 9) RAPID EXIT FROM THE PARLOR; 10) RELUCTANCE TO USE WATER BOWLS AND METALLIC FEEDERS; 11) ALTERED CONSUMMATORY BEHAVIOR (LAPPING FROM THE WATERING DEVICE)."

THE TEXAS AGRIBUSINESS ELECTRIC COUNCIL, KNOWN As TABEC, DISTRIBUTED A MANUAL ENTITLED STRAY VOLTAGE IN DAIRIES. THIS MANUAL WAS PRODUCED AT THE TEXAS A + M UNIVERSITY'S DEPARTMENT OF AGRICULTURAL ENGINEERING AND WAS INCORPORATED INTO THE IDAHO POWER COMPANY'S "GUIDELINES FOR THE INVESTIGATION AND RESOLUTION OF STRAY VOLTAGE PROBLEMS". IN THE TABEC MANUAL IT LISTS THE SYMPTOMS OF STRAY VOLTAGE AS 1) UNEVEN MILKOUT 2) COWS EXTREMELY NERVOUS IN THE PARLOR 3) COWS RELUCTANT TO ENTER THE PARLOR 4) INCREASED MASITIS (THEY DO NOT BLAME STRAY VOLTAGE FOR THE MASTITIS DIRECTLY BUT ON THE UNEVEN MILKOUTS. THEY REPORT A FAILURE OF TREATMENTS IN

THE UDDER DUE TO DILUTION OF THE MEDICAMENT BY THE EXCESS MILK IN THE UDDER) 5) REDUCED FEED INTAKE IN THE PARLOR 6) RELUCTANCE TO DRINK WATER 7) INCREASED MANURE DEPOSITION IN THE PARLOR 8) LOWERED MILK PRODUCTION.

IN A MAILER TO CUSTOMERS IN THE LATE 1990's, WISCONSIN PUBLIC SERVICE CORPORATION (ONE OF THE LARGEST SUPPLIERS OF ELECTRICITY IN WISCONSIN) LISTED THE SYMPTOMS OF STRAY VOLTAGE AS 1) NERVOUS COWS, 2) REDUCED FEED AND WATER INTAKES, 3) INCREASED INCIDENCE OF MASTITIS, 4) HIGHER SOMATIC CELL COUNTS, 5) DECREASED MILK PRODUCTION. THE ARTICLE ALSO STATES THAT SOME STUDIES INDICATE THAT, UNDER CERTAIN CIRCUMSTANCES, RELATIVELY LOW LEVELS OF STRAY VOLTAGE, BELOW THOSE WHICH WOULD ORDINARILY BE FELT BY PEOPLE, CAN HAVE AN ADVERSE EFFECT ON ANIMALS WITH CONTINUED EXPOSURE."

TO THE AFOREMENTIONED SYMPTOMOTOLOGY WE MUST ADD 1) POOR HAIR COATS, 2) POOR REPRODUCTIVE PERFORMANCE, :3) POOR FEET AND LEGS, 4) POOR RESPONSE TO TREATMENTS FOR INFECTIOUS AND METABOLIC DISEASES, 5) DIFFICULTY IN RAISING CALVES 6) POOR WEIGHT GAINS, 7) ABNORMAL POSTURING IN STANCHOINS AND TIE STALLS, 8) HIGHER THAN EXPECTED DEATH RATES, 9) HIGHER THAN NORMAL CULL RATES 10) DEPRESSED IMMUNE SYSTEMS HIGHER THAN EXPECTED INCIDENCE OF ABORTIONS.

THESE SYMPTOMS ARE THE ONES COMMONLY REPORTED BY VETERINARIANS WHO WORK WITH HERDS IMPACTED BY STRAY VOLTAGE. WHERE MITIGATION IS ACHIEVED, ALOT OF THE SYMPTOMS DISAPPEAR.

Appendix G

This is a very interesting article which was submitted to the Wisconsin Agriculturist magazine and should have been published; it was not. The reason is simple enough. Since electric utilities contribute financially to the magazine, it was clearly in the Agriculturist's best interest to censor itself in order that the funds would not dry up. Of course this can't be proven and therefore no one can accuse the utilities of employing blackmail tactics. Their inexhaustible resources are a big part of what makes them dangerous. The author of the piece is an electrical engineer with on axe to grind.

The 'Missing Wire' on Utility Electric Lines

An electrical engineer says utility distribution lines are poorly engineered and increasingly dangerous.

By Kurt Gutknecht

Most people know electricity arrives on a wire. Most people also assume it also leaves on a wire.

They're wrong on the last assumption. In most of the Unites States, electricity is indiscriminately dumped into the environment on its return trip, a practice that even alarmed Thomas Edison.

Until recently, dairy farmers were about the only ones to notice because their cows sickened and died from all the electricity in the environment. To most electrical engineers who work for utilities, however, the practice is considered "normal" and harmless.

Not to Don Zipse, however. The balding, bespectacled independent electrical engineer from Pennsylvania is one of the few professionals in the industry who wants to end the practice by putting up another wire on utility poles so electricity can return safely.

Zipse says the electrical distribution system used in every state except California is missing a wire. He says it's one of the biggest and most expensive engineering errors in history and one of the most dangerous. So far, however, most of his colleagues don't think it's a problem.

According to Zipse, utilities avoided the expense of putting up an additional "ground" wire, a safety feature, and have instead relied on a single wire to both carry return (neutral) electricity and to serve as a green ground wire. In theory, electricity should flow off the return wire (over the ground wire) into the earth only during a lightning strike or other dangerous surge of current or short circuit.

In practice, electricity often flows continuously into the earth from the most widely used distribution system, which is known as a "multigrounded neutral system." As electric loads have increased, so have levels of electricity spilling from utility lines, he says.

It gets worse. In another cost cutting maneuver, the utilities use customers' electric lines as relief valves for dangerous surges of electricity, such as during lightning strikes. However, the method also lets return current flow continuously into customers' lines.

More and more customers are complaining about the consequences of these practices, says Zipse.

"Dumping" electricity

Utilities and regulatory agencies have contended that releasing return electricity from their lines into the earth, which results in electrical pollution often known as "stray voltage," was necessary to maintain the safety of the electrical distribution system. For several decades, farmers questioned why the utility was allowed to "dump" electricity on their land without their permission. That issue prompted the attorney general of Michigan to bring legal action against a utility. The case was recently dismissed largely because the technical standards didn't prohibit the practice.

How much electricity can a utility dump into your back yard, house or school? In practical terms, as much as they want. The limits are established by Rule 92D in the National Electrical Safety Code (NESC, which states that there can't be any "objectionable flow" of electricity into the earth.

And who determines what's "objectionable"? The utility, who (as might be imagined) seldom find anything objectionable about the practice.

Complaints are now surfacing in many other parts of the country, including cities on the East Coast where utilities have warned children not to play outside lest they be injured by electricity from utility lines. Zipse has written several technical articles about urban residents shocked in their showers and swimming pools by electricity streaming from utility lines.

Ironically, Zipse says it's common practice in commercial installations to install a wire strictly for grounding purposes. The retrofit for thousands of miles of utility distribution lines would be expensive, but Zipse says it's time to start addressing the problem. A recent article published in the Journal of the Institute of Electrical and Electronics Engineers showed that adding another wire effectively reduced so called "stray voltage." So far, however, the utilities haven't made any effort to apply the technique.

The failure to use another wire has saved utilities "millions and millions of dollars," writes Zipse, and shows an "indifference and insensitivity to the safety of humans and animals." Zipse now serves as an expert witness on behalf of farmers with "stray voltage" complaints against utilities.

So far, however, he hasn't received much support from his professional colleagues. Zipse is a Life Fellow of the Institute of Electrical and Electronic Engineers and serves on the grounding committee for the National Electrical Safety Code (NESC), which establishes national standards for utility systems.

The revamped utility lines would essentially incorporate the familiar green grounding wire that has been required in household wiring since 1950s. Farmers often complain about utility recommendations to install a four¬ wire system on farms when utility lines don't meet the same standard.

The practice of connecting the utility's return wire to the return wires of their customers apparently originated in 1932 when a Chicago utility asked researchers to study surge protection for its distribution lines. The practice, which is known as connecting the primary and secondary neutrals together, has since become standard.

The connection amounted to an "invasion of the customer's wiring system," writes Zipse, allowing current from the utility to "flow unimpeded into and over the customers' homes into the homeowner's ground rod, through the metallic water piping, into their showers and bathtubs, into the farm and into the dairy barns and milking parlors without the owners' knowledge or consent."

Zipse says electrical engineers who don't work for a utility are surprised to learn that utility return electricity is allowed to flow onto customers' lines, Zipse proposed a rule change that would let customers install a four wire system from their facilities to the utility's transformer (at the customer's expense), which would provide customers with more protection from return electricity from the utility line. Both the NESC and the National Electrical Code,, which sets wiring standards beyond the service point, rejected the proposal.

Transformers that let electricity flow into customers' lines are much less expensive than the double bushing transformers required for commercial uses, thereby adding to utility profits, Zipse says.

Regulations now require that the electrical system be grounded at five points, which provides several paths for the uncontrolled flow of electricity. Zipse recommends confining "all continuously flowing current" to an insulated wire and connecting the neutral wire to the earth at only one place.

Not in California

The problems are inherent to the multigrounded neutral system used in all states except California, says Zipse. In many areas of California, most electrical distribution lines are configured in a different manner than the rest of the country. And where lines are configured the same, they incorporate features to prevent the flow of current into the ground.

Utilities claim the system used in California is unsafe. Critics of the present distribution system scoff at this assertion, noting that the state ranked as the sixth largest economy in the world would be unlikely to prosper if the system's safety was doubtful. And some analysts say the decline of dairying in Wisconsin and its ascendancy in California directly reflect the different levels of electrical pollution in the two states.

Several California dairy farmers who were recruited to move to Wisconsin left the state after a few years after trying unsuccessfully to cope with "stray voltage." California dairy farmers who have expanded to South Dakota and several other states have similar complaints about "stray voltage."

The cases of "stray voltage" in California are caused by faulty wiring in buildings or equipment, not by electricity routinely dumped into the ground by utilities.

Until 1992, utility lines in rural areas in Wisconsin were required to have four connections to earth per mile to allow electricity from utility lines to flow into the earth. The Wisconsin Public Service Commission (PSC) then mandated installation of a wire at every pole about nine per mile, supposedly to reduce the risks associated with lightning strikes, although the agency never cited any proof that lightning had increased. Critics say the additional wires, which are known as down grounds, simply funneled more electricity into the earth from overloaded neutral lines.

"The earth is not a sponge. It does not absorb electricity," Zipse says. "This is trespassing. They didn't tell you they were going to use your property." Zipse says the assertion by representatives of Wisconsin PSC that only a multigrounded system is legal in Wisconsin is "a complete fabrication."

Zipse doesn't think utilities will be quick to respond. "It took 21 years for the National Electrical Code panel to finally wake up" and require changes in the way electric ranges were wired, he says. Since he's been talking about the issue for three years, it could mean a decade or more before utilities adopt the changes.

But it could occur sooner than that. "Too many people are starting to be hurt," he says. He also wonders whether coroners recognize the subtle signs associated with many forms of electrocution and may therefore be underestimating deaths caused by electrical pollution.

Thomas Edison, who patented the first electrical distribution system, warned against using the earth as a return path for current, Zipse says. The utilities have forgotten why the present system is configured as it is, having been told for generations that there wasn't any alternative.

"It was all done to save money. They [the utilities] don't understand their own history," he says.

In 1994, the Public Utilities Commission in California issued a resolution that resulted in Rule 33.2, which stated that "Ground or earth shall not be used as a normal return or circuit conductor.... The neutral or any other conductor is permitted to be grounded only for the purposes of stabilization and protection."

The rule essentially prevented utilities from dumping electricity into the earth. The utilities in California have complied with the rule.

Appendix H

The graph shows the drastic changes in the level of stray current on my farm pre and post-reconfiguration. The flat line just above the date 10/26 shows no current whatsoever after the work was completed. However, WPS didn't eliminate the stray current; they just shunted it away from me and on to other farms, merely shifting the problem instead of eradicating it. Following the graph is a list of my creditors at the time of the trial. As a result of WPS' current damaging my cows and their milk-producing capacity, I had nothing with which to pay the bills and was falling behind an additional $50,000 per month.

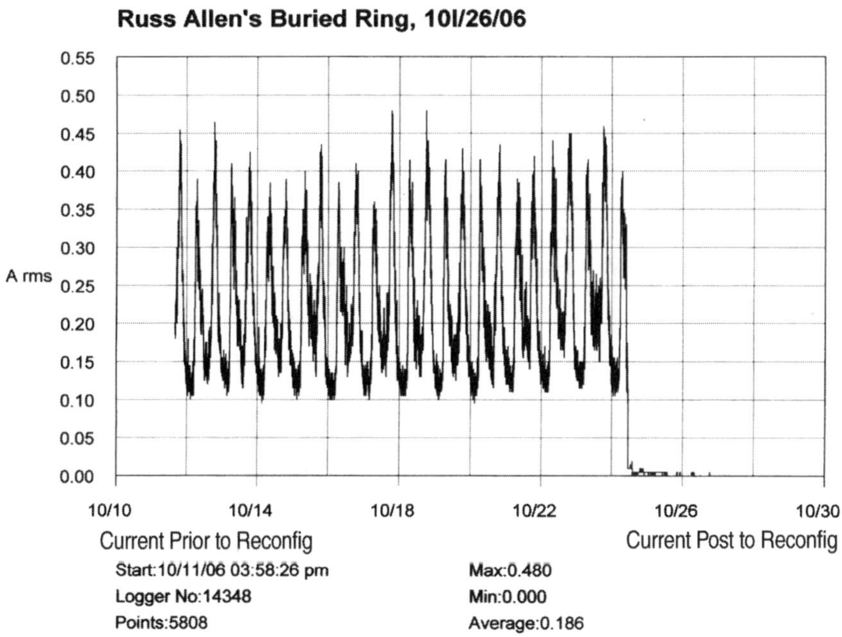

ALL BILLS NOT PAID UP TO DATE	
PROGRESSIVE COOP	$ 6,797.99
COOPERATIVE SERVICES	$ 88,814.56
MONSANTO	$ 9,902.77
REINKE EQUIPMENT	$ 2,582.32
DEAN DAVISTER	$ 914.00
KNOWLES BRANDT	$ 1,639.58
SEYMOUR DAIRY	$ 9,638.78
ANIMART	$ 2,619.57
VALLEY VET CLINIC	$ 4,911.78
RK PRODUCTS	$ 296.00
HYDROCLEAN (WAIT FOR STATEMENT)	$ 355.25
SERVICE MOTORS (WAIT FOR STATEMENT)	$ 57.24
FARM PLAN (PAID ONLY MINIMUM)	$ 4,441.60
CONCEPT ELECTRIC	$ 16,274.56
CONCEPT SERVICES	$ 7,416.32
CHRISTIANSON ELECTRIC	$ 371.10
SCHAEFFERS	$ 1,182.55
GREEN BAY ORTHOPEDICS - CARLOS	$ 1,313.00
ST. VINCENT HOSP - CARLOS	$ 3,452.10
GB EMERGENCY PHYS - CARLOS	$ 863.00
BELLIN HEALTH	$ 89.00
CARDOLOGY OF GREEN BAY	$ 129.00
RURAL -CROP INSURANCE	$ 5,001.00
TOTAL BILLS NOT PAID	**$ 174,310.73**

Appendix I

One of the ways utilities evade blame for electrical poisoning of the earth is by commissioning scientific studies "proving" that earth currents and stray voltage a) either do not exist, or b) if they do, exist at levels so low as to have no effect on animals or people. Since the utilities fund the vast majority of the studies and retain professors on their payroll as year-round witnesses in court cases, the results of these studies predictably support all utility claims, no matter how preposterous. One of the more interesting pieces of research was a Minnesota field study of stray current on nineteen dairy farms in that state. The study is noteworthy not for its results but because of its unscientific nature and the harsh dissent from its findings of two eminent members of the advisory panel, Drs. Abraham Liboff and Charles Polk. I personally spoke to Dr. Liboff and asked him if the results of the study were accurate. Below are some of his statements, followed by a progress report written by Dr. Polk.

> *"The power companies were only too happy not to give farmers a fair hearing."*
> *"I don't like Reinemann's research. It's unimaginative."*
> *"Reinemann knew ahead of time the results he would get."*
> *"None of the things I wanted to include were in the field research study."*
> *"Reinemann keeps doing the same old things he's done for years, which show nothing."*
> *"Reinemann should've looked for changes in biochemistry; all he looked at was if the cow's tail switched or its eyes moved."*
> *"The general attitude of the researchers on the Minnesota study seemed to be that farmers don't know how to manage their farms. And that's just wrong."*
> *"The research was one-sided. Charley Polk refused to sign the final draft. I backed him up. This indicated their prejudice."*
> *"The study was political, in terms of the state legislature. The farmers had a problem and we didn't solve it."*
> *"Another problem was that the research money suddenly dried up. I think the utility got to the legislature."*
> *"Where we tried to do this scientifically, they made it political."*
> *"Working with cow variables—housing, feed, etc. is an insane way of conducting earth current experiments. Research money is spent in the*

wrong way. It's grossly incorrect and wrong to call it research."
"I was frustrated by the study. We didn't get answers we should have. We couldn't help the farmers."

Final Report of the Science Advisors to the Minnesota Public Utilities Commission:

Research Findings and Recommendations Regarding Claims of Possible Effects of Currents in the Earth on Dairy Cow Health and Milk Production

July 31,1998
Minnesota Public Utilities Commission
121 7th Place East, Suite 350
St. Paul, Minnesota 55101-2147

Science Advisors[1]
Roger W. Staehle, Ph.D. *(Chairman)*
Adjunct Professor
Chemical Engineering and Materials Science
University of Minnesota
Minneapolis, Minnesota

Larry E. Anderson, Ph.D.
Staff Scientist
Bioelectromagnetics Group
Environmental and Health Science
Battelle Pacific Northwest Laboratories
Richland, Washington

Harold E. Dziuk, D.V.M., Ph.D.
Professor Emeritus
College of Veterinary Medicine
University of Minnesota
McGregor, Minnesota

David Hird, D.V.M., Ph.D.
Professor
Department of Medicine and Epidemiology
School of Veterinary MedicineUniversity of California
Davis, California

Abraham R. Liboff, Ph.D.
Professor
Department of Physics
Oakland University
Rochester, Michigan

Charles Polk, Ph.D.
Professor Emeritus
Department of Electrical Engineering
University of Rhode Island
Kingston, Rhode Island

Jimmie L. Richardson, Ph.D.
Professor
Department of Soil Science
North Dakota State University
Fargo, North Dakota

LaVerne E. Stetson
Agricultural Engineer
U.S. Department of Agriculture
Agricultural Research Service
Department of Biological Systems Engineering
University of Nebraska
Lincoln, Nebraska

Minnesota Public Utilities Commission

Patricia J. Hobert, Ph.D.
Research Director
Liaison, PUC/Science Advisors

Riley C. Hendrickson
Field Study Coordinator

[1]Alex Furo, Electrical Engineer and Consultant, was a Science Advisor from December 1994 until April 1998.

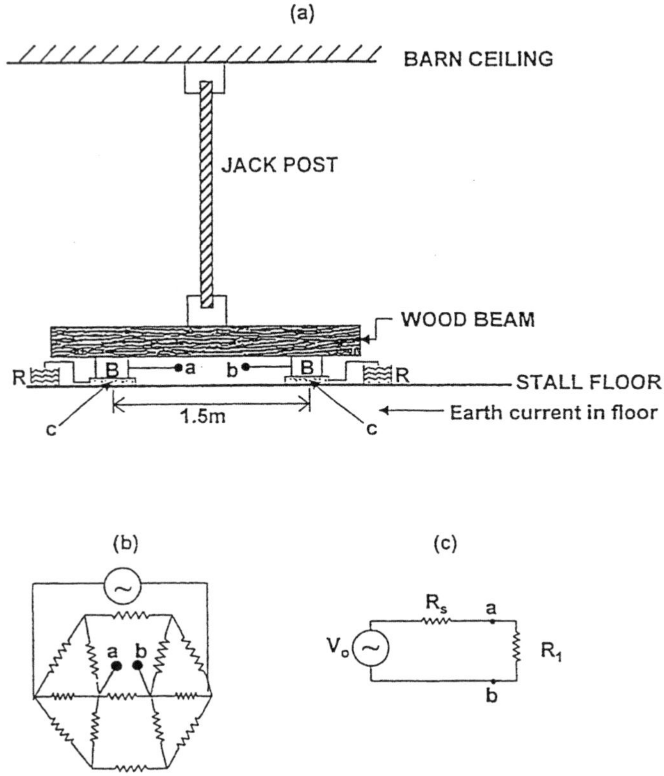

Fig. 1. Measurement of step voltage (SV). (**a**) Physical arrangement: Terminals a, b are connected to different values of R_1 for measurement of source resistance R_s. Source of the earth current is not known, but the source resistance necessarily includes the contact resistance between metal contact blocks B and the floor through saturated cotton squares C. For details see text. For 12 h continuous recording of SV terminals a, b are connected to a 500 Ω resistor and a high impedance voltmeter (output stored in computer memory). (**b**) Schematic representation of earth current paths (multiple sources replaced by a single source). (**c**) Thevenin equivalent circuit of (**b**) with added load resistor R_1.

selected for measurements [Hendrickson and Patoch, 1998a,b]. The physical arrangements and equivalent circuit for this measurement are illustrated by Figure 1. For SV measurements, connection to the stall floor was made after removal of packed bedding by using 6.25 × 6.25 × 0.8 cm square aluminum blocks (B) with terminal screws, simulating hooves. Front and rear hoof blocks were positioned approximately along the centerline of the stall and separated by 1.5 m. Electrical connection to the floor was enhanced using 7 × 7 cm cotton squares (C) kept wet with a saturated saltwater solution using a wick (veterinary umbilical tie) and salt water reservoir. The front and rear hoof blocks were spanned with a wood beam, which was depressed at the center by using a jack post to the barn ceiling. In barns that used rubber mats in the stalls, measurements were made with the mats in place.

The source resistance R_s was measured near the start and end of each 12 h measurement period and calculated by the expression $R_s = R_1[(V_0/V_1) - 1]$ where V_0 is the open circuit voltage and V_1 is the voltage across the load resistor R_1. An average source resistance was determined using three values of $R_1 = 500$, 1000, and 1500 Ω. Source resistances ranged from 256 Ω to more than 20 kΩ [Hendrickson and Patoch, 1998a]. Beginning and ending comparisons of source resistance indicated adequate control of source resistance changes due to variable contact wetness and pressure. Simulating the impedance of a cow between front and rear hooves, the low frequency (essentially 60 Hz) SV during the recording period was measured across a 500 Ω resistor. Although average and peak values of this voltage were recorded, the average value for the one hour of lowest electricity use, V-lu, will be examined in our discussion, since it and soil resistivity was the electrical quantities most significantly related to milk production (Table 1 and Figure 2).

Secondary Neutral-to-Earth Voltage

Secondary neutral-to-earth voltage (Vne) was measured between the neutral bus at the barn panel and a reference ground rod located more than 100 m from any ground connection of the farm electrical system or the supply line [Hendrickson and Patoch, 1998a]. Average values for each farm during "1-hour low electric use" are shown in Table 2 and Figure 5.

Soil resistivity

Soil resistivity (ρ) was measured using a commercial ("Vibroground") 4-electrode "Wenner array" [Grant and West, 1965] with a spacing of 5 feet (= 1.524 m) between adjacent electrodes. Thus, the current (outermost) electrodes were 15 feet apart and the potential (innermost) electrodes were spaced 5 feet. Since measurements were taken only with this single electrode spacing ($a = 5$ ft), the calculated ρ value is an apparent resistivity that assumes an earth which is electrically homogeneous, both laterally and in depth. Assuming a two layer earth with a top layer of resistivity ρ_1 bounded at the depth h by an infinite layer of resistivity ρ_2, it can be shown [Grant and West,

TABLE 1. Discrimination Between LP and HP Herds: Correlation with Milk/Cow/Day

Risk Factor	LP herds (Avg.)	HP herds (Avg.)	*P* value[a]	Correlation coefficient (r)	*P* value
Use of mats in stalls (17 stanchion and tie stall barns only)	0 of 9	5 of 8	0.0048	Not done	Not done
Net energy in feed stuffs (percentage)	31.6	36.9	0.0059	0.635	0.0047
Cow vaccination score (13 types of vaccinations possible)	3.1	5.9	0.0109	0.722	0.0005
Comfort score–cows[b]	3.1	2.1	0.0134	−0.772	0.0002
Maximum AC voltage between milk line and cow hoofs	0.63	0.25	0.0334	0.175	0.4869
Stall length (inches)	64.0	73.7	0.0446	0.452	0.1209
Dry matter intake (pounds/cow/day)	44.3	48.8	0.0490	0.564	0.0147
Soil resistivity in barnyard (ohm-m)	89.604	29.62	0.0506	−0.233	0.3381
Comfort score—calves	3.3	2.4	0.0543	−0.219	0.3832
Soil resistivity in the field (Ohm-m)	97.264	28.30	0.0670	−0.474	0.0406
Average step voltage, low electrical use (volt)	0.0070	0.00167	0.0727	−0.471	0.0418
Comfort score—heifers	1.2	2.4	0.0778	−0.348	0.1714
Magnetic field at the cow's back during milking (milligauss)	0.26	1.24	0.0779	0.299	0.2437
Average AC volts between milk line and cow hoofs, low electricity use	0.0662	0.0282	0.0832	−0.195	0.4380

Reproduced with permission from Staehle et al. [1998]

Data is ordered from the lowest *P* value for the comparison of averages to the highest *P* value. Nineteen herds were included for all herd parameters except those indicated. LP, low milk production herds (45.9 lb/cow/day average); HP, high producing herds (67.2 lb/cow/day average).

[a] *P* value calculations were based on a two-sample *t*-test for all variables except for use of mats in stalls, for which a chi square test for independent variables was employed.

[b] Cow comfort ratings were performed by the veterinarian on the research team using the following scoring system: 1 = clean, dry, well-bedded; good ventilation. 3 = adequate cleanliness, mostly dry, with some stale air. 5 = dirty, wet, with little or no bedding and stagnant air.

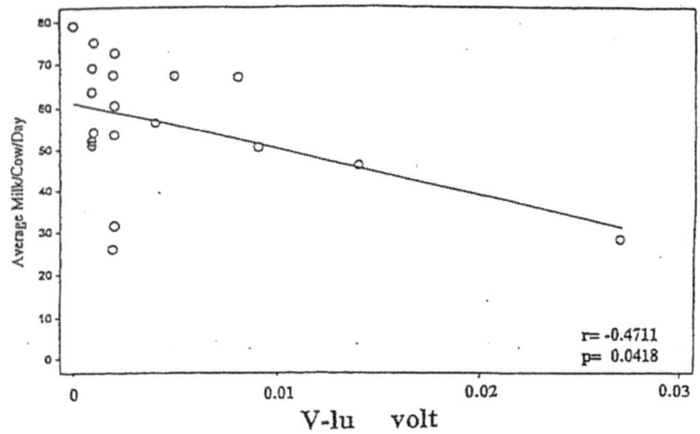

Fig. 2. Milk per cow per day (m/c/d) vs. V-lu: SV average during low use hour. The Pearson correlation coefficient r and its level of significance P on this and the other scatter plots (Figures 3, 5, 6) were evaluated as in Mould [1989] pp. 175–177: P is the probability that there is no linear association between the abscissa and ordinate values. All 19 test farms are included.

TABLE 2. Data for Each Test Farm

Farm identification	m/c/d lbs	ρ Ω-m	V-lu Volt	Vne Volt	NEF %	CVS	CC
A	53.92	36.768	0.001	0.077	29.65	5	2
B	74.64	28.918	0.001	0.084	38.94	5	1.5
C*	66.5	25.086	0.005	0.818	36.95	6	3
D	66.7	8.618	0.008	0.629	36.62	5	2.333
E*	78.64	22.6	0	0.547	40.98	7	2
F	31.34	19.916	0.002	0.412	27.13	5	4
G	49.91	170.438	0.009	0.547	n	0	3
H	25.67	36.002	0.002	0.165	28.28	0	4
I*	69.14	19.152	0.001	0.132	39.17	7	2.67
J	55.71	185.756	0.004	0.738	32.11	5	2.5
K	46.08	141.71	0.014	1.328	33.38	0	4
L	53.25	21.258	0.002	0.251	36.39	5	2.33
M*	63.54	35.81	0.001	0.358	37.41	5	2.67
N*	67.43	48.26	0.002	0.813	35.47	5	2
O	51.91	11.3	0.001	0.228	25.80	5	2
P	60	16.854	0.002	0.372	37.03	5	2.33
Q	27.85	321.724	0.027	0.795	32.87	0	n
P1	72.63	21.256	0.002		36.55	8	1
P2	50.64	55.918	0.001		32.08	6	3.67

Abbreviations: m/c/d = milk per cow per day; V-lu = step voltage during low electricity use; Vne = neutral to earth voltage during low electricity use; NEF = net energy in feed stuffs, CVS = cow vaccination score (13 best, 0 worst); CC = Cow comfort score (1 best, 5 worst). n = no data.

*Farm using rubber mats in stalls. Farms P1 and P2 are "parlor farms."

1965] that $\rho \approx \rho_1$ if $(a/h) < 0.3$. For the 5-foot electrode spacing used, this would require a depth $h > 16$ ft. However, from data given by Grant and West (their Figure 14-5), it can also be shown that for a more likely farm situation of $h = a = 5$ ft, the ratio of the apparent ρ, recorded in the field study, to ρ_1 is given by $0.65 < (\rho/\rho_1) < 1.5$ if $(1/2)\rho_1 < \rho_2 < \infty$. This ρ was measured "in the field" between 90 and 400 m from the barn, as well as in the barnyard. The field values showed significant correlation with milk production

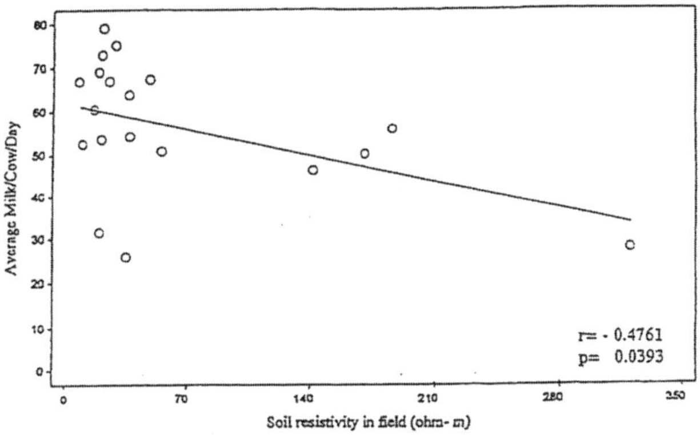

Fig. 3. Scatter plot of m/c/d vs. soil resistivity in field ρ. All 19 test farms are included.

(Table 1 and Figure 3) and will be considered in more detail in the analysis below.

RESULTS OF THE FIELD STUDY

The large amount of data gathered by the field study was subjected to statistical analysis by Dr. Will Marsh of the Department of Clinical and Population Sciences at the College of Veterinary Medicine, University of Minnesota. The most salient results of this analysis are summarized in Table 1, which is reproduced from Staehle et al. [1998]. It indicates those variables, which exhibited the most significant discrimination (lowest *P*-values) between HP and LP. The *P* value calculations were based on a two-sample *t*-test for all variables except for use of mats in stalls, for which a chi square test for independent variables was employed. Cow comfort ratings were assigned by the veterinarian on the field study team, *R.* Schell, DVM, using a scoring system ranging from 1 = clean, dry, well-bedded, good ventilation to 5 = dirty, wet, with little or no bedding and stagnant air. Data on the table are ordered from lowest *P* value for the comparison of averages to highest *P* value. Nineteen herds are included for all variables except for the use of mats in stalls, which is not applicable to the two farms with parlor facilities. Table 1 also gives the correlation coefficient r between each variable and milk/cow/day (m/c/d) and its *P*-value, which depends on the scatter about the assumed linear relation to m/c/d as explained, for example, in Mould [1989].

Table 1 shows that many variables, normally considered by veterinarians in the dairy industry, are strongly correlated with milk production. Particularly notable are net energy in feed stuffs, comfort scores, and vaccination history. However, Table 1 also shows that average SV during the one-hour low electricity use period on the farms (V-lu) and soil resistivity in the field ($\rho_F = \rho$) are significantly, but negatively correlated with m/c/d. This relation is further confirmed by the fact that use of mats in stalls, which provides electrical insulation, indicated highly significant discrimination between (LP) and (HP). Figure 2 shows considerable scatter of m/c/d when SV is less than 0.01 V, but a linear relation between m/c/d and SV above that value. In view of the hypothesis discussed above that V-lu $\geq$ 10 mV is necessary for a possible effect on cow physiology, it is of interest to note that when one considers only V-lu values above 9 mV in Figure 2, the correlation coefficient between m/c/d and V-lu becomes 0.994. However, in view of the small number (3) of farms over 9 mV, the *P*-value becomes 0.069. SVs of 3 mV or less that were measured on 15 of the 19 farms have apparently no relation to m/c/d. Figures 2 and 3 and Table 2 show that for the three LP farms where SV during low electricity use (V-lu) was 9 mV or larger, the value of m/c/d decreased nearly linearly with soil resistivity.

We were intrigued by the fact that both V-lu and ρ_F indicate trends in the same direction, particularly since "intuitive" examination would possibly suggest that high ρ should lead to small earth currents and therefore to low SV, opposite to what was observed (Table 1 and scatter plots of Figures 2 and 3). Closer examination, however, will show that the observed relation is what one should expect due to the characteristics of electrical distribution to and on farms and the construction of dairy barns.

CHARACTERISTICS OF RURAL ELECTRICITY DISTRIBUTION IN THE U.S.

To relate the observations to the electricity distribution system, it is necessary to indicate some of its characteristics. The usual electrical connections to farms are illustrated by Figure 4. The primary, high voltage side (7.2 kV line to ground in Minnesota) is usually a single phase line that can be as short as 100 feet or as long as several miles. It originates on one of the phases of a 3-phase Y-connected system. The grounded neutral conductor of the single-phase line is bonded to the neutral of the more or less distant 3-phase line, which comes from a substation. At the transformer, which is usually located on a pole at the entrance to the farm, the high voltage is stepped down to 240 V. The center of the low voltage winding (as well as the transformer casing) is grounded; 120 V loads on the farm are connected to one or the other half of the secondary side, preferably (but rarely in practice) in such a way that loads are "balanced" resulting in zero net current through the central "neutral" wire. Heavy machinery is normally connected directly across the 240 V supply.

If the 120 V farm loads are not equally distributed among the two 120 V circuits, substantial current may flow in the neutral wire. Since it has finite resistances (R_N), different points along its length will be at different potential. Generally, the distribution panels at each building entrance on the farm are separately grounded as indicated by P_1 and P_2 in Figure 4. If current flows in the neutral wire with resistances R_N, points P_1, P_2, and P_T (the transformer

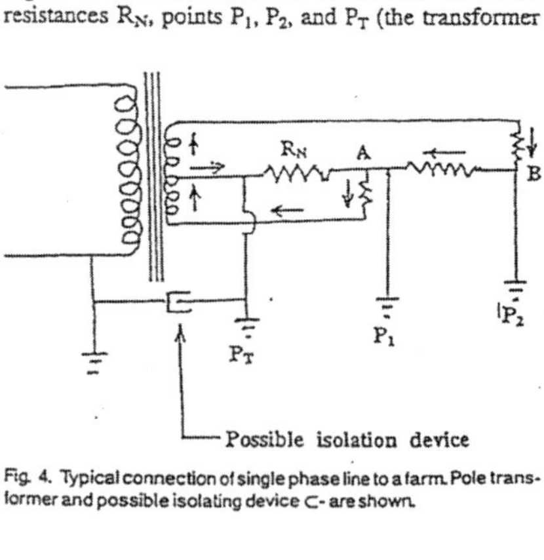

Fig. 4. Typical connection of single phase line to a farm. Pole transformer and possible isolating device C- are shown.

ground) will be at different potentials. This will cause current flow through the earth or preferentially through the grounded metal structures (e.g., water pipes) which exist in almost all barns. Sometimes metal housings of equipment, which should always be grounded, are also electrically connected to the neutral wire (a code violation!) and thus provide additional ground connections at still different potentials. Most of the undesirable ground current on the farm will flow on electrically connected metal structures and will cause potential differences, for example, between a metal water trough and soil. However, if points A and B in Figure 4 (for example, panels at the farm home and in the barn) are at different potentials, some current will also flow through the soil. The magnitude of that current will depend on ground resistivity, on the potential difference along the neutral wire, and particularly on the resistance of the various ground connections.

In addition to the on-farm sources of earth current, off-farm sources are also possible. These are of at least two types: those related to the primary feed line, which is grounded on the primary side of the transformer in Figure 4 and those that are completely, or at least partially, independent of the farm electrical supply. Considering first the sources related to the individual farm supply, we note that the primary neutral is grounded at least every 400 m along the distribution line and sometimes more frequently. Since the neutral wire on the primary side may be of relatively small size, a substantial fraction of the neutral current may flow in the soil. In a survey of 48 utility distribution systems in Minnesota [Hendrickson et al., 1995] mean earth currents for different utilities varied between 0.14 A and 25.74 A, with an average of 4.46 A for all utilities. If the exceptionally high value of 25.74 A for one utility is excluded, the average for the remaining systems is 3.98 A.

For safety reasons, for example to avoid fire or other damage by excessive current flow in case of an accidental short circuit or lightning strike, the primary and secondary neutral conductors at the transformer are normally rigidly connected. This will, however, cause return current of the primary feed to flow through the secondary neutral and over ground connected metal structures on the farm, both under normal operation and even if the farm circuit is disconnected at the main switch. It also makes the combination of primary and secondary grounds a terminal for any possible earth current circuit whose other terminal may be the ground connection of a distant transmission line or other source. Since potential differences due to the current flowing through the conductive path between primary and secondary have been found to cause undesirable "stray voltage" problems, isolation devices, as indicated in Figure 4, are used in some installations. These ground isolation devices are normally open, but to provide safety, for example during a lightning strike, they close when a preset threshold voltage between line and farm ground is exceeded. Several different types are in use and their frequency characteristics vary, so that while all devices provide isolation at 60 Hz under normal conditions, some provide no isolation at any time for higher frequencies, and therefore for transients [Jenkins et al., 1995].

Since all ground rod or other ground connections have finite resistance and since any earth current will place the earth at some non-zero value of potential (see Appendix), any measured neutral-to-earth voltage V_{ne} will indicate earth currents. On farms without isolators, these currents could originate on the primary side and be due to the earth return current on the utility line leading to the farm. Measured values of ground rod resistance vary between a few and several hundred Ohms, with an average of about 50 Ω; the values are weakly correlated ($r = 0.64$) with soil resistivity [Hendrickson et al., 1995].

We consider next the possible off-farm sources of earth current that are not directly related to the distribution line to the particular farm. It is apparent that the multiple ground connections on the farm (at one or several different potential levels) can provide a terminal for an earth "circuit" whose other terminal may be the different earth potential due to some distant transmission or distribution line or other source, such as a distant current-carrying water or oil pipe line. Approximating such a distant source and a local ground rod as a two-point source-sink system, we note from Eqn. A-(8) in the Appendix that even a current of 1 mA from a distant source will cause a ground surface potential difference of 24 mV across 1.5 m when the local ground rod is at a distance of 2 m and ρ is 400 Ωm. Although a completely rigorous analysis of earth currents due to transmission or distribution lines would have to include magnetic field induction effects [Meliopoulos, 1988; Olsen, 1996], they would give not more than very minor corrections to surface potentials when only very short segments (< 400 m) of 60 Hz distribution lines between grounding points are considered.

ANALYSIS OF MEASURED ELECTRICAL VARIABLES

In the survey summarized in Tables 1 and 2 and Figures 2 and 3, the correlation of milk production (m/c/d) with step voltage V-lu, as well as with earth resistivity ρ, was significant. However, as indicated in Table 2, five of the test farms used rubber mats in stalls. Since these mats provided substantial, but unknown resistance between concrete floors and the measurement points, the V-lu data from these farms are not useful for any analysis designed to establish possible relations to earth currents. Two additional farms were "parlor farms" where animals may move rather freely to different locations in the barn or may spend most time outside. These seven farms, six with V-lu ≤ 2 mV and one with V-lu = 5 mV, were therefore excluded from analysis exploring relations between the different electrical variables. For the remaining twelve farms the step voltage during low electricity use, V-lu is compared with the neutral to earth voltage, Vne in Figure 5, and with soil resistivity ρ in Figure 6. While the scatter plot of Figure 5 suggests a possible linear relation between V-lu and Vne, the correlation coefficient is only 0.693, compared with $r = 0.850$ for correlation between V-lu and ρ in Figure 6. The suggested association within each variable pair is r^2 [Sheskin, 1997] or, respectively, 48% and 72%. (More precisely, r^2 represents the proportion of variances in one variable that can be accounted for by variance in the other variable). It is apparent that neutral-to-earth voltage, measured at the distribution panel in the barn is not as strongly correlated with V-lu as is soil resistivity measured in the field. We were at first surprised to find the strong V-lu to ρ correlation, since resistivity of the concrete barn floor is likely to be different than resistivity measured in the field (it was not possible to drive electrodes for a Wenner array into the barn floors). However, it is shown in the Appendix (Eqn. A-4) that for the usual barn floor construction—less than 8 inches (≈ 20 cm) of concrete and crushed rock—the surface potential depends on the resistivity

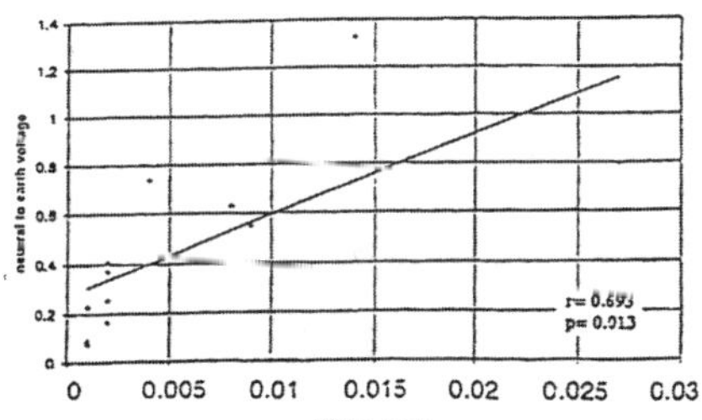

Fig. 5. Scatter plot of neutral-to-earth voltage V-ne vs. step voltage V-lu. Two parlor farms and five farms with rubber mats in stalls are excluded.

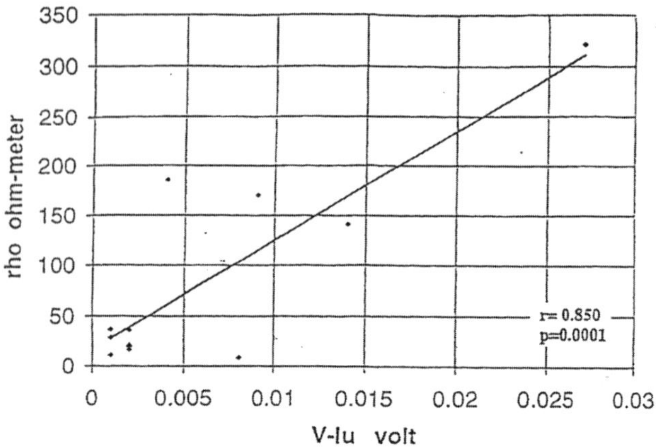

Fig. 6. Scatter plot of soil resistivity in field ρ vs. step voltage V-lu.

of the underlying earth, which is likely to be the same that measured in the field.

Both Figures 5 and 6, as well as Figures 2 and 3, indicate considerable clustering of points below V-lu values of 2 mV. Examination of Table 2 shows that seven (78%) of the nine HP farms with m/c/d > 60 lb have V-lu ≤ 2 mV and the other two (D and C) have V-lu ≤ 8mV. Also 70% (7 out of 10) of the LP have V-lu ≤ 4 mV, but 30% of the LP have V-lu ≥ 9 mV. Thus, while most farms (16 out of 19 or 84%) in the sample had V-lu ≤ 8 mV, the highest V-lu (≥ 9 mV) were only found among the LP. The farm with the highest V-lu (= 27 mV) also had the highest ρ (= 322 Ωm).

CONCLUSIONS

1. As expected, many different environmental conditions that are completely unrelated to the electrical distribution system can and do affect milk production on dairy farms. Nevertheless, in the investigated sample of 19 dairy farms, the step voltage during low electricity use on the farm V-lu was 4.2 times higher on average in the 10 below-average producing farms (LP) than in the nine farms (HP) with above average milk production.
2. The Minnesota field study did not find "credible scientific evidence to verify the specific claim that currents in the earth or associated electrical parameters... are causes of poor health and milk production in dairy cows" [Staehle et al., 1998], but neither did it find any evidence that currents in the earth from on- or off-farm sources are not the source of the small step voltages which were measured or that these SVs cannot affect cow health.
3. The three farms with V-lu > 9 mV contributed significantly to the difference in m/c/d between LP and HP. If they were to be excluded, the difference in averages of m/c/d between LP and HP would be smaller (19.3 instead of 21.3) (compare Table 2 and Figure 2). All three were among the 10 LP farms selected at random from the 10% of Minnesota dairy farms with the lowest milk production.
4. The inverse relation between V-lu > 9 mV and m/c/d shown in Figure 2 involves the farms identified as G, K, and Q in Table 2. Examination of Table 2 shows that net energy in feed stuffs (NEF) is relatively high for farms K and Q (35.2% average compared with 31.6% for all LP farms) and thus is unlikely to contribute to low m/c/d. However, the inverse relation between V-lu > 9 mV and m/c/d could be accidental, because all three farms have zero (the lowest possible) cow vaccination score (CVS) and two (G and K) have also poor cow comfort scores (3.5 average—worse than the 3.1 value for all LP farms). The only other farm (H) in the sample with CVS = 0 has also exceptionally low m/c/d despite small V-lu (= 0.002 V). The possibility that V-lu > 9 mV may contribute to low m/c/d can nevertheless not be excluded, particularly

since neither CVS nor CC can explain the linear decrease of m/c/d with increasing V-lu among farms G, K, and Q (Figure 2).

5. When farms using electrically insulating rubber mats in stalls, as well as two "parlor farms" with relatively freely moving animals are excluded from the sample, the weak, but statistically significant correlations between milk production and either step voltage ($r = -0.471$) or soil resistivity ρ ($r = -0.476$) suggested by Figures 2 and 3, appear as a strong correlation ($r = 0.850$) between V-lu and ρ for the remaining 12 farms (Figure 6).
6. Step voltage (V-lu) is also correlated ($r = 0.693$) with the voltage measured between secondary neutral and ground (Vne) as shown in Figure 5, but not as strongly as with ρ. A perfect linear relation between V-lu and V_{ne} could only be expected if the conducting paths between the secondary neutral and various grounding points in the barn were identical in all farms; this is clearly not the case. A perfect linear relation between V-lu and ρ could only be expected—assuming a single source-to-sink path—if the ratio of current to the square of the distance to the nearest grounding point were identical for all farms (Eqn. A-8).
7. The amplitude of earth currents near primary distribution lines appears to depend primarily on neutral return current, the size of neutral wires, and resistance as well as location of electrical system-to-earth connections. With a predetermined level of earth current, one can then expect a strong correlation between step voltages and soil resistivity (compare Eqn. A-8).
8. At present it is not known whether and how the long-term exposure to step voltages above 9 mV can affect health and/or milk production of dairy cows. This can only be established by laboratory experiments where major variables known to affect animal health can be controlled.
9. In view of the results discussed above, it presently appears to be desirable to measure soil resistivity when a dairy farmer reports possible electricity-related cow health and production problems that are not explainable by other variables (e.g., absence of vaccinations). A very detailed investigation of the on- and off-farm electrical distribution systems is then indicated if ρ of the uppermost earrth layer (as was measured for the present report) is larger than about 150 Ωm.

ACKNOWLEDGMENT

The author is indebted to Mr. Riley Hendrickson, who performed the electrical measurements in the field study and with whom this paper was discussed in great detail. Also acknowledged are the many contributions to the design of the Minnesota field study by the author's other colleagues as "Science Advisors" to the Minnesota Public Utilities Commission: L.E. Anderson, H.E. Dziuk, A. Furo, D. Hird, A.R. Liboff, J.L. Richardson, R.W. Staehle, and L.E. Stetson and by the liaison with the PUC and PUC Research Director Patricia Hoben. Credit is also due to R.C. Hendrickson's colleagues on the team that performed the measurements during the summer of 1997, J. W. Patoch, R. Schell DVM and W. Bickner, as well as to Dr. Will Marsh who did the statistical analysis of the data shown in Table 1 and Figures 2 and 3. The author alone is responsible for the analysis presented here and the conclusions drawn.

REFERENCES

Anderson LE, Dziuk E, Furo AJ, Liboff AL, Polk C, Richardson JL, Staehle RW, Stetson LE, Hoben PJ, Hendrickson RC. 1996. Progress report of the science advisors to the Minnesota Public Utilities Commission: proposed research for evaluating possible electrical causes for poor health and production in dairy cows. St. Paul, MN 55101-2147: Minnesota Public Utilities Commission.

Burchard JF, Nguyen DH, Richard L, Block E. 1996. Biological effects of electric and magnetic fields on productivity of dairy cows. J Dairy Sci 79:1549–1554.

Burchard JF, Nguyen DH, Richard L, Young SN, Heyes MP, Block E. 1998. Effects of electromagnetic fields on the levels of biogenic amine metabolites, quindinic acid and β-endorphin in the cerebrospinal fluid of dairy cows. Neurochem Res 23:1527–1531.

Burchard JF, Nguyen DH, Block E. 1999. Macro- and trace element concentrations in blood plasma and cerebrospinal fluid of dairy cows exposed to electric and magnetic fields. Bioelectromagnetics 20:358–364.

Dawson TW, Stuchly MA. 1998. High-resolution organ dosimetry for human exposure to low-frequency magnetic fields. IEEE Trans Magnetics 34:708–710.

Dawson TW, Caputa C, Stuchly MA. 1997. High-resolution organ dosimetry for human exposure to low-frequency electric fields. IEEE Trans Power Delivery 13:366–373.

Grant FS, West GF. 1965. Interpretation theory in applied geophysics, Chapter 14-4. New York, Toronto, London: McGraw-Hill Book Company.

Hendrickson RC, Michaud M, Bierbaum A. 1995. Survey to determine the age and condition of electric distribution facilities in minnesota. Report 1: Analysis of overhead distribution feeder testing data. St. Paul, MN: Minnesota Public Utilities Commission.

Hendrickson R. 1997. On-farm research pilot field study—electrical measurements presentation for the science advisors to the Minnesota Public Utilities Commission, December 15, 1997. St. Paul MN: Minnesota P.U.C.

Hendrickson RC, Patoch JW. 1998a. Electric environment on 17 dairy farms. ASAE, 2950 Niles Rd., St. Joseph, MI 49085-9659: Proc 1998 ASAE Annual International Meeting.

Hendrickson RC, Patoch JW. 1998b. AC earth current on dairy farms. ASAE paper no 983002. ASAE, 2950 Niles Rd., St. Joseph, MI 49085-9659: Proc 1998 Annual International Meeting.

Jenkins D, Aneshansley DJ, Pellerin RA, Throop JA, Ludington C. 1995. Dynamic response of ground isolation devices used on livestock facilities. Paper 953622. International ASEE Annual Meeting. June 18–23, Chicago, IL.

Kaune WT, Forsythe WC. 1985. Current densities measured in human models exposed to 60-Hz electrical fields. Bioelectromagnetics 6:13–22.

Kaune WT, Forsythe WC. 1988. Current densities induced in swine and rat models by power-frequency electric fields. Bioelectromagnetics 9:1–24.

Lefcourt AM (editor in chief). 1991. Effects of electrical voltage/current on farm animals: how to detect and remedy problems. Washington, DC: U.S. Government Printing Office, U.S. Department of Agriculture, Agriculture Handbook No. 696.

Meliopoulos APS. 1988. Power system grounding and transients. New York and Basel: Marcel Dekker.

Mould RF. 1989. Introductory medical statistics, 2nd ed. Bristol and Philadelphia: Adam Hilger. p 175–177.

Olsen RG. 1996. Power transmission electromagnetics. URSI Radio Science Bulletin 177:29–37.

Reinemann DJ, Stetson LE, Reily JP, Laughlin NK. 1999. Behavioral measures of dairy cow sensitivity to short duration electrical currents. Trans ASEE 42(1):215–222.

Sheskin DJ. 1997. Handbook of parametric and nonparametric statistical procedures. p 546. Boca Raton: CRC Press.

Smythe WR. 1989. Static and dynamic electricity, 3rd ed. New York, London: Hemisphere Publishing Corp.

Staehle RW, Anderson LE, Dziuk HE, Hird D, Liboff AR, Polk C, Richardson JL, Stetson LVE, Hoben PJ, Hendrickson RC. 1998. Final report of the science advisors to the Minnesota Public Utilities Commission: research findings and recommendations regarding claims of possible health effects of currents in the earth on dairy cow health and milk production. St. Paul, MN: Minnesota Public Utilities Commission.

TERF. 1994. Dairy farm stray voltage summary-final report for the Minnesota Department of Public Services. Underwood, MN: The Electromagnetics Research Foundation, Inc.

Walleczek J. 1995. Magnetokinetic effects on radical pairs: a paradigm for magnetic field interactions with biological systems of lower than thermal energy. In: M. Blank, editor. Electromagnetic fields—biological interactions and mechanisms, Chapter 23. Washington, DC: American Chemical Society p 305–422.

APPENDIX

Surface Potential Difference Between Two Grounded Electrodes: Relation to Earth Current and Earth Resistivities

The problem is illustrated by Figure 7 and the analysis is based on Smythe [1989]. Current I is injected into a two-layer earth by a pair of electrodes, separated by a distance $2b$. The midpoint between the electrodes is located at $x=0$. The y-coordinate (positive, out of the plane of the paper) is perpendicular to the straight line between the electrodes, and the z-coordinate is the vertical distance into the soil, measured from the surface. The earth resistivities are ρ_1 for $0 \leq z < a$ and ρ_2 for $z \geq a$. Smythe [1989] showed that the potential at the surface $V = V_S$ is given by

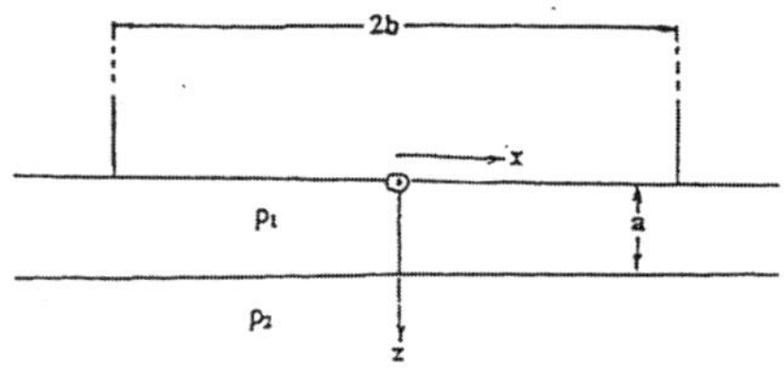

Fig. 7. Two layer earth with electrodes at $x = \pm b$.

$$V_S = \frac{\rho_1 I}{2\pi} \int_0^\infty \frac{1 - \beta e^{-2ka}}{1 + \beta e^{-2ka}} J_0(k\Gamma)dk \qquad \text{A-(1)}$$

where Γ = radial distance from either electrode in the x-y plane,

$$\beta = \frac{\rho_1 - \rho_2}{\rho_1 + \rho_2} \qquad \text{A-(2)}$$

and

$$k^2 = \frac{1}{Z}\frac{d^2 Z}{dz^2}. \qquad \text{A-(3)}$$

Z is the z-dependent part of the solution of Laplace's equation in cylindrical coordinates $V = R\Phi Z$, where R is a solution of Bessel's equation and Φ a sum of trigonometric functions. The value of the separation constant k follows from the boundary conditions of the present problem: continuity of V at $z=0$ and $z=a$, $V=0$ at $z=\infty$, continuity of z-directed current density $(1/\rho)(\delta V/\delta z)$ at $z=a$ and $(\delta V/\delta z)=0$ at $z=0$. For details see Smythe [1989, pp. 157, 180, 189, 192, and 263].

If the thickness of the layer near the surface is very small compared with the distance b or if $\beta \ll 1$, V_S becomes practically only a function of ρ_2 rather than of ρ_1 and ρ_2. If $ka \to 0$ as $k \to \infty$, it follows from A-(1) and A-(2) that

$$V_S = \frac{I}{2\pi}\rho_1 \frac{\rho_2}{\rho_1} \int_0^\infty J_0(k\Gamma)dk = \frac{I\rho_2}{2\pi\Gamma}. \qquad \text{A-(4)}$$

This result can also be obtained by a series expansion of the integral in A-(1) in powers of β.

To indicate the validity of A-(4) for the problem of measurements of V_S over a concrete floor, i.e., to show that V_S will depend on the conductivity ρ_2 of the substrate (which is essentially the conductivity measured "in the field"), we consider the following example: distance from the source or sink of current $\Gamma = 10$ m, thickness of concrete and underlying gravel (if any) $a = 0.2$ m. If the resistivity of the concrete floor (which is usually porous and contains some water) $\rho_1 = 2\rho_2$, where ρ_2 is the conductivity in the field ($a = 0$), $\beta = (1/3)$ and the error in V_S resulting from using A-(4) rather than A-(1) is not more than 4%.

For electrodes at $y = 0, x = b$ and $y = 0, x = -b$

$$\Gamma_+ = \left[(x-b)^2+y^2\right]^{1/2} \qquad \Gamma_- = \left[(x+b)^2+y^2\right]^{1/2}. \qquad \text{A-(5)}$$

Superposition of potentials due to electrodes at Γ_+ and Γ_- defined by A-(5) and now assuming a uniform earth conductivity $\rho = \rho_1 = \rho_2$, as for the measurement of "conductivity in the field," we obtain from A-(4) and A-(5)

$$V_S = \frac{\rho I}{2\pi}\left\{\left[(x-b)^2+y^2\right]^{-1/2} - \left[(x+b)^2+y^2\right]^{-1/2}\right\}. \qquad \text{A-(6)}$$

This expression is the basis for evaluation of ρ by the Wenner array where $2b = 3s$ and V_S is measured at $y = 0$ and $x = \pm(s/2)$. The potential difference ΔV_W between the points at $x = (s/2)$ and $x = -(s/2)$ is then

$$\Delta V_W = \frac{\rho I}{2\pi s}. \qquad \text{A-(7)}$$

Evaluating the open circuit voltage ΔV_{OC} between two points separated by a distance x_1 near one of the source or sink electrodes at $x = (b - \delta)$, where $\delta \ll b$ and $(x_1/2) \ll \delta$, one finds

$$\Delta V_{OC} \approx \frac{\rho I x_1}{2\pi\partial^2}. \qquad \text{A-(8)}$$

Charles Polk, Ph.D.
Science and Engineering Consultant
(Professor emeritus, University of Rhode Island)
53 Spring Hill Road
Kingston, RI, 02881

Review of "Second Progress Report on Dairy Cow Response to Electrical Environment" to the Minnesota Public Utilities Commission, dated January 31,1999.

This review follows the organization of the report.

Part I. Sensitivity Testing, and Comparison of Behavioral to Physiological Responses.

This part of the research was essentially concerned with testing cow response to acute, short period exposure to 60 Hz electric current (e.g. 0.5 seconds on, 2 seconds off for 1 minute). To this reviewer the most interesting results were:

(1) Cows are more sensitive to front-hoof to rear-hooves current exposure [F-R] than to muzzle to 4-hooves exposure (page 13 of report);

(2) While the mean reaction level to F-R exposure was 3 mA, the authors give from observations on 8 cows a standard deviation of 0.534 mA rms. This means (as illustrated by the graph on page 8 of the report) that 5 per cent of animals can be expected to have a reaction level at or below about 2 mA (using t-statistic with 7 degrees of freedom). Assuming a cow and contact resistance of 500 ohm, such a reaction level would correspond to a step potential of 1 Volt;

(3) The investigators found that human observation of cow behavior (detection of "flinch") is a more consistent indicator of current perception by the animals than the motion sensing equipment that was tested. They also found that blood cortisol levels are very variable and therefore are not a useful response indicator for response to current (at 1.5 times threshold level) (see graph on page12).

I agree with the authors of the report that no further studies should be conducted on this subject under the present contract.

Part II. Comparison of Treatments Applied During Milking.

Response to a 1 mA current from front to rear hooves during milking was compared to the stress produced by pulsation failure on the milking machine. Since this experiment gave a significant result (p=0.026) [page 16 of report] that could possibly attributed to unintended equipment malfunction, the authors planned a rerun during the week of February 3-5 [page 2 of report}. **It would be useful to explain in the final report what actually happened, but no repetition of this experiment should be attempted which would take time away from "Part III" of the research proram.**

Part III. Immune Function Response to Sub-Acute Voltages.

The copy of the report originally received by this reviewer did not contain pages 21 and 22 which show data on PMA chemiluminescence (index of phagocytic cell function), Con A-induced lymphocyte blastogenesis and PHA-induced blastogenesis. These pages were faxed to me by Dr Reinemann on March 3 and are only for the first replicate of the experiment. Data for the second replicate (apparently obtained during the period January 8 to 29) were not included in the report. Data on lymphocyte response to other mitogens during current exposure, and data on antibody and interleukin production for the first replicate of the study were also not included in the report.

This reviewer contacted Dr Reinemann, as well as Mr.Riley Hendrickson, in February to make them aware of his concerns about the "long duration" exposure system. The concerns arose after receiving, first from Dr Reinemann and then from the manufacturer, requested information on the electrical characteristics of the "conductive" mats used in the exposure system. Suggestions were also made via telephone to Dr Reinemann in February, and again on March 3 for possible continuous monitoring of cow exposure as discussed below.

The principal possible problems with the exposure system are:

(1) The separate "conductive" mats on which the front and hind legs of the cows are located have apparently very high electrical resistance. Unless the gap between the mats is large enough (as is apparently the case as per telephone transmitted information by Dr Reinemann) and is kept clean and dry at all times, one cannot be certain that the current (kept constant by high source resistance) actually flows through the animals and not through moisture or dirt in the gap, or through wet concrete below the mats (if the mats are not sufficiently electrically isolated from the concrete floor). **In any case, it will be essential to include in the final report on this contract a much more detailed description of the exposure system than given under "Test Stall and Procedures" on pages 16 and 17 of the present report. This description should include results of measurements taken to verify that current was actually flowing through the "exposed" animals during the entire exposure period (10 minutes on, 10 minutes off for two weeks).**

(2) Even if the exposure system is performing as intended, the contact area between each animal and the mats will change with hoof lifting and, particularly, when the animal lies down. Therefore the resistance of the mat-cow-mat path and the resulting voltage across the cow will change while the total current is kept constant by the high source resistance. Such voltage changes would be indicative of changes in cow position, as well as changes in current path through the cow (e.g. front to rear legs versus belly to chest). Since the voltage across the gap between mats (as measured by a high impedance meter) would be identical with the voltage across the cow (see circuit diagram below), it would provide information about the nature of the animal exposure if recorded continuously. Furthermore, if for some reason an electrical path other than through the cow is established (by moisture, dirt in the gap or other insulation failure), the voltage across the gap would be immediately substantially

reduced. **I therefore recommended to Dr Reinemann (hopefully in time for implementation before the third replicate of the immune function long period exposure study) that the voltage between mats be monitored continuously.** Only such monitoring would confirm that cows were actually exposed to current flow during the intended exposure period.

Circuit diagram to illustrate the discussion in "Part III" (Exposure system using "conductive" mats):

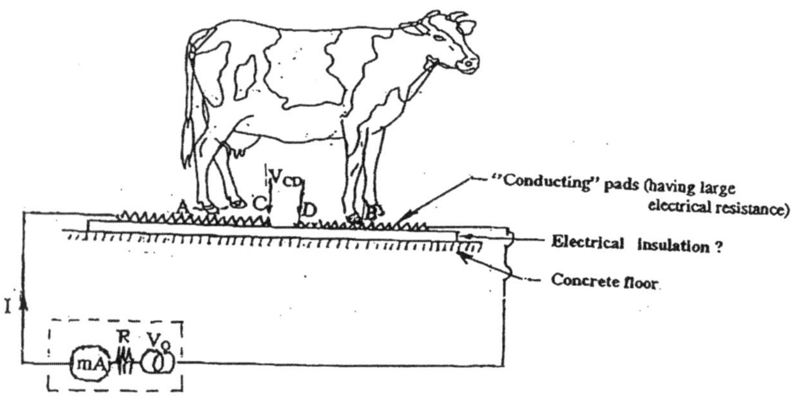

Only if no current flows from C to D is $V_{CD} = V_{AB}$
and all current I goes through animal (A to B)

Problem: How much current (what part of I, if any, flows through a path other than the cow?
V_{CD} should equal V_O if cow is not in place.

Review submitted on March 4, 1999.

Charles Polk, Ph.D.

Comments concerning **Regulatory Policy Implications** of the Final report of the Science Advisors and the Dairy Cow Response to Electrical Environment, Final Report
by Charles Polk, Ph.D.,
Professor emeritus of Electrical Engineering,
University of Rhode Island
53 Spring Hill Road
Kingston, RI 02881

(1) It would be very unfortunate if any regulatory policy decisions were to be based only on the three statements that are highlighted by numbers on page 38 of the Report under "General Findings", rather than on the entire report or at least on all the information summarized under "General Findings".

(2) That information includes (on page 38) that "the 19-farm field study did find significant differences between high and low producing herds in the levels of electrical step potentials and soil resistivity in the field". (Details are found on page 32, Table II, where the negative correlation coefficients between milk production and 'average step potential, low /electricity/ use' and 'soil resistivity in the field' are shown to be statistically significant below the 0.05 level).

(3) In view of (2) it is clear that the 19-farm field study suggests a possible connection between low milk production and the electrical variables of step potential and soil resistivity.

(4) The usefulness for regulatory policy of high frequency transient measurements made as part of the "field study" is limited, because such measurements between cow contact points were recorded only when they exceeded 0.5 Volt (see page 21 of "Final Report").

(5) The "general finding" no.2 on page 38 of the "Final Report" is "**At the present time** /my emphasis/ there is no basis for altering the PUC-approved standards by which electric utilities distribute power onto or in the vicinity of individual dairy farms". Do such written standards exist and is any special consideration given to the particular environment of dairy farms?

(6) The research described in "Part III Immune Function Response to Low-Level Electrical Current Exposure" does **<u>not</u> conclusively show that long-term exposure to 1 mA 60 Hz electrical current has no effect on immune function of dairy cattle.** Apart from the fact that a two-week experiment is to short to permit conclusions concerning long-term exposure, the data in the report did show a statistically significant difference in one important measure of immune cell response. "This result suggests that the difference in B-cell response between control and exposed animals cannot be discounted" (page 2 of peer review by Charles Polk dated July 12,1999) and "In the conclusions, the measurement of response to S.aureus (it measures a different end point, B lymphocytes) cannot be rejected just because it is not "entirely consistent" with other measures- it doesn't necessarily need to be" (Peer review by Larry S. Anderson dated 10/15/99). Two of the three reviewers of this research, L. Anderson and C. Polk also point out that the control cow that came down with a mastitis infection within 4 days could have impacted the results of the study.

(7) In summary, I suggest that careful examination of the data given in the "Final Report of the Science Advisors" and the subsequent report on the experimental attempt to simulate "long-term" current exposure can have the following implications for regulatory policy:

(I) The contribution of electrical problems ("step potential") to adverse effects on milk production cannot be excluded. Since this an extremely complex problem with many unknown variables, each reported case of apparent electrical effects on milk production or animal health deserves careful individual examination. **The data given in the Final Report clearly do not justify dismissal without careful investigation of any complaint by a dairy farmer showing some evidence that electrical problems impact milk production and/or animal health.**

(II) The "Field Study" described in the Final Report provides no information on whether the relatively high step potential differences (above 9 mV across 1.5 m) measured on 3 of the 19 farms originated from conditions on these farms, were caused by conditions on the utility operated power lines, or were due to some other source. **The report can therefore not be used to conclude that undesirable step potentials are not caused, at least in some cases, by conditions outside particular farms.** The fact that only step potentials during low electric use on the farms are significantly correlated with low milk production (Table II in Final Report) actually suggests off-farm sources.

(III) The correlation between higher "average step potential, low use" and higher "soil resistivity in the field" that is apparent from Table II suggests that higher step potentials are more likely to be found in areas with higher soil resistivity (my own analysis suggests above approximately 150 ohm-meter when measured with a "Wenner array" employing 5- foot electrode spacing as in the "Field Study"). **The implication for regulatory policy of this fact might be that in areas with high soil resistivity particular consideration should be given to all measures that would help to prevent high earth surface potential differences.**

(IV) The "Science Advisors" report leaves open many questions; the "Field Study" involved a small number of farms and the laboratory research was inconclusive concerning immune function response to long term current exposure. Therefore regulatory policy that affects electric power supply to dairy farms (as well as other systems that may have an impact on dairy farms and are under PUC control, such as construction, maintenance and location of oil pipe lines) should involve all available sources of information, including the results of research that may have become available after publication of the "Final Report ".

04/30/00

Appendix J

Concerning the Reinemann research that indicates there is no problem with electricity entering cows, the Wisconsin Department of Agriculture sponsored a conference entitled "Stray Voltage and Dairy Farms," wherein the following was carefully noted. Since electroshocks change an animal's blood chemistry, it's unfathomable how Reinemann can assert there is no problem. This is in direct contradiction to the testimony of Dr. Richard Schulte, who was asked about the matter during his testimony in the Gumz case, a suit for stray voltage damages against Northern States Power Company.

Excerpt from testimony of Richard Schulte, D.V.M.
In case of James Gumz, Grace Gumz, Michael Gumz and Susan Gumz
v.
Northern States Power Company

Q. Doctor, can electricity be a limiting factor on production in dairy herds?
A. It absolutely can.

Q. What happens over time to cows exposed to electricity that can limit their production?
A. Well, sometimes it doesn't have to be over time. You can get a real poor response to treatment of these animals and the feed and water intakes, you know, that can be – that can be a real immediate type of thing that can slow down the production.

Q. What's mastitis?
A. Oh, mastitis is basically infection in the udder of the cow, and it can really get a cow sick if the infection starts to overwhelm the cow. They can get hot hard swollen mammary tissue and run a high temperature and really start to feel pretty sick.

Q. In your experience, Doctor, does exposure to electricity sometimes create health problems with cows?
A. Oh, absolutely.

Q. What's a somatic cell count?
A. Somatic cell count or SCC is a measure of the number of white blood cells in the milk. And we are – the dairy industry is regulated by the State Department of Agriculture on how many white blood cells you can have in the milk. The white blood cells in the milk are an indication of infection, or the healthy status or unhealthy status of the udders. And, in the Grade A business, it used to be higher, but now we're – you're limited to 750,000. They – you would like to have a herd at, oh, 100 to 200 thousand. I would say the average probably in the State of Wisconson is somewhere in the neighborhood of about, I think I read this not too long ago, 324. If you took every pound of milk that's produced in Wisconson.

Q. Why is the somatic cell count important?
A. Well, the somatic cell count is indicative of the health of the cow's udder. And also if you start getting up above 750,000, this isn't – not deemed a food that you would want to eat.

Q. Will exposure to stray electricity sometimes result in higher somatic cell counts than would otherwise be present?
A. That has been my experience, yes.

Q. Can exposure to stray electricity have an impact on the disease status of a dairy herd?
A. Oh, absolutely. And that's probably the – one of the biggest complaints that I hear from other veterinarians, and that was the biggest complaint that I had. You felt like failure as a veterinarian. You go out and you treat this cow, and you're using good technique, good pharmaceuticals, and the guy calls the next day and says, this cow hasn't responded.

Appendix K

At one point I filed a complaint of trespass against WPS, meaning that electric current had invaded my property. By way of highlighting the utility's arrogance, the following (crazy) legal brief is the WPS response. In it they argued that, since they'd been poisoning my earth for the past 10 years and getting away with it, they should be permitted to do so indefinitely! As if this were not bad enough, they were aware of the problem from the first day they reconfigured the electrical distribution system, in the mid-seventies. Their own engineer, Steve Schott, admitted this. His statement during deposition follows the WPS brief.

From WPS Brief

Russell Allen
Vs.
Wisconson Public Service Corporation

C. WPS would have an easement for any alleged trespass from earth currents.

> WPS has maintained a distribution system adjacent to plaintiff's property continuously for more than 10 years. This system has resulted in current flowing through plaintiff's property on a continuous basis for more than 10 years. Given these facts, even assuming that earth currents could otherwise give rise to a claim for trespass, WPS would have an easement by prescription to continue to permit its earth currents to flow under plaintiff's property.
>
> The Wisconsin statute that governs utilities' acquisition of prescriptive rights is directly on point. It provides in pertinent part that "[c]ontinues use of rights in real estate of another for at least 10 years by a domestic corporation organized to…transmit…electric current to the public or for public purposes…establishes the prescriptive right to continue the use…" Wis. Stat. § 893.28(2). If, as plaintiff asserts, earth currents generated by WPS has generated earth currents that have flowed through plaintiff's property in the course of transmitting electricity to the public, and that it has done so for more than 10 years. Therefore, even if earth currents are a

potential basis for a trespass claim, and even if WPS is not privileged to generate those earth currents, WPS has obtained a prescriptive easement to do so.

Excerpted from Deposition of Steven Schott

Russell Allen

v.

Wisconson Public Service Corporation

February 18, 2002

Q. When that distribution system was installed to the Allen farm in 1975, would it be fair to say that WPS knew that it would be using the earth as part of its return path?

A. Yes.

Appendix L

Here are two letters I wrote regarding the stray current problem on my property and the nuisance it created. The first is to the CEO of WPS Corporation, while the second is to the Ag/Extension agent in Madison. Following my letters is a Foley & Lardner brief to Judge McKay, written pretrial, rejecting all of my proposals for relief. The proposals were those that Don Zipse wrote and clarified on the witness stand during his testimony. And yet WPS acceded to proposal number four when, under my legal pressure, it reconfigured the system in October of 2006, driving the current onto other farms.

Russell Allen
Allen Dairy Farms
5206 Little Apple Rd.
De Pere, WI 54115

Mr. Larry Weyers, Chairman & CEO
Wisconsin Public Service Corporation
700 N. Adams street
P. 0. Box 19001
Green Bay, WI 54307 9001

RE: Continued Trespasses and Nuisances on My Real Estate in Brown County

Dear Mr. Weyers:

On July 17, 2002, I delivered to your legal representative, Trevor Will, and your employee Michael Moore, my letter advising you to terminate your trespasses onto my real estate. To date nothing has been done to stop the flow of electricity onto my premises, and you continue to use my property for the path of your electricity. This has now been demonstrated by several tests. Most notably, when your system was shut down, all of' the electricity flowing onto our farm property of any substantial magnitude disappeared. The only source for this electricity is the WPS distribution line.You are not authorized, and I specifically forbid you to continue further trespasses of your electrical current and/or electrical energy onto my property. This constitutes a nuisance, and it is obvious that you are intentionally continuing this nuisance. This must terminate immediately.

I attach a copy of a graph taken of a copper wire laid on the ground of my property which shows the level of current coming onto the property, indicating that even since my letter advising you to stop your trespass and nuisance, the flow of current from WPS's distribution system continues This Must stop.

Since I have won my lawsuit, WPS continues to use my property for the path of your electricity. This must stop.

Very truly yours,
Russell Allen

5206 Little Apple Road
De Pere, WI 54115

UW Extension
ANRE Program Leader
Ric, Klemme
633 Extension Building
432 North Lake Street
Madison, WI 53706 1498

November 1, 2002

Dear Mr. Klemme:

As you may be aware "Stray Voltage' has been a very contentious Issue on Wisconsin's Dairy Farms, including my own. I recently held an informational field day on the subject which was attended by 150 farmers, Agribusinesses and concerned citizens. I am enclosing a video of that field day for your enlightment as to what some non university experts have determined about the subject of stray voltage and earth currents rents. I would hope the University of Wisconsin and soon to be former Dairy State would evaluate this ongoing crisis to our declining dairy industry with an open unbiased mind, before it's too late. I realize the University has done research on this topic, but the campus based research does not support what hundreds of farms across the state and county are seeing regarding the threshold affecting cows. It is time a farmer based advisory committee be assembled to discuss future research directions regarding stray voltage. Thank you for taking the time to read this letter and view the video.

Sincerely,
Russ Allen

Reconfigure the Single-Phase System

Mr. Zipse's fourth proposal is for the Court to order WPS to reconfigure the distribution line serving the Allen farm (Mystery Hills 242) 80 that those customers on the sin vie phase service to plaintiff's south and east are transferred to the Wesmark substation. This would be done by disconnecting the single phase line that begins at the Allen farm from the Mystery Hills 242 feeder and extending the southern end of that single phase line approximately 300 feet to connect to a single phase line from the Wesmark substation. This proposal is inappropriate and unwarranted.

Mr. Zipse conceded at his deposition and at trial that his proposal would simply redirect allegedly harmful astray currents from Mr. Allen's property to other landowners in the area including dairy farmers located south and east of the Allen farm.

Appendix M

This letter, from Scott Lawrence to Trevor Will after the verdict in my case, shows that we did not want another court case and hoped that WPS would institute a fix of its own volition. But, as the utility continued its foot-dragging, I had no choice but to take further legal action. Three and a half years after the verdict—from a new, non-biased judge—I finally got results. Also appended here are two newspaper articles reporting the upholding of my jury award by an appellate court.

Stray Voltage Award Of $1.75m Upheld
The Capital Times :: FRONT :: 8A
Tuesday, February 15, 2005

By Matt Pommer
The Capital Times

A $1.75 million award to a former Brown County dairy farmer in a stray voltage case was upheld today by the 3rd District Court of Appeals.

The court also remanded part of the case to the Circuit Court to determine whether the award had included potential future damages. The appeals court noted that Russell Allen, who won the award, has left the dairy business.

Wisconsin Public Service Co. had challenged the size of the award, questioned payment of nuisance damages, and asked whether Allen had taken reasonable steps to discover the problems associated with unintended electrical current on his property. The appeals court rejected the three pleas.

Allen had counter-sued, seeking treble damages, and had sought an injunction to order the company to eliminate stray voltage from the farm. The court rejected the claim for treble damages and said the injunction issue was related to the question of whether the damage award included amounts for future damages.

Allen started experiencing problems on his dairy farm in 1976 when he moved his cows to a new barn, but did not discover the stray voltage problem until 21 years later. That year he installed a Dairyland Isolater and the herd's health improved and milk production increased.

But by 1999 the herd's health worsened again and Allen had an electrician install a copper ground ring around the farm to prevent current from coming onto the farm. He also installed a surface ring of wire around the farm, but there was no improvement, the appeals court noted.

In a near monthlong trial, a jury found that Wisconsin Public Service was negligent in its distribution of electricity. The jury awarded $750,000 in damages and $1 million for Allen's inconvenience, annoyance and loss of use and enjoyment of his property.

"We conclude the evidence support the amount of the jury's award for nuisance damages," the appeals court said in its decision

Appendix N

Here are three letters from Minnesota farmers who underwent the same stray current ordeal at the hands of electric utilities. If only our state and national governments, the EPA, paid as much attention to the problem of electrical pollution as they do to air and water pollution, Americans could be saved a world of grief.

Response to Final Report of the Science Advisors to the MPUC

From: **Harold Johnson, Swanville, MN 56382**

What a waste of time and money! I thought we would get some answers! But it looks like the electrical utilities got to the team and it shows in the report. I am very disappointed that the MPUC did not follow the legislative guidelines. Some of the advisors could not handle the way the testing was done and quit. The others were swayed by the utilities. Reidamann's report was a joke!

I had a formal complaint against my utility, but the PUC decided to dismiss it. I had good proof that ground currents affect cows. My neighbor build a house and the electrical utility put in an underground cable to the house. The insulation got damaged when installed. The wire burned off completely when it rained for 2 days. During that time, our dairy cows dropped 800 pounds in milk production a day! And the calves that were born at that time had lots of problems. Some milkings the cows are find and the next milking they kick off the milkers all the time. One day they eat everything and the next they leave a lot of food.

If I could afford an attorney, I would prove that the utilities are lying about how much electrical current they are putting into the ground. It is the biggest cover up that is going on today! The major problem is that there is so much electricity used on the same distribution system. And that will only get worse unless something is done. I'm sure that the electricity is affecting the human immune system and causing the high rate of cancer!

It seems strange that the utilities were involved in the meetings but the farmers were not. Especially when the farmers are the ones affected by the problem! Patricia Hoben was not going to listen or help the farmer. I think we need to get the Attorney General involved to get this straightened out.

Respectfully submitted,
Harold Johnson.
PS – I'm 70 years old and was born on this place.

March 15, 2000

My Response to Final Science Advisor Report to Minnesota Pollution Control Agency

From: **Arthur J. Borgerding**
St. Cloud, MN 56303

To Whom It May Concern:

This is my final report on my problem with ground current on my dairy farm that was located in Freeport, Minnesota, which no longer exists.

We started experiencing ground current on our farm in 1981. The problems we experienced were: 1) Restless cow, 2) crippled cows, 3) cows not drinking enough water, 4) calves born crippled, and 5) good fresh cows coming into the barn would only last six weeks. They then would be so crippled up they had to be sole for little or nothing or they would die.

Northern States Power (NSP) was our electric supplier. Our farm was the fist farm west of Freeport, Minnesota. Our electricity came from town. The line to our farm was not adequate to carry the load. It had three different size wires on the line plus many inadequate splices in it. The ground wire on the transformer pole had so much current flowing into the ground that you could light a 25 watt light bulb. That was flowing through my barn and killing my cows, robbing me of production and ruining my health. I could feel the current in my body. Based on the health of my cows, I'm sure they could feel it as well.

Fr 16 years, we were fighting this problem with NSP. We would cut NSP's ground wires and they it would get better. NSP would come with the sheriff and hook them up, and it would get worse again.

They would send their NSP engineers out to do testing. They finally said we had stray voltage. We then installed an isolation transformer, which helped some, but not enough. NSP put in blockers at the poles, which again helped very little. The only thing that really helped was to cut the transformer ground wires. We had engineers there from all over, including the University of Minnesota. Our veterinarians would check our feed and do herd health and couldn't find a problem.

In 1989, we hired an engineer from Nebraska. He spent three days on our farm doing tests. His conclusion was that we had ground current. He said our problem was coming from the inadequate line running to our farm and still NSP said that this was not their problem.

We kept fighting. We worked with the University of Minnesota engineers, state senators, state representatives, and whoever would listen to us. Finally, in July 1991, NSP agreed to put in a new line to our farm from across the railroad tracks south of our farm. They hooked the two neighbors to the east of us to that line as well. The results were tremendous. This happened in July 1991. The cows started drinking more water, production increased and the cows were not as restless. It was great! We kept real close records of our milk production. Every month it went up. It was fun doing chores again!

It lasted from July 1991 (when the new lines were built) until March 1993, 1½ years later. That is when everything started falling apart again. We didn't notice it at first, but in March 1993, NSP put new insulators on the top wire of every pole and grounded every pole. That is when it got so bad in the barn again. The same problems we had prior to July 1991 came back—only worse.

Everything was really bad in the barn again. We had six cows that were so bad that they needed help getting up. The calves were born crippled and some were born blind. The cows would put their mouths in the water bowls and would pull them right back out again. They were getting shocked every time they drank water. My health problems were also coming back again. I could feel it in my ankles, my knees and my hips. By the time I was done milking, my eyes would feel like they were going to pop out of their sockets, and I would have tears in my eyes.

We started fighting NSP again. They said they could reroute the current past our farm to see if it would make a difference but they never did. Then, NSP engineers told me my problem was ground current and there was no state standards set on ground current, so they didn't have to do anything about it. We fought and struggled as much as we could. At one point, we just couldn't handle it any more, so I called an engineer at NSP. I left a message stating that if I didn't hear from him by 4:00 p.m. that afternoon, I was going to go across the tracks and cut the ground wires on the poles. It was 20 degrees below 0 and windy. I was mad! I didn't hear from them, so I cut seven ground wires. There was 1,400 volts going into the ground at ever pole. You could hold the wires a ¼" apart and they would buzz like crazy. They would arc like a welder. There couldn't have been many amps there, or I would be dead. Everything got much better in the barn again. I could feel the difference immediately.

Approximately 40 cows were left in the barn, of which five were severely crippled. On Christmas night, one was in such misery (she couldn't get up anymore). She was moaning so badly from pain. I couldn't watch her suffer any longer so I had to shoot her. That was our Christmas gift from NSP.

In January 1995, we decided to give up. We called the auctioneer to set up a date. We decided to have an auction on February 16, 1995, my wife's 62nd birthday.

Several days before the auction, we sold four cows that were unable to get up before I cut the ground wires. We received a decent price for them. If I wouldn't have cut those ground wires, most likely these four cows would have been dead. Normally, we should have had 150 head of cattle, but we only had 64 left to sell at the auction. We were out of business due to ground current.

We couldn't sell our farm as a dairy farm, so we sold the land to out neighbors. We sold the house and four acres separate. We also sold separately 19 acres, which were in city limits. I was born and raised on that farm. I was the third generation to farm this land. I farmed and worked this land all my life. By the time we were all done, we paid all out bills and barely had enough money left to make a down payment on a house.

Nellie, my wife, works as a bridal consultant in a bridal shop four days per week. Randy, our youngest son who farmed with us, has a job here in St. Cloud and still lives with us and helps pay the bills. I work as a janitor at a car dealership. I work over 50 hours per week so we can make our $700/month house payments. I'm 68 years old and I don't know where the house payments are going to come from when I can't work anymore.

When NSP's engineers were at our farm doing all their testing, they would brag about the good times they had at all their homes over the weekend and going on their hunting and/or fishing trips. My wife and I, at the age of 67 and 68, are struggling to make our house payments thanks to ground currents. We should be enjoying our retirement! At one time, one of the engineers told me, "We can't fix this problem because if it works for you, then we'll have to fix it for others and that would cost them a lot of money."

Our neighbors to the west of what used to be out farm were also dairy farmers. Their electricity came from the same lines as ours did – from across the track. They were also forced out of the dairy business and had the same problems we did. This happened after we quit farming.

When are you people going to wake up and do something about this problem? How many more farmers will be forced out of business?

Yours truly,

Arthur J. Borgerding

John Jackson's letter to Public Utilities Commision

Public Utilities Commission
121 Seventh Place East, Suite 350
St. Paul, Minnesota 55101-2147

Dear Sirs:

I am informed that you will receive comments on the Science Advisor's research regarding stray voltage.

I was a member of that group representing the Farmers Union and had personally fought the stray voltage war on my own farm. and sad to say, lost both the battle and the war, resulting in the dispersal of my herd and ultimately of the farm operation. There is no question in my mind that the cause of our problem was transient voltage. The problem was, we were never able to conclusively identify the source, and the people who were trying to assist us kept coming back to "ground currents".

My reason for commenting on the report is because of great unhappiness with the conclusions reached.

I believe that the only reasonable conclusion to the study would be this:

Using the techniques and methodology described, and considering the time of the year when the various measurements were taken, that no statistically significant evidence of ground currents or other transient voltage affecting herd health was detected.

The remarks about vaccination, sanitation, and other pious platitudes expressed or implied are an insult to those of us who exhausted every one of these avenues and our problems with sometimes detectable voltage, especially during the winter months, remain unsolved.

This letter is not a criticism of the work done by Riley Hendrickson and the follow up by Patricia Hoban, it is only hat the conclusion as printed are indefensible.

Your truly,
John A. Jackson

INDEX

T

U

W

Z